VISION AND
MOTOR CONTRO

ADVANCES
IN
PSYCHOLOGY

85

Editors:

G. E. STELMACH
P. A. VROON

N·H

NORTH-HOLLAND
AMSTERDAM • LONDON • NEW YORK • TOKYO

STANDARD LOAN

UNLESS RECALLED BY ANOTHER READER
THIS ITEM MAY BE BORROWED FOR

FOUR WEEKS

To renew, telephone:
01243 816089 (Bishop Otter)
01243 812099 (Bognor Regis)

VISION AND
MOTOR CONTROL

Edited by

Luc PROTEAU
Department of Physical Education
University of Montreal
Montreal, Quebec, Canada

Digby ELLIOTT
Motor Behaviour Laboratory
Department of Physical Education
McMaster University
Hamilton, Ontario, Canada

N·H

1992

NORTH-HOLLAND
AMSTERDAM • LONDON • NEW YORK • TOKYO

NORTH-HOLLAND
ELSEVIER SCIENCE PUBLISHERS B.V.
Sara Burgerhartstraat 25
P.O. Box 211, 1000 AE Amsterdam, The Netherlands

Distributors for the United States and Canada:
ELSEVIER SCIENCE PUBLISHING COMPANY, INC.
655 Avenue of the Americas
New York, N.Y. 10010, U.S.A.

Library of Congress Cataloging-in-Publication Data

Vision and motor control/ edited by Luc Proteau, Digby Elliott.
     p.   cm. -- (Advances in psychology ; 85)
   Includes indexes.
   ISBN 0-444-88816-0
   1. Perceptual-motor processes. 2. Motion perception (Vision)
 3. Visual perception. 4. Kinesiology.   I. Proteau, Luc.
 II. Elliott, Digby. III. Series: Advances in psychology (Amsterdam,
 Netherlands) ; 85.
 BF295.V57 1992
 152.3'5--dc20   91-42377
                                                            CIP

ISBN: 0 444 88816 0

Printed in The Netherlands

Preface

Since the classic studies of Woodworth (1899) almost 100 years ago, the role of vision in the control of movement has been an important research topic in experimental psychology. While many early studies were concerned with the relative importance of vision and kinesthesis and/or the time it takes to use visual information, recent theoretical and technical developments have stimulated scientists to ask questions about "how" different sources of visual information contribute to motor control in different contexts. In this volume, we present a collection of articles that provide a broad coverage of the current research and theory on vision and human motor learning and control.

Many of the contributors are colleagues that we have met over the years at the meetings and conferences concerned with human movement. They represent a wide range of affiliation and background including kinesiology, physical education, neurophysiology, cognitive psychology and neuropsychology. Thus the topic of vision and motor control is addressed from a number of different perspectives. In general, each author sets an empirical and theoretical framework for their topic, and then discusses current work from their own laboratory, and how it fits into the larger context.

The volume consists of four main sections and an epilogue. The first section deals with the role of vision in manual aiming. Manual aiming studies have been historically important in the study of speed-accuracy relationships in goal-directed movement. Because many of the specific speed-accuracy models of motor control involve visual feedback processing, the first chapter by Les Carlton deals with the issue of visual processing time. The next chapter by Digby Elliott deals with research concerning the amount and type of visual information that is important for precise aiming. Richard Carson then discusses the contribution of visual feedback processing in determining speed-accuracy asymmetries between the right and the left hands. In the next chapter Luc Proteau exposes the myth that people become less dependent on visual information with repeated practice, setting the stage for a chapter by Gordon Redding and Benjamin Wallace on the interaction of visual and motor subsystems in motor control. In this chapter they discuss how visual rearrangement techniques can be used to examine these coordinative linkages. Then Richard Abrams discusses the coordination and role of eye movements in manual aiming. Finally, Malcolm Cohen and Robert Welch provide insight into visual-motor control in altered gravity.

The second section consists of four chapters concerned with prehension and gesturing. Heather Carnahan provides a link with the previous section by discussing the coordination of eye, head and limb movements in both aiming and prehension. Marc Jeannerod and Ron Marteniuk then review how perturbation techniques have been used to study the interaction between the reaching and grasping components of this type of movement, while Barb Sivak and Christie MacKenzie focus on the role of central and peripheral vision in prehension. Finally Eric Roy and Craig Hall discuss how we can gain a better understanding of the mechanisms involved in the normal visual control of movement by examining deficits resulting from cerebral brain damage.

In the next section, we have three chapters concerned with the spatial timing of movements. Reinoud Bootsma and Lieke Peper introduce this section with a discussion of how predictive visual information can be used to regulate action in activities such as catching and hitting. Michelle Fleury, Chantal Bard, Marie Gagnon and Normand Teasdale then provide a comprehensive review of research concerned with vision and coincidence-anticipation. In the next chapter Dave Goodman and Dario Liebermann review their work on free-fall landings, and visual information about time to contact.

In the final section, the focus is on gross motor activities that involve maintenance of posture and locomotion. Janet Starkes, Cindy Riach and Beverley Clarke lead off this section discussing the role of vision in postural sway in normal adults, children and Parkinson's disease. Next John Corlett discusses the role of vision in goal-directed locomotion. Jean Pailhous and Mireille Bonnard cover locomotor automatism and visual information providing a physiological framework for behavioral evidence. Finally, Daniel Mestre reviews the ecological approach to the perception of self-motion in locomotion.

In the epilogue, Dan Weeks and Robert Proctor attempt to place the various chapters into a larger framework of visual information processing research. As well, they suggest some future directions for scientists interested in vision and motor control.

While the volume does not exhaust the topics of interest to students of vision and motor control, we believe it provides a broad sampling of the issues of current interest. Although in several chapters the authors speculate about the neural mechanisms underlying visual-motor control, the level of analysis in most chapters is behavioral. Given this content, the book should fill the role of a resource text for behavioral scientists already working in the field. As well, the volume should provide a good starting point for a new researcher. Readings from this volume would be appropriate for graduate and senior undergraduate students in psychology, kinesiology and physical education.

Acknowledgement

This project took nearly 18 months to complete. This is a long period of time. However, it is quite short if we compare it to how long it would have taken us to do it all by ourselves. We are indebted to many people. We would, first, like to thank our contributors. We asked a lot from them ranging from rewriting complete sections of their chapters and redrawing figures to simply adding a reference or two, and proofreading. They never complained (at least, not to us!) and did a superb job. Any glitches that can (and probably will) still be found in this book are our sole responsibility. We would also like to thank Lucie Senneville who had to retype a few chapters and was in charge of the Author Index. We really appreciated her help and her smile when we lost the final version of a chapter and when she did not complain about doing it again. We also appreciate Barbara Pollack's contribution to the Subject Index and the library search to retrace figures from other sources. A good graduate student has always been and will always be of invaluable help. Lastly but not the least, we would like to thank the members of the McMaster graduate seminar (Craig, Lydia, Romeo, Tim, Tina) who helped us review the first draft of many chapters. Tim Lee also helped us in the first stage of this endeavor when we were trying to build a list of potential contributors. We thank all of them for their time and effort.

Digby Elliott   Luc Proteau
Hamilton, Ontario   Montréal, Québec

Permissions

We thank the following authors/publishers for allowing us to reproduce/redraw previously published figures and tables.

American Physiological Society. Figure 1.4 (Georgopoulos et al., copyright 1981); Figure 18.4 (Talbott & Brookhart, copyright 1980).

American Psychological Association. Figure 1.3 (Carlton, copyright 1981), Figure 6.1 (Abrams et al., copyright 1990); Figure 17.2 (Warren et al., copyright 1988); Figure 17.3 and 17.4 (Warren et al., copyright 1991).

Canadian Psychological Association. Figure 2.4 (Elliott & Calvert, copyright 1990); Figure 9.4 (Marteniuk et al., copyright 1987).

Elsevier Science Publishing Company Ltd. Figure 18.2 (Pailhous et al., copyright 1990); Figure 18.3 (Young, copyright 1988).

Heldref Publications. Figure 2.2 and 2.3 (Elliott et al., copyright 1990); Figure 5.5 (Redding & Wallace, copyright 1990); Figure 8.1 and 8.2 (Carnahan & Marteniuk, in press).

Karger AG. Figure 18.1 (Kanaya et al., copyright 1985).

Permagon Press. Figure 10.2, 10.3, 10.4, and 10.6 (Sivak & Mackenzie, copy 1990); Figure 18.5 (Cappozzo, copyright 1981).

Pion Ltd. Table 18.1 (Rock & Smith, copyright 1986).

Quarterly Journal of Experimental Psychology. Figure 2.1 (Elliott & Madalena, copyright 1987); Figure 4.2 (Proteau & Cournoyer, copyright 1990); Figure 4.4 (Proteau et al., in press).

Springer-Verlag GmbH & Co. Figure 9.1 and 9.2 (Pélisson et al., copyright 1986); Figure 9.3, 9.5, and 9.7 (Paulignan et al., copyright 1991); Figure 9.5, 9.6, and 9.8 (Paulignan et al., in press).

Taylor and Francis. Figure 1.2 (Whiting et al., copyright 1970).

The Company of Biologists Ltd. Figure 18.7 (Frost, copyright 1978).

List of Contents

Preface   v

Acknowledgement   vii

Permissions   ix

List of Contents   xi

List of Contributors   xv

Part One   Manual Aiming

Chapter 1   **Visual processing time and the control of movement**
Les G. Carlton   3

Chapter 2   **Intermittent versus continuous control of manual aiming movements**
Digby Elliott   33

Chapter 3   **Visual feedback processing and manual asymmetries: An evolving perspective**
Richard G. Carson   49

Chapter 4   **On the specificity of learning and the role of visual information for movement control**
Luc Proteau   67

Chapter 5   **Adaptive eye-hand coordination: Implications of prism adaptation for perceptual-motor organization**
Gordon M. Redding and Benjamin Wallace   105

Chapter 6   **Coordination of eye and hand for aimed limb movements**
Richard A. Abrams   129

Chapter 7   **Visual-motor control in altered gravity**
Malcolm M. Cohen and Robert B. Welch   153

Part Two   Prehension and Gesturing

Chapter 8   **Eye, head and hand coordination during
                    manual aiming**
                    Heather Carnahan   179

Chapter 9   **Functional characteristics of prehension:
                    From data to artificial neural networks**
                    Marc Jeannerod and Ronald G. Marteniuk   197

Chapter 10   **The contributions of peripheral vision
                    and central vision to prehension**
                    Barbara Sivak and Christine L. MacKenzie   233

Chapter 11   **Limb apraxia: A process approach**
                    Eric A. Roy and Craig Hall   261

Part Three   Spatial-Temporal Anticipation

Chapter 12   **Predictive visual information sources for
                    the regulation of action with special emphasis
                    on catching and hitting**
                    Reinoud J. Bootsma and C. (Lieke) E. Peper   285

Chapter 13   **Coincidence-anticipation timing:
                    The perceptual-motor interface**
                    Michelle Fleury, Chantal Bard, Marie Gagnon,
                    and Normand Teasdale   315

Chapter 14   **Time to contact as a determiner of action:
                    Vision and motor control**
                    David Goodman and Dario G. Liebermann   335

Part Four   Posture and Locomotion

Chapter 15   **The effect of eye closure on postural sway:
                    Converging evidence from children
                    and a Parkinson patient**
                    Janet Starkes, Cindy Riach and, Beverley Clarke   353

Chapter 16   **The role of vision in the planning and guidance
                    of locomotion through the environment**
                    John Corlett   375

Chapter 17   **Locomotor automatism and visual feedback**
              Jean Pailhous and Mireille Bonnard   399

Chapter 18   **Visual perception of self-motion**
              Daniel R. Mestre   421

Chapter 19   **The visual control of movement**
              Daniel J. Weeks and Robert W. Proctor   441

Author Index   455

Subject Index   473

List of Contributors

RICHARD A. ABRAMS
Department of Psychology, Washington University, St. Louis, MO, 63130

CHANTAL BARD
Laboratoire de Performance Motrice Humaine, Université Laval,
Sainte-Foy, Québec, G1K 7P4

MIREILLE BONNARD
Université d'Aix-Marseille II, Faculté de médecine, Cognition & Mouvement,
URA CNRS 1166 IBHOP, Traverse Charles Susini, 13388 Marseille, France

REINOUD J. BOOTSMA
Department of Psychology, Faculty of Human Movement Sciences,
Free University, Van der Boechorststraat 9, 1981 BT Amsterdam,
The Netherlands

LES G. CARLTON
Department of Kinesiology, University of Illinois at Urbana-Champaign,
Urbana, IL, 61801

HEATHER CARNAHAN
Department of Physical Therapy, Elborn College,
University of Western Ontario, London, Ontario, N6G 1H1

RICHARD G. CARSON
School of Kinesiology, Simon Fraser University,
Burnaby, British Columbia, V5A 1S6

BEVERLEY CLARKE
Department of Neurology, McMaster University Medical Centre,
Hamilton, Ontario, L8N 3Z5

MALCOLM M. COHEN
NASA-Ames Research Center, Moffett Field, CA, 94035-1000

JOHN CORLETT
Faculty of Human Kinetics, University of Windsor,
Windsor, Ontario, N9B 3P4

DIGBY ELLIOTT
Motor Behaviour Laboratory, Department of Physical Education,
McMaster University, Hamilton, Ontario, L8S 4K1

MICHELLE FLEURY
Laboratoire de Performance Motrice Humaine, Université Laval,
Sainte-Foy, Québec, G1K 7P4

MARIE GAGNON
Laboratoire de Performance Motrice Humaine, Université Laval,
Sainte-Foy, Québec, G1K 7P4

DAVID GOODMAN
School of Kinesiology, Simon Fraser University,
Burnaby, British Columbia, V5A 1S6

CRAIG HALL
Faculty of Kinesiology, University of Western Ontario,
London, Ontario, N6G 1H1

MARC JEANNEROD
Vision et Motricité, INSERM U94, 69500 Bron, France

DARIO G. LIEBERMANN
Center for Research and Sports Medicine Sciences, Wingate Institute,
Netanya, 42902, Israel

CHRISTINE L. MACKENZIE
Department of Kinesiology, University of Waterloo,
Waterloo, Ontario, N2L 3G1

RONALD G. MARTENIUK
Department of Kinesiology, University of Waterloo,
Waterloo, Ontario, N2L 3G1

DANIEL R. MESTRE
Université d'Aix-Marseille II, Faculté de médecine, Cognition & Mouvement,
URA CNRS 1166, IBHOP, Traverse Charles Susini, 13388 Marseille, France

JEAN PAILHOUS
Université d'Aix-Marseille II, Faculté de médecine, Cognition & Mouvement,
URA CNRS 1166, IBHOP, Traverse Charles Susini, 13388 Marseille, France

C. (LIEKE) E. PEPER
Department of Psychology, Faculty of Human Movement Sciences,
Free University, Van der Boechorststraat 9, 1981 BT Amsterdam,
The Netherlands

ROBERT W. PROCTOR
Department of Psychological Sciences, Purdue University,
West Lafayette, IN, 47907-1364

LUC PROTEAU
Département d'éducation physique, Université de Montréal,
P.O. Box Office 6128, Station A, Montréal, Québec, H3C 3J7

GORDON M. REDDING
Department of Psychology, Illinois State University, Normal, IL 61761

CINDY RIACH
Department of Physical Education, McMaster University,
Hamilton, Ontario, L8S 4K1

ERIC A. ROY
Department of Kinesiology, University of Waterloo,
Waterloo, Ontario, N2L 3G1

BARBARA SIVAK
Department of Kinesiology, University of Waterloo,
Waterloo, Ontario, N2L 3G1

JANET STARKES
Department of Physical Education, McMaster University,
Hamilton, Ontario, L8S 4K1

NORMAND TEASDALE
Laboratoire de Performance Motrice Humaine, Université Laval,
Sainte-Foy, Québec, G1K 7P4

BENJAMIN WALLACE
Department of Psychology, Cleveland State University,
Cleveland, OH 44115

DANIEL J. WEEKS
Department of Physical Education, Lakehead University,
Thunder Bay, Ontario, P7B 5E1

ROBERT B. WELCH
NASA-Ames Research Center, Moffett Field, CA, 94035-1000

Part 1:
Manual aiming

VISION AND MOTOR CONTROL
L. Proteau and D. Elliott (Editors)
© 1992 Elsevier Science Publishers B.V. All rights reserved.

CHAPTER 1

VISUAL PROCESSING TIME
AND THE CONTROL OF MOVEMENT

LES G. CARLTON

Department of Kinesiology, University of Illinois at Urbana-Champaign, Urbana, IL, 61801

Visual information is critical for performing a variety of motor activities, especially when the performer's movements must coincide with a changing environment, such as in catching a ball, or in motor activities requiring precise movements of the hand to a target. The study of "manual aiming" activities, such as pointing at a target, relocating a body segment in space, or reaching for an object, has been directly linked to vision and movement control since the seminal work of Woodworth in 1899. One of the outgrowths of this work is the realization that vision plays a number of roles in producing skilled movements. For example, it is evident that vision facilitates performance in manual aiming by specifying the target position, by indicating the position of the moving limb, and/or by providing error information regarding the discrepancy between the target and limb position.

The actual performance benefit derived from visual information is a function of a number of factors (Carlton, 1987), with one of the most important being temporal processing delays. The duration of the temporal delay between picking up visual information and using this information to trigger or guide our movements is critical for precise manual aiming, and is also important for a broad range of other activities. Survival in high speed - low altitude fighter aircraft maneuvers, as well as more mundane activities such as pedestrian and automobile travel, sport performance, and the catching of living prey by animals, is also influenced by visual processing delays.

Reaction time paradigms have provided one method for studying processing latencies. A visual stimulus, often a low intensity light, is presented at an unpredictable time and the delay between light onset and the initiation of movement by the subject is measured. This delay has been studied in an attempt to understand stages of information processing dating back to Donders (1868/1969). Using a simple reaction time paradigm, the delay in responding to a visual stimulus is usually on the order of 190 to 210 ms in humans. There is reason to believe, however, that the reaction time approach to studying processing delays may not be appropriate when the visual information represents feedback from an ongoing movement, or when the time of the visual event is reasonably certain.

This chapter will be concerned with visual processing delays associated with the control of ongoing movements. The activities covered will mainly be aimed or pointing movements of the hand and arm, but includes some work on whole body actions. Oculomotor responses will not be examined. The chapter contains four primary sections. The first section provides a review of some of the classic studies examining manual aiming and visual processing time. This is followed by recent work suggesting that visual feedback information may be used in aimed movements with latencies much shorter than previously supposed. Recent evidence is then presented supporting rapid visual processing when visual stimuli in the environment are changed, such as movement of the target in aimed movements or a sudden change in the flight path of a ball in striking activities. The last section provides a synthesis of research on visual processing time for the control of movement. It is suggested that the visuomotor delay is not constant but is a function of the interaction of a number of factors.

Historical Review

In this section some of the original work on visual processing time is reviewed. The time period covered is from just before 1900 up to the mid 1970s. While the studies cited do not represent an exhaustive list of work, they are among the most influential on current thinking and illustrate the progression of research over time. It will become evident that visual processing time and the tradeoff between movement speed and accuracy are related, and it is to this work that we now turn.

Woodworth and Vince

The initial work on the accuracy of movements by Woodworth in 1899 provided significant insights into the control of aiming movements, and set the tone for much of the work that followed on visual processing delays as well as the relation between movement speed and accuracy. Woodworth was interested in how movement speed influenced movement accuracy, and realized that spatial accuracy was at least partially a function of visual feedback received during movements. This, and knowing that movement accuracy decreases as speed increases, led to the hypothesis that short duration movements would not benefit as much from visual feedback as movements made with longer durations.

The experimental procedure that Woodworth used was somewhat crude in comparisons to today's standards, but the basic strategy used to estimate visual feedback processing time is still used in current research studies (e.g., Zelaznik, Hawkins, & Kisselburgh, 1983). Woodworth had subjects draw lines on a sheet of paper using a pencil, while the paper rolled at a constant speed. Reciprocal (back and forth) movements were used, with subjects either attempting to come as close as possible to a fixed target or trying to match the spatial endpoint of the previous movement. This procedure allowed for the measurement of spatial accuracy as well as giving some insights as to the kinematic characteristics of the movements. Movement duration was varied by having subjects pace their movements to the beat of a metronome, and vision was manipulated by having subjects open or close their eyes during the movements.

The basic tenet of this paradigm is that movements made with vision should be more accurate than those made without vision as long as there is sufficient time to process the visual feedback. This argument is at least partly related to assumptions about how aimed movements are controlled, and it was Woodworth's belief that aiming movements were made up of two phases. The first, or initial adjustment phase, moved the hand toward the target and the second, or current control phase, allowed for the correction of errors, resulting in a high degree of spatial accuracy. While, in principle, error corrections could be mediated by a number of sensory information sources, it seemed to Woodworth that visual feedback of the moving limb was the most salient in this task.

Thus, Woodworth sought to determine the shortest movement time where vision facilitated movement accuracy. Just as he had expected, movements with the shortest durations were no more accurate when the eyes were open than when they were closed (see Figure 1.1). As the average time to complete each movement in the sequence increased, spatial errors decreased when the eyes were open but remained approximately constant when the eyes were closed. Woodworth's interpretation of the results was that movements made at a rate of 140 / minute or greater were equally accurate with or without vision, leading Woodworth to conclude that the time to process visual feedback for the control of movement was about 450 ms.

While Woodworth's contribution is now well recognized, research on visual feedback processing delays, and, more generally, speed-accuracy tradeoffs, was not taken up again in any significant way until the late 1940s. In 1948 Vince, also using reciprocal movements and manipulating visual

L.G. Carlton

feedback, re-examined the importance of visually based corrections in aimed movements. Instead of subjects keeping their eyes closed during a series of movements, subjects only closed their eyes during the "down" stroke toward the target and then opened them during the return to the start line ("up" stroke). The time for the return stroke was fixed at 100 ms by the apparatus. Due to difficulty in rapidly opening and closing the eyes, only slow rates (20 to 100 strokes per minute) were completed without vision.

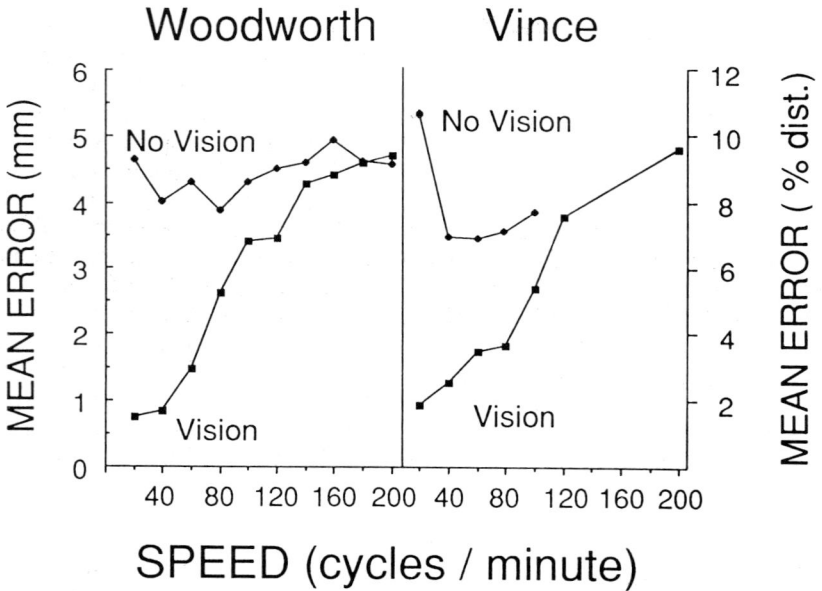

Figure 1.1. The left side of the Figure provides the results from Woodworth (1899). Subjects attempted to match the length of the immediately preceding stroke with either the eyes open or closed. The data presented is from four subjects using their right hand (Tables 1 through 4, Woodworth, 1899). The right hand side of the Figure represents the results from a similar experiment by Vince (1948, Experiment 2, Table 2). Subjects closed their eyes on half of the cycle, when the hand was moving down toward the target. Because of problems encountered when subjects opened and closed their eyes rapidly, the no vision condition was only performed at movement rates of 100 cycles / minute or less. The errors are measured as a percentage of the goal movement distance.

The results, also presented in Figure 1.1, show that subjects were always more accurate with vision, but the shortest down stroke time was 500 ms (100 strokes per minute; 600 ms cycle time minus 100 ms return time). Vince observed that errors were relatively constant in the eyes-closed condition at about 7.5% of the movement distance. With the eyes open, errors increased as movement time decreased (see Figure 1.1). Reasoning that visual information was not used if spatial errors in the eyes-open condition were larger than 7.5%, Vince concluded that movements shorter than 400 ms in duration could not be visually controlled.

The visual processing time estimates by Woodworth and Vince were quite long and were at least partly the result of the experimental methods used and the way the results were interpreted. Woodworth (1899) had subjects produce reciprocal movements, and movement accuracy was only measured at one end of the cycle, with the reversing motion being made to a stop. If subjects only used visual feedback from the forward stroke, and if movements back and forth were made at equal rates, Woodworth's visual processing estimate of 450 ms could be halved to around 225 ms! Vince also overestimated visual processing time by assuming that spatial errors are constant at all movement times with the eyes closed. While some studies (e.g., Woodworth, 1899; Zelaznik et al., 1983, Experiment 3), have demonstrated this phenomenon over a limited set of conditions, spatial errors usually increase as movement time decreases regardless of the visual feedback condition (Carlton & Carlton, 1991). This makes it difficult to make projections about what the eyes-closed error rates would have been at short movement times in Vince's experiment. Similar spatial errors for eyes-open and eyes-closed conditions may have only occurred at the fastest movement rates.

Fitts' Law and the Iterative Model

Woodworth and Vince were interested in how movement characteristics such as speed and amplitude are related to movement accuracy. While each made important contributions to our understanding of the relation between speed and accuracy, a formal description of this relation awaited the work of Fitts (1954). Using reciprocal movements and manipulating movement amplitude and target size, Fitts observed that the time to complete movements was linearly related to the logarithm of twice the movement distance divided by the width of the target. Fitts originally demonstrated this relation using three different tasks. Subsequent studies using a variety of movement tasks and subject populations, showed that this relation was highly reliable and generalizable (see Hancock & Newell, 1985; Meyer, Smith, Kornblum, Abrams, & Wright, 1990 for a discussion), so much so, that it ultimately became known as Fitts' Law.

Fitts' experiments, and his interpretation of the speed-accuracy relation, did not involve visual function or visual feedback processing time. Other investigators (Crossman & Goodeve, 1963/1983; Keele, 1968), however, argued that Fitts' Law could be derived from an iterative correction model. In this model, aiming movements are made up of a number of submovements that are under feedback control. Each submovement is assumed to be of constant duration, and the total movement time is, therefore, equal to the number of submovements multiplied by the submovement duration. While the feedback information used might be either visual or kinesthetic (e.g., Crossman & Goodeve, 1963/1983), Keele (1968) suggested that spatial accuracy was a function of visual feedback information, based partly on the finding that movement accuracy decreases when visual feedback is withheld (Vince, 1948; Woodworth, 1899). Estimates of visual feedback processing time of 400 ms, however, made vision an unlikely candidate for controlling submovements, or explaining Fitts' law when movement times are relatively short (under 400 ms).

Keele and Posner

About the same time as Keele's influential review on the control of skilled movements (Keele, 1968), Keele and Posner (1968) were questioning the long estimates of visual feedback processing time obtained by Woodworth (1899) and Vince (1948). It was difficult to understand why the visual processing delay should be so long, when choice visual reaction times were typically only 250 ms. They speculated that one of the reasons for the long processing estimates of Woodworth and Vince was that reciprocal rather than discrete movements were used. In reciprocal movements as much as 75% of the total cycle time can be taken-up in reversing the movement for the next response. This, they argued, resulted in an overestimation of visual processing time.

To overcome this limitation, Keele and Posner (1968) had subjects produce discrete movements to a target, attempting to control movement time through practice and knowledge of results. Instead of measuring the actual end points of the movements as had Woodworth and Vince, movements were made to a small circular target, and accuracy was assessed from the percentage of target misses. Vision was manipulated by controlling the only light source in the room, a small wattage bulb over the apparatus. On half the trials the light was extinguished as soon as a hand-held stylus left the start position. Four designated movement times were used ranging from 150 to 450 ms, in 100 ms increments.

The basic procedures were similar to those used by Woodworth with three notable exceptions. First, discrete rather than reciprocal movements were used; second, vision was manipulated unpredictably; and, third, spatial accuracy was measured indirectly as the percentage of response errors. The results from the Keele and Posner study indicated that movements were equally accurate at the fastest movement condition with or without visual feedback, but movements with longer durations had a smaller percentage of target misses when subjects could see their movements. Subjects did not match the criterion movement times exactly. The mean movement times produced for the vision conditions were 190, 267, 357, and 441 ms. Keele and Posner argued that the time to process visual feedback for controlling movements was between 190 and 260 ms because movement accuracy was aided by having visual feedback information available when the mean movement time was 260 ms or greater. Two years later, Beggs and Howarth (1970) provided additional evidence for visual processing times under 300 ms.

The estimate of 190 to 260 ms obtained by Keele and Posner was a significant reduction from earlier estimates, and compared favorably with visual choice reaction times. Keele (1968) used these results to argue that even fairly rapid movements could be controlled by visual feedback, providing support for the iterative correction model. Assuming that the time to process visual feedback information is 200 ms and that individual submovement durations are also 200 ms, Keele showed that the iterative correction model could accurately predict the slope of the logarithmic speed-accuracy function for discrete responses (data from Fitts & Peterson, 1964).

Early Evidence for Faster Processing Times

The estimates of visual feedback processing time provided by Keele and Posner (1968) and Beggs and Howarth (1970) were considerably shorter than those by Woodworth and Vince, but were still longer than typical simple reaction times. At about the same time, evidence was accumulating that visual information could be processed and used for the control of movement with latencies even shorter than simple reaction times. The evidence came from three sets of data. Only one of the studies involved aiming movements, and in many respects the studies where unrelated to each other. While these studies provided the initial suggestion of rapid visual processing for motor control, in some cases the authors discounted this evidence.

Slater-Hammel. The late 1940s and 1950s saw increased interest in processing limitations and their impact on motor control. Craik's ideas on intermittency and information processing (Craik, 1948) paved the way for future work on stages of processing (e.g., Broadbent, 1958; Welford,1952) and the psychological refractory period (e.g., Davis, 1959). Slater-Hammel, being influenced by this work, sought to examine temporal delays in "transit" reactions (Slater-Hammel, 1960). Transit reaction was the term Slater-Hammel used to describe reactions that occur in coincidence timing activities, where the performers task is to produce a response that coincides with another event. The task used by Slater-Hammel will serve as an example. Subjects watched a sweep hand of a clock and attempted to release a signal key, that was being held down with the fingers of the right hand, at the same time as the clock hand reached a target pointer. The transit reaction time, or what Slater-Hammel referred to as the "refractory threshold period", was estimated from catch trials, where the sweep hand stopped before it reached the target pointer. When this occurred, subjects were supposed to inhibit the key release.

The results of this experiment showed that when the sweep hand stopped 70 to 90 ms before reaching the target pointer, subjects almost always incorrectly initiated the key release. The ability to inhibit the key release increased when the sweep hand stopped more than 100 ms before the target, and at 170 ms subjects were almost always able to inhibit the key release. Slater-Hammel argued that the refractory threshold value should be defined as the catch trial interval were subjects correctly inhibited the key release on 50% of the trials. Using this criterion, 140 ms was obtained for the refractory estimate. Because the catch trial procedure caused subjects to delay the release time past the required target pointer on control trials by an average of 26 ms, this bias was added to the 140 ms value, yielding a final estimate of 166 ms. Slater-Hammel also measured simple reaction times of his subjects and obtained a mean value of 221 ms. Thus, transit reactions were 55 ms faster than the visual reaction time. These results demonstrated that the temporal delay for using visual information to influence a motor response could be shorter than typical visual reaction times.

Whiting, Gill, and Stephenson. Also interested in the intermittent nature of information processing, Whiting, Gill, and Stephenson (1970) attempted to isolate critical time intervals for using visual feedback in a ball catching task. Whiting et al. had subjects attempt to catch a ball, and experimentally manipulated the portion of the flight that the ball was visible. This was accomplished by placing a light inside a translucent ball and projecting the ball inside a totally dark room. The ball was made visible for the initial 100, 150, 200, 250, or 300 ms of ball flight and the light was then extinguished prior to the catching attempt. There was also a condition where the ball remained illuminated during the entire trajectory of approximately 400 ms.

The results indicated that the percentage of catches increased as the illumination period increased. These results are in contrast to what would be predicted if a typical visual reaction time delay of 200 ms was operating. Figure 1.2 displays the actual catching performance from Whiting et al. (1970), and the predicted performance assuming a 200 ms processing delay. The predicted curve indicates that catching performance should improve when the ball can be seen for longer periods of time, until 200 ms before the catch.

Vision during the last 200 ms of flight would not improve performance because of the processing delay, and, therefore, performance should level off.

Whiting et al. attempted to account for these findings by suggesting that subjects used varying strategies for catching depending on the visual manipulation. They suggested that when the illumination time was short (300 ms or less), subjects may have moved the hand forward and caught with a snatching movement. In contrast, when the entire flight path was visible subjects may have moved the hand backward, adding up to 100 ms of viewing time for the full vision condition. There was, however, little evidence for this post-hoc explanation. The task was performed in an entirely dark room and the experimenter, presumably, was unable to see the strategies used by the subjects.

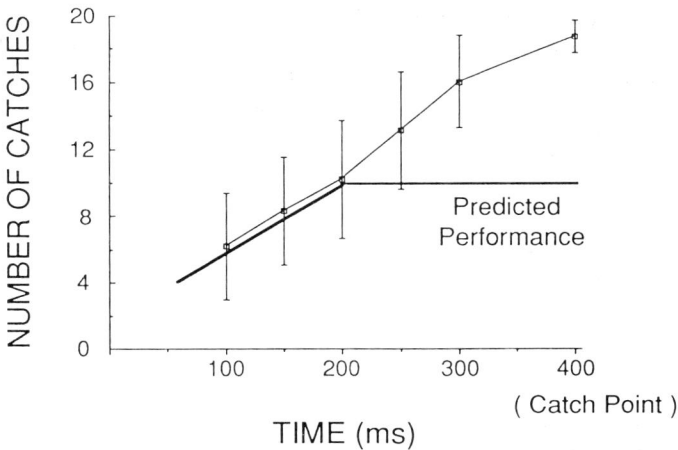

Figure 1.2. The results from Whiting, Gill, and Stephenson (1970, Table 1) demonstrating that the number of ball catches increased when subjects could see the ball for longer periods of time, all the way up to the catch point. The error bars represent 1 standard deviation. The thick line represents expected performance changes if subjects were not able to use visual information from the last 200 ms of ball flight. The predicted performance level is arbitrary and is plotted to correspond to the actual performance for this experiment. (Reprinted with permission).

Even if the explanation of Whiting et al. (1970) is accepted, the data still provide evidence for rapid visual processing. Post hoc statistical tests showed that seeing the first 300 ms of flight resulted in a statistically greater number of catches than when the subjects only saw the first 250 ms of the trajectory. This corresponds to a time of 100 to 150 ms before the catch, and still provides evidence of rapid visual processing.

Beggs and Howarth. At about the same time as Whiting et al. were studying the intermittent nature of visual processing in catching activities, Beggs and Howarth (e.g., Beggs & Howarth, 1970, 1972; Beggs, Sakstein, & Howarth, 1974), were examining the intermittent nature of visual processing in manual aiming. One of their experiments (Beggs & Howarth, 1970) examined time delays in processing visual feedback using a somewhat different method than previous investigators. Subjects moved a hand held stylus from a position near their shoulder, attempting to strike a vertical line on a wall about 61 cm in front of them. Rather than eliminate visual input at the start of a movement (e.g. Keele & Posner, 1968) or for a series of movements (e.g., Woodworth, 1899), Beggs and Howarth eliminated vision when the hand was at various distances from the target. The initial part of the trajectory was performed with vision, and the room lights were extinguished as the hand approached the target. Beggs and Howarth reasoned that the accuracy in hitting a target should depend on whether visual information is eliminated farther from the target than the distance corresponding to a corrective reaction time. That is, if vision is removed when the hand is less than one corrective reaction time from the target, its removal should have little effect on aiming accuracy. Eight different velocities were used ranging from 184 to 12 strokes per minute, and for three of the intermediate speeds, 144, 125, and 100 strokes per minute, eliminating vision at distances of 16.5, 14.5, and 9 inches (41.9, 36,8, 22.9 cm) or closer, respectively, had no effect on accuracy. These distances corresponded to times of 295, 290 and 285 ms, which lead Beggs and Howarth to conclude that the time delay in using visual information was 290 ms.

Some of the data from the Beggs and Howarth (1970) experiment, however, suggest that subjects used visual information much more rapidly. When subjects moved at a rate of 85 strokes per minute, error rates decreased with longer periods of vision, up to the point where vision was eliminated at slightly less than 3 inches (7.6 cm) from the target. This distance corresponded to a time of 165 ms. While Beggs and Howarth considered the corrective reaction time estimate of 165 ms to be an anomalous result, it is possible that short processing time estimates emerge only when there is an optimal combination of time remaining to impact and the pick up of critical error information.

Rapid Visual Processing: Feedback in Manual Aiming

The late 1960s and the 1970s were marked by renewed interest in the study of motor learning and control in psychology and merging interests with the physiology of motor control (e.g., Stelmach, 1976). A major issue was the role of closed-loop feedback processes (e.g., Adams, 1971, 1977) and centrally stored motor commands or motor programs (e.g., Schmidt, 1975), for the control of movement. One of the primary arguments used by proponents of motor program theory was that feedback processing delays are too long for feedback to be used to control rapid motor actions (e.g., Schmidt, 1976). Estimates of processing delays were important, therefore, because these estimates provided a time limitation on closed-loop feedback control.

A number of experimental strategies have been used to provide evidence for rapid processing of visual feedback during manual aiming. For purposes of this discussion, these will be grouped into two general methods. One method incorporates the recording of response dynamics. If vision influences movement accuracy, it follows that the processing latency must be shorter than the response duration. Therefore, it should be possible to directly estimate visual processing time as the time interval between visual feedback becoming available and changes in response dynamics related to the visual feedback. A second method, similar to that used by Woodworth (1899), Vince (1948), and Keele and Posner (1968), is to measure movement accuracy while manipulating visual feedback. In most studies the visual manipulation is produced at the start of the movement (exceptions include Beaubaton & Hay, 1986; Elliott & Allard, 1985) and the shortest response duration where the vision manipulation influences accuracy is the processing estimate. This second method is indirect in that there is no attempt to directly determine the point in time when movements were influenced by the feedback manipulation.

Processing Estimates Derived from Changes in Response Dynamics

The visual processing estimates of Woodworth (1899), Vince (1948), and Keele and Posner (1968), assumed that increased spatial accuracy was a function of visual information picked up at the initiation of the movement. That is, if visual processing time is estimated as the shortest movement time where vision improves aiming accuracy, the visual information used must come from the very start of the movement. Feedback control models used to explain the way vision is used to guide movements, however, have argued that movements are made up of two (e.g., Woodworth, 1899) or possibly more submovements (e.g., Keele, 1968; Meyer et al., 1990), with visual feedback from the first submovement used to "home" in on the target or make movement amendments. This suggests that the first submovement must be nearly completed before visual feedback would aid movement accuracy. In support of this, there is evidence that movement kinematics associated with the initial portions of even short duration movements are poorly related to response outcome (Carlton, Newell, & Carlton, 1984).

Carlton (1981) argued that the procedure used by Keele and Posner (1968) overestimated visual processing time in two ways. First, if the visual information used to make response amendments did not become available until some portion of the response had been completed, the time needed to obtain the error information should to be subtracted from the processing estimate. Second, a finite period of time must elapse between the onset of the amendment and contact with the target area, and this time should also be subtracted from the estimate. Using a visual barrier to block vision of the initial portion of movements, Carlton (1981) found that visual feedback of the last 25% of the movement distance was most critical for reducing spatial errors (cf. Bard, Hay, & Fleury, 1984). Movement kinematics were then obtained using high speed cinematography techniques under conditions where the initial 75% or 93% of the movement was not visible. Visual processing time was estimated as the time between the hand becoming visible and the time of a visually based discrete correction, available from the response kinematics.

Figure 1.3 shows an example velocity and acceleration pattern of the stylus. The movement was initiated at time 0.0, first accelerating and then decelerating as the hand approached the target. At 250 ms into the movement the hand and stylus were in a position above and short of the target. By this time the stylus was visible to the subject and had nearly stopped, its velocity approaching 0.0. At 290 ms into the movement the stylus accelerated toward the target and made contact 60 ms later. Subtracting the time that the stylus became visible (169 ms) from the time of the correction (290 ms), provides an estimate of visual processing time for this trial (121 ms). The mean error correction time using this procedure was 135 ms and this estimate was independent of whether the initial 75% or 93% of the movement was produced without visual feedback.

The assumption that kinematic changes near the target reduced error and were based on visual feedback was important, and there was good reason to believe that it was correct. Feedback control models of aimed movements (Crossman & Goodeve, 1963/1983; Keele, 1968) postulated multiple submovements identifiable by discrete kinematic changes based on feedback information. Carlton (1979) had previously demonstrated that aimed hand movements under moderate index of difficulty conditions (see Fitts, 1954, for a discussion of index of difficulty) were characterized by discrete kinematic changes when the hand was near the target. It was also known that spatial accuracy at moderate index of difficulties was dependent on visual feedback. Jeannerod (1988) has pointed out that secondary movements near the target need not be triggered by vision, but can occur in the absence of visual feedback. However, the low error rates (mean < 5%) produced by subjects (Carlton, 1981), even though the task had a fairly high index of difficulty (5.65 bits), indicates that vision was used to facilitate movement accuracy.

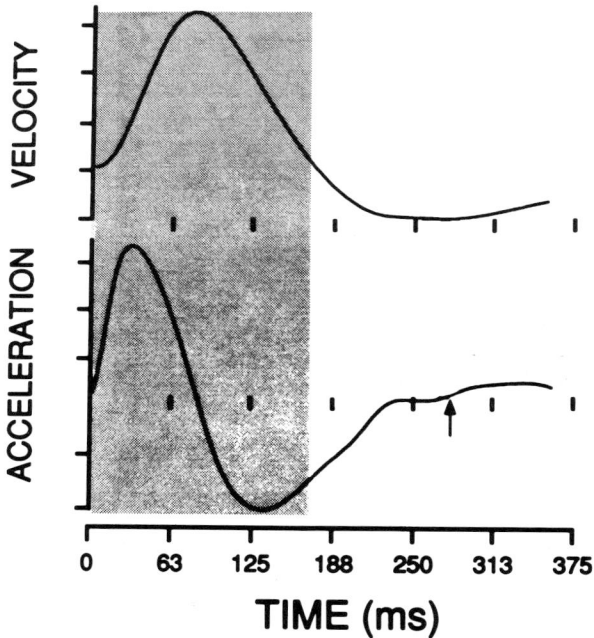

Figure 1.3. The velocity and acceleration profile of the stylus during an aimed movement where the initial 93% of the movement was made without visual feedback (from Carlton, 1981, Figure 5). The shaded area represents the portion of the movement that the subject could not see, and the arrow marks the time of the movement correction (290 ms). (Reprinted with permission).

A similar processing time estimate was obtained by Cordo (1987). Subjects produced isometric torque on a handle that resulted in the movement of a cursor on a graphic screen. This technique, while not producing movement of the hand as in reaching, has the advantage of being able to easily manipulate feedback by controlling when the image on the graphic screen is seen. In addition, the response kinetics are directly available from the torque transducer used to control the cursor.

In Experiment 1 vision was eliminated at the start of torque production or was available throughout the response. Visual feedback was found to decrease errors in torque production 233 ms after response initiation. The torque adjustments were demonstrated to be visually mediated, because responses produced without visual feedback did not show this characteristic. In order to obtain an estimate for visual processing time, Cordo argued that the initial portion of the force response, during which visual feedback was not useful for making error corrections, should be subtracted from the 233 ms time of adjustment. This time was estimated from a second experiment where only the initial portion of the torque trajectory was seen. The results of this experiment indicated that visually based corrections did not occur when less than 60 ms of the initial torque output was seen. Longer periods of initial vision (61 to 120 ms) resulted in error reducing response corrections. The difference between the mean time of correction (233 ms) and the 61 to 120 ms needed to obtain functional visual feedback, produced a visual processing estimate of approximately 110 to 170 ms.

Estimates Derived from Changes in Movement Accuracy

Keele and Posner (1968) found that visual processing estimates could be lowered by a small change in experimental procedure - using discrete rather than reciprocal movements. In the past 10 years researchers, using other procedural changes, have provided evidence suggesting that visual processing time can be as short as 100 ms. Visual manipulations are made either by eliminating, perturbing or distorting visual feedback and examining errors in aiming accuracy. Depending on the visual manipulation, the accuracy measure may reflect response bias, variability, or some general measure of overall accuracy (e.g. absolute error, RMS error, radial error, dispersion).

Vision-No Vision Manipulations. Zelaznik et al. (1983) manipulated vision and measured error rates in an attempt to replicate and extend the findings of Keele and Posner (1968). Zelaznik et al. argued that the long estimates of visual processing time by Keele and Posner (1968) might have been due to two methodological factors. One factor was that visual feedback availability was uncertain in Keele and Posner's experiment. Subjects did not know in advance of movement initiation whether the light illuminating the apparatus would remain on or be extinguished. Keele and Posner (1968) had designed their experiment this way deliberately so that subjects could not use different strategies under the two conditions. Zelaznik et al. also hypothesized that the use of a specific movement time goal by Keele and Posner might cause subjects to focus more on timing accuracy than on spatial accuracy.

Zelaznik et al. manipulated visual feedback certainty and movement time constraint as part of a systematic set of four experiments examining visual processing time. As in Keele and Posner's experiments, subjects used a hand-held stylus to make aimed movements to a target and vision (lights on or off) was manipulated at movement initiation. Instead of measuring spatial accuracy using a hit or miss measure with a fixed size target, as Keele and Posner had done, movements were made to a point target and the errors for individual trials were measured. This measurement technique allowed for the separation of end point errors with respect to the principal direction of movement and errors perpendicular to this plane (designated as distance and direction errors, respectively).

The results from these experiments indicated that movements completed with vision were more accurate with movement times as low as 100 ms (Experiment 1). Knowing that the lights would be on resulted in increased spatial accuracy in comparison to trials where the lights were on, but randomly presented with trials where the lights were off. However, trials with the lights on were always more accurate than when the lights were off, independent of feedback certainty or the movement time goal manipulation. Unfortunately, specific comparisons at each of the movement times were not conducted, as Keele and Posner (1968) had done. Zelaznik et al. noted that every subject produced less error in the vision condition as compared to the no vision condition when feedback was certain. While this was true when errors were averaged over all movement times, it was not true at *each* of the movement times. In the certain visual feedback condition, the mean distance error was smaller in the no vision condition at a movement time of 140 ms and the mean direction error was also smaller for the no vision condition at a movement time of 160 ms (from Zelaznik et al., 1983, Tables 1 and 2; Preset movement time conditions). Under uncertain visual feedback conditions, spatial errors were lower *without* vision with movement times as long as 200 ms for distance errors (Preset movement time), and 180 ms for direction errors (Prompted movement time).

Inconsistent benefits of vision (Carlton & Carlton, 1991) have also been observed in our laboratory. Using a wide range of movement conditions, we have seen lower spatial error scores for movements produced without vision with movement times as long as 225 ms, even though subjects knew whether or not vision would be available. Why visual feedback does not facilitate movement accuracy for some conditions with relatively long movement times is not known. It is possible that it is a statistical phenomena. Given a large number of conditions, and the rather small reduction in errors attributable to vision when movement durations are short, it is possible that, due to probability, mean spatial accuracy for a group of trials will occasionally be greater without vision.

Visual Feedback Distortion. Another strategy that has been used to estimate visual processing time is to distort rather than eliminate vision.

Subjects typically produce movements under various visual manipulation conditions, movement duration is varied, and spatial end-point accuracy is measured. Visual information has been distorted by having subjects look through prisms, which distorts both hand and target position, (e.g. Elliott & Allard, 1985; Smith & Bowen, 1980), or by delaying visual feedback using video technology (Smith & Bowen, 1980), which only distorts feedback of the hand because the target does not move.

Both feedback delay and prism distortion techniques were used by Smith and Bowen (1980). Using a television camera, monitor, and a two-surfaced mirror, Smith and Bowen were able to position a reflected television image of their aiming apparatus so that it lined up exactly with the actual image. Thus, subjects in the experiment viewed their hand indirectly, and made aimed movements using a hand-held stylus. Using a video disk unit, the image from the television camera was presented either in real time or it was delayed by 66 ms. This manipulation resulted in subjects performing under a normal visual feedback condition and a delayed feedback condition. In a third condition the entire visual field was displaced by combining the real time video image with a pair of prism goggles worn by the subject.

It was anticipated that movements of short duration would not be influenced by delaying the visual feedback because the target position was not displaced, and there would not be sufficient time for the delayed visual feedback to influence accuracy. As movement duration increased, however, the visual feedback delay should cause subjects to overshoot the target. The opposite results were predicted when the visual field was displaced by prisms. With short movement durations spatial error should be large because the target is displaced and insufficient time is available to use feedback information to guide the movement to the correct spatial location. The data from this experiment closely paralleled these predictions, with one exception. Subjects tended to overshoot the target when visual feedback was delayed, even when mean movement durations were as short as 164 ms. Estimating the processing time as the shortest movement duration where vision influences spatial accuracy, leads to a visual processing estimate of 164 ms. Because visual feedback was not actually available until 66 ms into the movement, Smith and Bowen argued that the correct visual processing estimate should be about 100 ms (movement duration [164 ms] minus the feedback delay [66 ms]). While the visual representation of the hand did not move until 66 ms after movement initiation, subjects did receive feedback immediately from kinesthetic inputs. As a result, there was a mismatch between the visual and kinesthetic inputs at the initiation of the movement and this may have lead to the overshooting seen with short movement durations. Taking this into consideration, the more conservative estimate (164 ms) seems more appropriate.

Mixed Support for Rapid Visual Processing. The studies by Zelaznik et al. (1983) and Smith and Bowen (1980) provide the strongest *indirect* evidence for rapid visual processing in aimed movements, while others have provided mixed support. Elliott and Allard (1985, Experiment 2), for example, found that subjects could use visual feedback to reduce errors while wearing prism goggles when the movement duration averaged about 170 ms. A subsequent experiment (Experiment 3) was conducted where the lights on / off manipulation occurred 80 ms after movement initiation. Subjects performed with and without prisms and always saw the first 80 ms of the movement. The results (Elliott & Allard, 1985) revealed no differences in spatial error between the lights on and lights off conditions when the visual manipulation time was 140 ms or less, but smaller errors were produced when the lights were on when the vision manipulation time was approximately 170 ms. Interestingly, when subjects performed the experiment without prisms, movements produced with vision had smaller spatial errors for the shortest vision manipulation times (less than 141 ms and 141-200 ms) but larger spatial errors when the vision condition had 200 to 300 ms of visual feedback.

Inconsistent increases in accuracy with vision were also found by Beaubaton and Hay (1986). For very short duration movements (110-150 ms) constant error of target end-points was reduced with complete vision or when movement of the hand was visible for the last 50% of the movement distance, as compared to when only the end point of the movements were known (knowledge of results). Because movements with an approximate mean time of 130 ms are more accurate with vision of the last 50% of the response, it could be argued that the visual processing time estimate should be about 65 ms. This estimate is based on the assumption that the mean time between vision becoming available and the time of movement completion was 65 ms, or half the total movement time. Not all of the evidence supported rapid visual processing, however. When the movement duration ranged between 190 and 230 ms subjects provided only with end-point information were as accurate as when they could see the entire movement.

While the above studies suggest that visual processing latencies can be quite short, the inconsistent benefit of vision across movement times is troubling. Inconsistent findings have occurred in a number of experiments (Beaubaton & Hay, 1986; Carlton & Carlton, 1991; Elliott & Allard, 1985; Zelaznik et al., 1983) and is not isolated to any one accuracy score (e.g. absolute error, RMS error, dispersion), or any one laboratory. In addition, other studies (e.g., Christina, 1970; Wallace & Newell, 1983) have not found evidence of short visual processing times using aimed movements. Because differences in spatial errors between vision and no vision conditions are small at short movement durations, typically ranging from 1 to 3 mm, the non-beneficial effect of vision could be due to statistical chance. Alternatively, subjects may use different strategies under different visual conditions, and these strategies may change with the time available to use vision.

Rapid Visual Processing: Changes in the Environment

Evidence is mounting that visual information can also be used rapidly to control movements when there are changes in the external environment. In contrast to the studies in the previous section examining the processing of response produced feedback during aimed movements, this section focuses on visual processing delays associated with changing environmental conditions. This task grouping is similar to the distinction that has been made between the role of vision as feedback and feedforward (e.g., Beaubaton & Hay, 1986), and the open-closed task classification of Poulton (1957). Support for short visuo-motor delays to a changing environment comes from three lines of investigation. These include studies examining the control of arm trajectories when the target is displaced, studies examining temporal delays in responding to perturbed flight trajectories in ball games, and studies estimating visuo-motor delays in naturally occurring tasks involving time-to-contact.

Changing Spatial Movement Goals

One strategy for examining the control of aimed responses, and the role that visual information about the position of the target plays in these movements, is to eliminate or move the target prior to, at, or some time after movement initiation. The first situation, changing the goal of the movement before movement initiation, is usually done in experiments studying the psychological refractory period. The psychological refractory period represents an increase in the length of time taken to respond to the second of two stimuli presented closely in time (Telford, 1931). This situation typically results in processing times longer than normal reaction times, especially when the second stimulus occurs before the movement corresponding to the first stimulus is initiated (e.g., Davis, 1959). Although these delays do not always occur (Georgopoulos, Kalaska, & Massey, 1981; Soechting & Lacquaniti, 1983), latencies to the second stimuli have not been reported to be shorter than simple reaction times.

When the second stimulus occurs near or after the start of movement, there is some evidence that the reaction to the second stimulus can be shorter than typical reaction times. Pélisson, Prablanc, Goodale, and Jeannerod (1986) as well as Prablanc and Pélisson (1990) have argued that updated information about target position derived from retinal and extraretinal signals is *immediately* used to fine-tune targeted hand movements. In these experiments subjects pointed at targets located 30, 40 or 50 cm from a "home" position. The subject's head was fixed, and the initial movement to the target was an eye movement. On some trials the target was displaced to a new position 10%

farther to the right when the first saccade reached maximum velocity. The end of the first saccade, and the observation of the new target position, occurred at about the same time as the initiation of the hand movement.

Analysis of constant errors indicated that subjects shifted their pointing responses to the right when the target was moved. Movement times, however, were not increased in comparison to trials where the target was not displaced, and there were no indications of movement corrections from the kinematic profiles of the arm trajectory. Pélisson et al. used these results to argue that visual information was used to update the hand trajectory *immediately*. While it is a given that there is some processing delay due to neural conduction of impulses, the electro-chemical delay in muscle activation, and the excitation-mechanical delay in force production, the lack of any clear changes in the response kinematics makes it impossible to estimate what the processing time was. One way to make the size of the kinematic change larger would be to increase the size of the second target jump, but this manipulation leads to a normal reaction time (Prablanc & Pélisson, 1990).

Recent evidence from an experiment with cats (Alstermark, Gorska, Lundberg, & Pettersson, 1990) indicates that movement trajectories can be amended in the direction of a displaced target with a latency of approximately 100 ms. After training the cat to reach for food in a lighted tube, the illumination of the initial tube was turned off shortly after movement initiation on some trials, and a second tube was illuminated. The cats were able to modify the movement pattern and reach toward the second tube 100 ms after the target was changed.

Similar results have recently been reported with humans subjects as well (Paulignan, MacKenzie, Marteniuk, & Jeannerod, 1990). Paulignan et al. found that shifting the location of an object to be grasped at the onset of movement shortened the time to peak acceleration, even though the time of peak acceleration in control trials was only about 100 ms. Although Paulignan et al. used these data to suggest that visual information could be used to abort the initial movement in 100 ms, the amplitude of peak acceleration did not decrease, which is inconsistent with an early reduction of acceleration toward the target. A corrective acceleration to the new target did not take place until 180 ms after the target was moved.

The 100 ms processing estimate by Paulignan et al. is considerably faster than those obtained by Georgopoulos et al. (1981) and Soechting and Lacquaniti (1983) using a similar paradigm with rhesus monkeys and humans, respectively. Both Georgopoulos et al. (1981) and Soechting and Lacquaniti (1983) concluded that reactions to perturbed targets were as long as typical reaction times, but a close examination of the data, particularly those from

Georgopoulos et al., suggests shorter processing delays. Georgopoulos et al. (1981) used the time of direction reversal to estimate the time of movement amendment even though it was clear that changes in the velocity profiles occurred much earlier. Figure 1.4 (redrawn from Georgopoulos et al., Figure 8) shows the mean velocity profile when the visual stimulus was not perturbed, and superimposed over this, mean velocity profiles when the target was changed near the time of movement initiation. It appears from these data that the velocity pattern deviated from the control profile approximately 150 ms after the target was moved to a new location. The movement reversal, defined as the movement away from the initial target toward the second target, did not occur until after a typical reaction time (approximately 250 ms). Like the results of Paulignan et al. (1990) the initial response kinematics were adjusted quickly, but corrective movements to the new target took considerably longer.

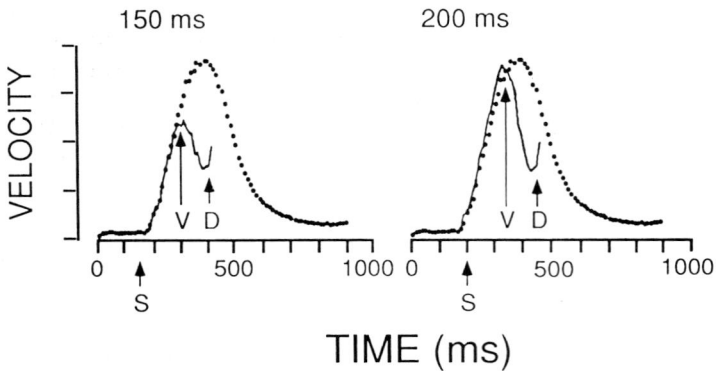

Figure 1.4. The results from Georgopoulos, Kalaska, and Massey (1981). The dotted line represents the hand velocity curve when a single target was presented. The curves represented by solid lines are the averaged velocity profiles when a second target was presented 150 or 200 ms after the initial target position was displayed. The time of the initial stimulus is time zero (0 ms); "S" marks the time of the second stimulus; "V" is the point where the velocity profile begins to deviate from the single target velocity pattern; and, "D" is the time when the hand changed movement direction. (Reproduced with permission.

Other studies that have used visual cues near movement initiation to specify a change in the movement goal (e.g., Carlton & Carlton, 1987; Henry

& Harrison, 1961) have not reported particularly fast processing time estimates. Carlton and Carlton (1987) did observe some short latency changes in electromyographic activity of the initial movement on some trials, but this did not occur consistently over trials.

Perturbing Projectiles in Ball Games

McLeod (1987) questioned the generalizability of rapid processing estimates observed with reaching and aiming activities to whole body actions. To examine this, McLeod observed the temporal delay between perturbing the path of a pitched ball and movement amendments of batsmen in the sport of cricket. It was assumed that highly skilled performers, having had years of practice at matching the flight trajectory of the ball to movements of the bat, would be able to respond to changes in the flight trajectory with minimal latency. In this experiment the ball was projected to a point in front of the batsmen so that it bounced 100 to 400 ms before being hit. Small diameter dowels were placed in the ball bounce area, parallel to the flight of the pitched ball. On trials where the ball hit one of the dowels, the ball bounced "inside" toward the batsman or "outside" away from the batsman. The time to process the visual information from the unexpected change in the path of the ball was estimated as the time between ball bounce and the change in the normal swing path of the bat as measured with the aid of high speed cinematograph. Using this procedure McLeod found no evidence of changes in the spatial characteristics of the swing in times under 200 ms.

There are a number of reasons why more rapid corrections might not have been produced by subjects. The first, as McLeod suggests, is that processing delays to changes in external stimuli are only as fast as normal reaction times. A second possible explanation is related to the task and the corrections required in the experiment. When the ball hit one of the dowels it bounced either in-toward or out-away from the performer. Because the bat was moving down along the flight path, a change in bat direction was required to compensate for the change in ball trajectory. This modification would necessitate a large reorganization of the motor command, leading to a long delay before the corrective action could begin (Schmidt, 1976). A third possibility is that because of the large mass of the bat, its inertia was hard to overcome, leading to long correction delays. A further possibility is that the movement recording technique, filming at 72, 83 or 200 frames / s and making the assessment of response changes from position records, may have led to an overestimation of the time of correction.

In order to demonstrate that short visuo-motor delays can occur in whole body actions, we (Carlton, Carlton, & Kim, 1991) have conducted a similar experiment using skilled tennis players. The court surface was altered by placing either smooth tape or a rough textured surface on selected portions of the court so that approximately 5% of the balls bounced unexpectedly fast (in the smooth tape condition) or unexpectedly slow (in the rough surface condition). The required correction was to accelerate or decelerate the swing, depending on the condition, and hit the ball over the net and into the court. The racquet was instrumented with a three dimensional accelerometer allowing for a comparison of the acceleration profiles for normal trials where the ball bounced on the regular court surface, and trials where the ball was perturbed.

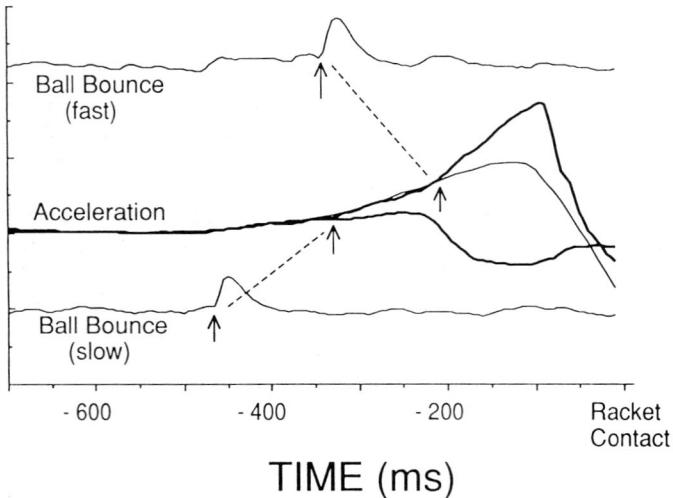

Figure 1.5. The results from Carlton, Carlton, and Kim (1991). The acceleration curves produced when the ball bounced unexpectedly fast or slow are compared to the average control acceleration pattern, that is, when the ball bounced on the normal court surface (dotted line). The time of ball bounce and the time of the change in the acceleration curves are marked with an arrow, and are linked together with a dashed line.

Figure 1.5 shows two example trials from one subject, one trial where the ball hit a smooth surface and bounced faster than normal and one trial where the ball hit a rough surface and bounced slower than normal. The acceleration profiles along the major direction of racquet movement for these experimental trials were compared to the acceleration pattern when the ball landed on the normal court surface. The acceleration patterns were very similar until a short time after the ball contacted the ground. The fast ball bounce resulted in a rapid increase in acceleration and the slow bounce resulted in a reduction in acceleration. Preliminary results from this experiment (n=2) indicate that subjects are able to process the visual information and amend these responses with a mean time of 150 to 190 ms depending on the subject and the condition.

Estimates of visuo-motor delay and τ

David Lee (Lee, 1976; Lee & Reddish, 1981) has argued that vision is used to determine the time-to-contact between an actor and objects in the environment through the optic variable τ, where τ is the inverse of the rate of dilation of the object on the retina. Lee, Young, Reddish, Lough, and Clayton (1983) argued that movements are not geared directly to τ, but to a time one visuo-motor delay before time-to-contact. Lee et al. (1983) had subjects jump up and strike balls that were dropped from different heights and showed that the subject's knee and elbow angle patterns were very similar when plotted against the value of τ at short times earlier than time-to-contact. The visuo-motor delay was estimated to be between 50 and 135 ms depending on the subject. This range of times is similar to the estimated visuo-motor delay of 60 ms for diving gannets (Lee & Reddish, 1981), and 105 to 156 ms for the forehand drive in table tennis (Bootsma & Van Wieringen, 1990). The experiments by Lee et al. (1983) and Bootsma and Van Wieringen (1990) did not directly manipulate visual information, and it has not been demonstrated that disruption of visual information just prior to the estimated visuo-motor delay alters performance patterns. If changes in the performance pattern did occur, this would provide a useful validation of the technique used to estimate the visuo-motor delay.

While a number of experiments have provided evidence that movements can be altered in times less than a simple visual reaction time, others have not. It is somewhat surprising that the spread of processing estimates is so great for different experiments, and for different individuals within an experiment. Using similar techniques and tasks, Lee et al. (1983) and Bootsma and Van Wieringen (1990) estimated processing delays ranging from 50 to 156 ms. Some of the factors that may influence the visual processing time estimate will be explored in the following section.

A Constant Visual Processing Delay-
or a Multi-Factor Phenomenon?

Over the last 10 years there has been increasing evidence that visual processing times can be shorter than simple visual reaction times, independent of whether the visual information is in the form of response produced feedback or changes in the external environment. Estimates of visual processing time, however, have been far from consistent across experiments. This is partly due to the different measurement techniques that have been used, but, more importantly, it is because visual processing time for movement control is not constant, just as simple reaction times are not constant over a range of stimulus, response, and subject variations. While one specific research focus has been to determine what the *minimum* processing time is for the control of movement, more generally, the factors that determine visual processing delays are at issue.

What determines visual processing latencies? Is the visuo-motor delay associated with perception-action coupling in naturally occurring activities (e.g., Bootsma & Van Wieringen, 1990) fundamentally different from visual processing delays associated with response amendments to target perturbations (e.g., Paulignan et al., 1990)? Are the benefits of vision in rapid aiming tasks (e.g., Zelaznik et al., 1983) attributable to the same visuo-motor processes that allow for rapid discrete corrections in aimed movements to a target (e.g., Carlton, 1981)? A number of studies have attempted to determine temporal processing delays for a particular task or activity, but there has been little attempt to isolate variables that influence processing delays.

One exception is the hypothesis that correction latencies are a function of the type of error that is produced. Schmidt (1976) purposed that errors in response selection, where the performer selects an incorrect response for accomplishing the goal of the task, leads to the reprogramming of the movement and results in a processing delay of at least one reaction time. Errors in response execution, where the correct response is selected but there is an error in its execution, can be corrected more rapidly. This distinction, however, does not account for the findings from a number of experiments on visual processing including studies on manual aiming (e.g., Carlton, 1981; Zelaznik et al., 1983) and the prehension experiment by Paulignan et al. (1990) where the object to be grasped moved at response initiation.

In contrast to suggestions that correction latencies can be attributed to a single characteristic (e.g., Schmidt, 1976), a variety of factors may interact to determine both the way that vision is used and how rapidly it can be processed for movement control. These factors include characteristics of the task and the nature of movement amendments, if any, that are required, and may also include characteristics of the performer as well. For example, a number of experiments examining visual processing have used highly practiced subjects (e.g., Bootsma & Van Wieringen, 1990; Carlton & Carlton, 1991; McLeod, 1987) and the assumption is that there is a direct and rapid perception-action coupling that develops with practice. Less practiced subjects might be expected to have longer visual processing delays.

While there have been few systematic attempts to determine the factors that contribute to rapid visual processing, a number of speculations can be made. Visual information may be used with short latencies in aimed movements, for example, because there is redundant information about limb position from other sources. Kinesthetic inputs as well as information about motor commands in the form of an "efference copy" provide information about limb movement (McCloskey, 1981), and information about the type of response adjustment that might be required. By knowing the time when visual information will be available, and by knowing the probable response adjustment that needs to be made, visual information may be used rapidly. This explanation does not account for short processing latencies found with changes in the external environment (e.g., Paulignan et al., 1990). But even under these conditions, the temporal placement of the response modification stimulus is often known. This is in contrast to reaction time tasks where the time of stimulus presentation is typically uncertain.

Another factor that could account for the variety of processing time estimates that have been observed is the size of the error that must be corrected. Large errors may require a fundamental reorganization of the movement to accomplish the task goal, whereas small adjustments may only require lower level neural processes. This has been demonstrated for the correction of limb perturbations (Carlton, 1983), and recent evidence suggests that it may be the case for visual processing as well (Prablanc & Pélisson, 1990). Displacing the target position during aimed hand movements, Prablanc and Pélisson (1990) found that response correction latencies were long - approximating a reaction time - when the target was displaced more than 15 to 20 % of the movement distance. They suggested that small adjustments could be accomplished more rapidly using quick internal loops to "fine-tune" the motor output. This distinction could explain the rather long processing estimate by McLeod (1987) where response modifications were often large, and the short visuo-motor delays associated with tasks requiring movements to be coupled to time-to-contact.

The nature of the required correction will also influence the processing delay (e.g. Carlton & Carlton, 1987; Quinn & Sherwood, 1983). Corrections that require additional activation of an already active muscle occur more quickly than corrections requiring deactivation of an active muscle and the activation of a new muscle (Carlton & Carlton, 1987). Similarly, further accelerating an already accelerating movement takes less time than slowing it down (Carlton et al., 1991).

At present, the evidence that visual processing delays vary systematically as a function of these variables is only circumstantial. In fact, it is not clear whether the various experimental manipulations used to study visual processing delays are examining a single visual-motor control process. It is probably incorrect to assume that there is a single visual processing latency for the control of movement, but that visual processing delays depend on the specific interaction of the information available and the nature of the response correction that is required. Current work in our laboratory is examining how action constraints influence visual processing and motor control.

References

Adams, J.A. (1971). A Closed-loop theory of motor learning. *Journal of Motor Behavior, 3*, 111-150.

Adams, J.A. (1977). Feedback theory of how joint receptors regulate the timing and positioning of a limb. *Psychological Review, 84*, 504-523.

Alstermark, B., Gorska, T., Lundberg, A., & Pettersson, L.-G. (1990). Integration in descending motor pathways controlling the forelimb in the cat. 16. Visually guided switching of target-reaching. *Experimental Brain Research, 80*, 1-11.

Bard, C., Hay, L., & Fleury, M. (1984). Role of peripheral vision in the directional control of rapid aiming movements. *Canadian Journal of Psychology, 39*, 151-161.

Beaubaton, D., & Hay, L. (1986). Contribution of visual information to feedforward and feedback processes in rapid pointing movements. *Human Movement Science, 5*, 19-34.

Beggs, W.D.A., & Howarth, C.I. (1970). Movement control in man in a repetitive motor task. *Nature, 221*, 752-753.

Beggs, W.D.A., & Howarth, C.I. (1972). The accuracy of aiming at a target: Some further evidence for a theory of intermittent control. *Acta Psychologica, 36*, 171-177.

Beggs, W.D.A., Sakstein, R., & Howarth, C.I. (1974). The generality of a theory of intermittent control of accurate movements. *Ergonomics, 17*, 757-768.

Bootsma, R.J., & Van Wieringen, P.C.W. (1990). Timing an attacking forehand drive in table tennis. *Journal of Experimental Psychology: Human Perception and Performance, 16*, 21-29.

Broadbent, D.E. (1958). *Perception and communication.* London: Pergamon Press.

Carlton, L.G. (1979). Control processes in the production of discrete aiming responses. *Journal of Human Movement Studies, 5*, 115-124.

Carlton, L.G. (1981). Processing visual feedback information for movement control. *Journal of Experimental Psychology: Human Perception and Performance, 7*, 1019-1030.

Carlton, L.G. (1987). The speed and accuracy of movements as a function of constraints to action. In L. S. Mark, J. Warm, & R. L. Huston (Eds.),

Ergonomics and human factors: Recent research (pp. 103-109). New York: Springer-Verlag.

Carlton, L.G., & Carlton, M.J. (1987). Response amendment latencies during discrete arm movements. *Journal of Motor Behavior, 19*, 227-239.

Carlton, L.G., & Carlton, M.J. (1991). [Vision and the speed-accuracy relation]. Unpublished raw data.

Carlton, L.G., Carlton, M.J., & Kim, K.H. (1991). [Visual processing time with changing environmental conditions]. Unpublished raw data.

Carlton, M.J. (1983). Amending movements: The relationship between degree of mechanical disturbances and outcome accuracy. *Journal of Motor Behavior, 15*, 39-62.

Carlton, M.J., Newell, K.M., & Carlton, L.G. (1984). Predicting individual discrete response outcomes from kinematic characteristics. *Journal of Human Movement Studies, 10*, 63-82.

Christina, R.W. (1970). Minimum visual feedback processing time for amendment of an incorrect movement. *Perceptual and Motor Skills, 31*, 991-994.

Cordo, P.J. (1987). Mechanisms controlling accurate changes in elbow torque in humans. *Journal of Neuroscience, 7*, 432-442.

Craik, K.J.W. (1948). The theory of the human operator in control systems: II. Man as an element in a control system. *British Journal of Psychology, 38*, 142-148.

Crossman, E.R.F.W., & Goodeve, P.J. (1983). Feedback control of hand-movement and Fitts' Law. Paper presented at the meeting of the Experimental Psychology Society, Oxford, July 1963. Published in *Quarterly Journal of Experimental Psychology, 35A*, 251-278.

Davis, R. (1959). The role of "attention" in the psychological refractory period. *Quarterly Journal of Experimental Psychology, 11*, 211-220.

Donders, F.C. (1969). On the speed of mental processes. In W.G. Koster (Ed. & Trans.), *Attention and performance II.* Amsterdam: North-Holland. (Original work published 1868).

Elliott, D. & Allard, F. (1985). The utilization of visual feedback information during rapid pointing movements. *Quarterly Journal of Experimental Psychology, 37A*, 407-425.

Fitts, P.M. (1954). The information capacity of the human motor system in controlling the amplitude of movement. *Journal of Experimental Psychology, 47*, 381-391.

Fitts, P.M., & Peterson, J.R. (1964). Information capacity of discrete motor responses. *Journal of Experimental Psychology, 67*, 103-112.

Georgopoulos, A.P., Kalaska, J.F., & Massey, J.T. (1981). Spatial trajectories and reaction times of aimed movements: Effects of practice, uncertainty, and change in target location. *Journal of Neurophysiology, 46*, 725-743.

Hancock, P.A., & Newell, K.M. (1985). The movement speed-accuracy relationship in space-time. In H. Heuer, U. Kleinbeck, & K.H. Schmidt (Eds.), *Motor behavior: Programming control and acquisition* (pp. 153-188). Berlin: Springer-Verlag.

Henry, F.M., & Harrison, J.S. (1961). Refractoriness of a fast movement. *Perceptual and Motor Skills, 13*, 351-354.

Jeannerod, M. (1988). *The neural and behavioural organization of goal-directed movements.* Oxford: Clarendon Press.

Keele, S.W. (1968). Movement control in skilled motor performance. *Psychological Bulletin, 70*, 387-403.

Keele, S.W., & Posner, M.I. (1968). Processing of visual feedback in rapid movements. *Journal of Experimental Psychology, 77*, 155-158.

Lee, D.N. (1976). A theory of visual control of braking based on information about time-to-collision. *Perception, 5*, 437-459.

Lee, D.N., & Reddish, P.E. (1981). Plummeting gannets: A paradigm of ecological optics. *Nature, 293*, 293-294.

Lee, D.N., Young, D.S., Reddish, P.E., Lough, S., & Clayton, T.M.H. (1983). Visual timing in hitting an accelerating ball. *Quarterly Journal of Experimental Psychology, 35A*, 333-346.

McCloskey, D.I. (1981). Corollary discharges: Motor commands and perception. In V.B. Brooks (Ed.), *Handbook of Physiology, Vol. 2, The Nervous System* (pp. 1415-1447). Baltimore: American Physiological Society.

McLeod, P. (1987). Visual reaction time and high-speed ball games. *Perception, 16*, 49-59.

Meyer, D.E., Smith, J.E.K., Kornblum, S., Abrams, R.A., & Wright, C.E. (1990). Speed-Accuracy tradeoffs in aimed movements: Toward a theory of rapid voluntary action. In M. Jeannerod (Ed.), *Attention and performance XIII* (pp. 173-226). Hillsdale, NJ: Erlbaum.

Paulignan, Y., MacKenzie, C.L., Marteniuk, R.G., & Jeannerod, M. (1990). The coupling of arm and finger movements during prehension. *Experimental Brain Research, 79*, 431-435.

Pélisson, D., Prablanc, C., Goodale, M. A., & Jeannerod, M. (1986). Visual control of reaching movements without vision of the limb. II. Evidence of fast unconscious processes correcting the trajectory of the hand to the final position of a double-step stimulus. *Experimental Brain Research, 62*, 303-313.

Poulton, E. C. (1957). On prediction in skilled movements. *Psychological Bulletin, 54*, 467-478.

Prablanc, C., & Pélisson, D. (1990). Gaze saccade orienting and hand pointing are locked to their goal by quick internal loops. In M. Jeannerod (Ed.), *Attention and performance XIII* (pp. 653-676). Hillsdale, NJ: Erlbaum.

Quinn, J.T., & Sherwood, D.E. (1983). Time requirements of changes in program and parameter variables in rapid ongoing movements. *Journal of Motor Behavior, 15*, 163-178.

Schmidt, R.A. (1975). A schema theory of discrete motor skill learning. *Psychological Review, 82*, 225-260.

Schmidt, R.A. (1976). Control processes in motor skills. *Exercise and Sport Sciences Reviews, 4*, 229-261.

Slater-Hammel, A.T. (1960). Reliability, accuracy and refractoriness of a transit reaction. *Research Quarterly, 31*, 217-228.

Smith, W.M., Bowen, K.F. (1980). The effects of delayed and displaced visual feedback on motor control. *Journal of Motor Behavior, 12*, 91-101.

Soechting, J.F., & Lacquaniti, F. (1983). Modification of trajectory of a pointing movement in response to a change in target location. *Journal of Neurophysiology, 49*, 548-564.

Stelmach, G.E. (1976). *Motor control: Issues and trends*. New York: Academic Press.

Telford, C.W. (1931). The refractory phase of voluntary and associative responses. *Journal of Experimental Psychology, 14*, 1-36.

Vince, M.A. (1948). Corrective movements in a pursuit task. *Quarterly Journal of Experimental Psychology, 1*, 85-103.

Wallace, S.A., & Newell, K.M. (1983). Visual control of discrete aiming movements. *Quarterly Journal of Experimental Psychology, 35A*, 311-321.

Welford, A.T. (1952). The psychological refractory period and the timing of high-speed performance - A review and a theory. *British Journal of Psychology, 43*, 2-19.

Whiting, H.T.A., Gill, E.B., & Stephenson, J.M. (1970). Critical time intervals for taking in flight information in a ball-catching task. *Ergonomics, 13*, 265-272.

Woodworth, R.S. (1899). The accuracy of voluntary movement. *Psychological Review, 3*, (Monograph Supplement), 1-119.

Zelaznik, H.N., Hawkins, B., & Kisselburgh, L. (1983). Rapid visual feedback processing in single-aiming movements. *Journal of Motor Behavior, 15*, 217-236.

Acknowledgment

I would like to thank Mary Carlton, Paul Cordo and Karl Newell for helpful comments on an earlier draft of this manuscript.

VISION AND MOTOR CONTROL
L. Proteau and D. Elliott (Editors)

34

CHAPTER 2

INTERMITTENT VERSUS CONTINUOUS CONTROL OF MANUAL AIMING MOVEMENTS

DIGBY ELLIOTT

Motor Behaviour Laboratory, Department of Physical Education
McMaster University, Hamilton, Ontario, L8S 4K1

Over the course of a normal day, the average person makes hundreds of limb movements to objects and positions in space. If the movement requires spatial precision, vision will usually play an important role in both the preparation and control of the movement. While continuous visual guidance of the limb is best for optimal performance (e.g., "look the football into your hands"), it is also possible to perform some movements with reasonable accuracy when sampling task-related visual information intermittently. It is not difficult for example, to insert a tape into the tapedeck of your automobile while trading glimpses between the tape, tapedeck and the road. I have colleagues who have no trouble locating their bottle of beer on the bar and bringing it to their mouth, while devoting most of their visual attention to a hockey game on the television. The point is, that although vision is important for limb control, continuous visual monitoring of the limb and movement environment is not always necessary.

This chapter deals with the issue of visual intermittency and the control of arm and hand movements. Most of the research discussed comes from our own laboratory, where we have been attempting to determine the mechanisms and/or information processing events that might account for our ability to perform reasonably precise movements, while sampling the movement environment intermittently.

After developing a historical context for our work, I review the experiments that have led us to suggest that a brief visual representation of the movement environment may, at times, provide an adequate replacement for more direct visual information. I argue that this representation allows us, in ome situations, to move accurately to targets in space even when vision is occluded for a short period of time. This representation of the movement environment allows for a certain degree of movement continuity even when visual sampling of the movement environment must be intermittent.

Intermittent Visual Pickup

Historically, investigators have attempted to examine the role of vision in the control of limb movements by manipulating its presence/absence during the movement (e.g., Keele & Posner, 1968; Woodworth, 1899). In many studies, the major goal has been to determine the time it takes an individual to use visual feedback about the position of the limb in space in order to improve the accuracy of the movement (see Carlton this volume). While recent experiments indicate that visual feedback can be processed very rapidly (Carlton, 1981; Elliott & Allard, 1985; Smith & Bowen, 1980; Zelaznik, Hawkins, & Kisselburgh, 1983), as Thomson (1983) has pointed out, equally impressive is the degree of accuracy that can be achieved when vision of the limb and target are eliminated upon movement initiation.

When one considers typical closed-loop explanations of limb control (e.g., Beggs & Howarth, 1970; Stubbs, 1976), the degree of precision that can be achieved without vision is all the more surprising. Specifically, most closed-loop models of limb control posit that visual, kinesthetic and feedforward information about limb position must be compared to visual information about target position in order to detect and correct error inherent in the initial movement impulse (see Elliott, 1990 and Elliott, Calvert, Jaeger, & Jones, 1990 for discussion). Although non-visual sources of information about limb position remain when vision is occluded, the elimination of visual target information would seem to obviate this process, since there would be no basis for evaluating the effectiveness of the limb movement (i.e., its position relative to the target). This of course assumes that visual information about target position is eliminated when vision of the target is occluded. Our work indicates that this assumption does not appear to hold. Specifically, it appears that a visual representation of the movement environment, including the target, persists for a brief period of time following visual occlusion. This representation can be used to guide limb movements when direct visual contact with the environment is prevented. It may be this representation of the movement environment that allows a movement to proceed with a degree of continuity even when direct visual contact with the environment is intermittent.

Our first evidence for this type of representation came from a study in which we (Elliott & Madalena, 1987) used an experimental protocol developed by Thomson (1983) to examine the visual control of locomotion (see Corlett this volume for a review of locomotion work). In our experiments, we (Elliott & Madalena, 1987) had subjects aim with a stylus at targets a short distance away (25-35 cm) under fast and slow movement time instructions. As is typical, subjects aimed at the targets in a condition of full visual information (i.e., room lights on), and in a condition in which the room lights were extinguished upon movement initiation. Following Thomson's (1983) procedure, we also included conditions in which the room lights were extinguished prior to movement initiation and remained off until the movement was completed. In our first experiment, subjects indicated to the experimenter their readiness and then the lights were extinguished for 2, 5 or 10 s before they were given the signal to point. As is evident in Figure 2.1, subjects performed better when vision was available than when it was eliminated upon movement initiation, particularly when moving slowly. Of greater interest was the deterioration in performance if subjects were required to sit in the dark for as little as 2 s prior to movement initiation (Experiment 1). In a second experiment, we were able to replicate this 2 s delay effect employing a between-subjects design, suggesting that it is not the result of asymmetric transfer between vision conditions (see also Holding, 1968).

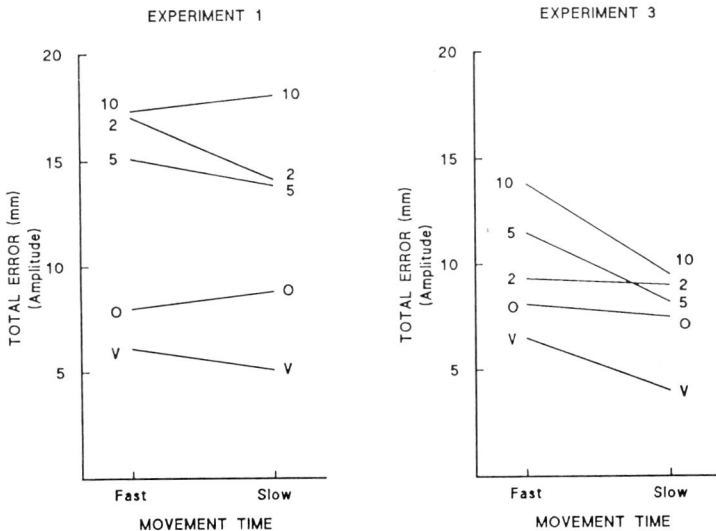

Figure 2.1. Total amplitude error (mm) in Experiment 1 (no target) and Experiment 3 (phosphorescent target) as a function of movement time (fast = 200-300 ms, slow = 400-500 ms) and vision condition (V = vision, O = no lights with O-s delay, 2 = no lights with 2-s delay, 5 = no lights with 5-s delay, 10 = no lights with 10-s delay; adapted from Elliott & Madalena, 1987).

Following the logic used by Thomson (1983) in his locomotion studies, we suggested that the relatively accurate aiming in the lights off upon movement initiation condition resulted from the persistence of target information over the duration of the movement. Spending 2 s in the dark prior to aiming however, allowed time for the target information to deteriorate resulting in decreased opportunity for on-line error reduction during the movement and therefore more inaccuracy.

In a third experiment (Elliott & Madalena, 1987) we were able to greatly reduce the 2 s delay effect by making the target visible when other visual information was eliminated (Figure 2.1). This procedure reduced the subjects' dependence on any mental representation of target position. Experiment 3 then not only supported our suggestion that a short-lived representation of the movement environment may be useful for the control of limb movements when vision is occluded, it also suggested that the representation may be visual in nature.[1]

In two subsequent experiments (Elliott, 1988), in which we manipulated vision of both the stylus and the target we were able to replicate our findings.[2] As well, we demonstrated that visual information about the location of the limb is most useful when there is also visual information about the target position available. Visual information about the target position can be direct, or contained in the brief representation identified in our earlier work (e.g., Elliott & Madalena, 1987; see Carlton, 1981 and Proteau & Cournoyer, 1990 for a discussion of the relative importance of limb and target information).

If a visual representation of the movement environment can be useful for motor control when it contains relatively accurate information about the location of a target, it follows that it will be detrimental to performance when it provides inaccurate or imprecise information. Armed with this logic, we (Elliott et al., 1990, Experiment 1) attempted to reverse the 2 s delay effect (Elliott & Madalena, 1987) by using a visual rearrangement procedure. Employing the same type of aiming task, subjects were required to complete 64 trials in which they aimed with a stylus to a single target 35 cm in front of them (movement time range: 400-500 ms). This practice was completed with

---

[1] It is interesting that for movements performed with the eyes closed, Vince (1948) actually showed increases in error with movement time when the movement times were exceedingly long. Perhaps in these situations a representation of the target's position in space had had an opportunity to decay before the subject reached the target area.

[2] The 2 s no vision delay effect has also been replicated in studies concerned primarily with learning (Elliott & Jaeger, 1988) and manual asymmetries in movement control (Roy & Elliott, 1989).

full vision, and was designed to provide the subjects with a fairly stable representation of the target position.

Following the practice, subjects were required to wear goggles containing a 25 diopter wedge prism with a variable base. It was at this point that our visual manipulations were introduced. Specifically, subjects pointed to the target under conditions of full vision or vision was eliminated 0, 2 or 10 s prior to movement initiation. The difference between this study and our other work (Elliott, 1988; Elliott & Madalena, 1987) was that prior to a trial the base of the prism was rotated randomly to one of 8 possible orientations. Thus the subject was always uncertain as to the direction in which his/her vision was displaced (approximately 15°). Given this situation the strategy adopted by subjects was to aim for the perceived target location.[3]

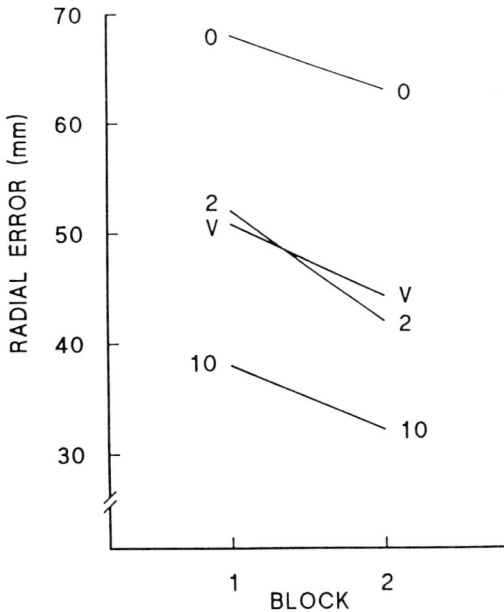

Figure 2.2. Radial error (mm) as a function of trial block and vision condition (adapted from Elliott et al., 1990).

---

[3]Even though the subjects had moved to the actual target location 64 times, the visual capture phenomenon (see Posner, Nissen, & Klein, 1976) is very compelling and subjects prepare their movement for the perceived, as opposed to actual, target location.

In the full vision condition, it was expected that subjects would visually detect their aiming error early in the movement, and at least to some degree correct the error (Elliott & Allard, 1985). In the three no vision conditions, it was expected that the degree of error would depend on the stability of the inaccurate representation of target position, since presumably as visual information became less salient, subjects would depend more and more on the representation developed for the movement during pre-experimental training. As is evident in Figure 2.2, this is exactly what happened. That is, the lights off upon movement initiation condition was associated with the most error with a large decrease in error following only a 2 s no vision delay.

In this study, there appeared to be further deterioration in the representation of the target position between 2 and 10 s suggesting that either target information may be maintained at several levels in the nervous system, and/or that at least some information in the representation decayed more slowly than we had anticipated it would (cf. Elliott, 1986, 1987; Elliott, Jones, & Gray, 1990). In any event, the study supported our claim that in the absence of direct visual information, subjects may use a visual representation of the environment to guide their limb movements. The representation may be useful or detrimental to performance depending on the quality of the information it contains.

Although some information about the movement environment persists for longer than 2 s after visual occlusion, our experiments to this point indicated that the majority of decay occurred very rapidly. This lead us to wonder if, what has been traditionally termed, iconic memory (Neisser, 1967) had a role to play in our target aiming effects. Because one of the characteristics of iconic memory is temporal integration (Coltheart, 1980), we (Elliott et al., 1990, Experiment 2) attempted to determine if the hypothesized representation subjects were using to guide their movements could be disrupted by a visual pattern mask. To achieve this end, we projected small targets (+) onto a screen in front of the subject (150 ms) and then either presented a visual pattern mask (200 ms) consisting of horizontal, vertical and diagonal lines, or a blank field. As in previous studies the subjects task was to hit the target with a hand held stylus. Subjects were more accurate in the no mask condition than when the target was followed immediately by the pattern mask. Presumably, this is because in the former situation target information was available to the subject longer (i.e., 150 ms plus duration of icon) than when the mask was introduced (150 ms). Thus it would appear that iconic memory may have some role to play in guiding movement when direct visual contact with the movement environment is prevented.

Encouraged by these findings, we decided to conduct an additional 'iconic memory' experiment employing a variation of Sperling's (1960) partial-full report procedure (Elliott et al., 1990, Experiment 3). In this study, we used a slide projector to present three potential targets to the subject. The target slide was presented for 60 ms, and the targets were located in a random position in each of three horizontally aligned rows (top, middle, bottom). At stimulus onset asynchronies of 0, 200, 400, 600, 800 or 1000 ms a high, medium or low pitched tone sounded designating the particular target for the trial. The subjects' task as always was to contact the target location (i.e., where the target had appeared) with the stylus. The experiment then was designed to determine the duration of the representation previously identified, since at longer stimulus onset asynchronies it was assumed that the representation would have decayed to a greater extent. As a control, we included a situation in which subjects were cued as to the target prior to a trial, so that the tone provided only the signal to point. In this situation, it was possible for the subject to prepare the movement upon target onset and simply wait for the tone before executing the movement.

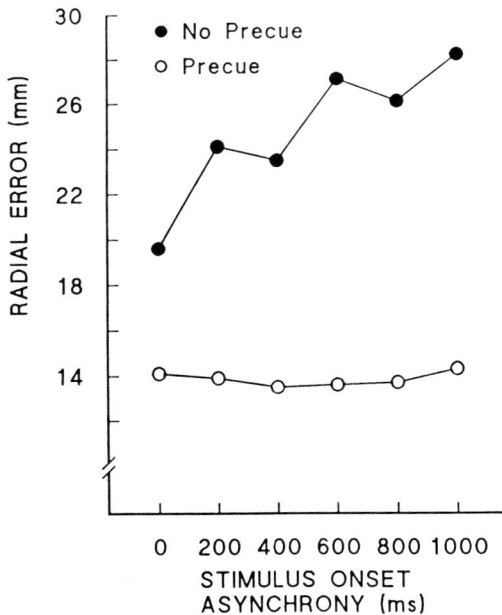

Figure 2.3. Radial error (mm) as a function of stimulus onset asynchrony (ms) and precue condition (adapted from Elliott et al., 1990).

Although the accuracy of subjects' pointing diminished with stimulus onset asynchrony, as is evident in Figure 2.3, there was no discrete point at which there were large increases in error. In fact even at 1000 ms target-aiming error seems to be increasing.[4] These findings indicate that as well as iconic memory (i.e., Elliott et al., 1990, Experiment 2), a longer more slowly decaying representation may be involved.

While our research to this point made it clear that subjects can use some sort of representation of a target position to guide their movements, we were uncertain as to whether the representation contains information about the whole movement environment, or only a single target location. In order to determine what subjects remember we decided to manipulate target uncertainty at the time of visual occlusion. In an initial experiment then, we (Elliott & Calvert, 1990) employed a reaction time protocol in which, following an auditory signal, subjects were required to make a rapid movement to a target either to the left or right of their midline. Subjects performed this task in both simple and choice reaction time conditions. In the former situation subjects were precued as to the target for that trial, while in choice conditions the frequency of the auditory signal specified the target. The target aiming was done under four different visual conditions: full vision, lights off upon movement initiation, lights off coincident with the auditory signal and lights off 2 s prior to the auditory signal. The idea was that in the simple situation subjects always had an opportunity to remember a specific target location, while in some choice conditions the target was known only after vision was eliminated. Presumably, in these situations the subject would need to encode the whole visual layout.[5]

The radial error findings in this study (see Figure 2.4) demonstrated the independence of uncertainty and vision condition. It would appear then that while there are some overall advantages associated with knowing the target in advance, subjects do have information available to them about more than one target for a brief period of time following visual occlusion. If this were not the case, one would expect error in the choice - lights off upon movement imperative condition to be as large as in the 2 s delay conditions.[6]

---

[4]Relatively long and variable reaction times and movement times probably tend to obscure stimulus onset asynchrony effects (see Elliott et al., 1990 for discussion).

[5]Another possibility is that subjects prepare two movements. We would have a stronger case for our line of thinking if we had used more than two targets.

[6]These basic findings were replicated in a second study in which we employed a cost-benefit paradigm (Elliott & Calvert, 1990).

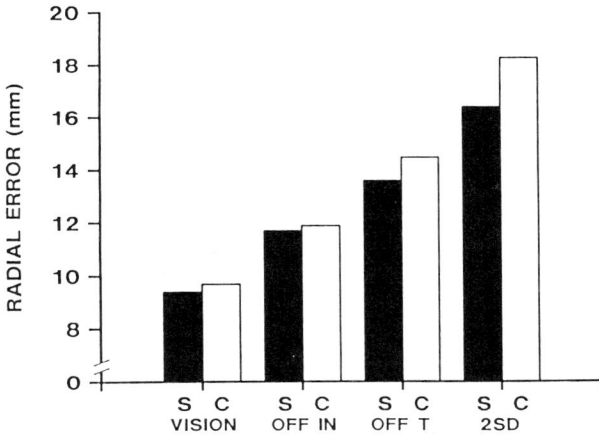

Figure 2.4. Radial error (mm) as a function of vision condition and response type (OFF IN = lights off upon movement initiation, OFF T = lights off upon tone, 2SD = 2 s no-vision delay; S = simple reaction time, C = choice reaction time; adapted from Elliott & Calvert, 1990).

One interesting aspect of this work was the importance of vision during the reaction time interval, even when the target was known in advance (i.e., simple reaction time situation). This finding is at odds with models that posit a large degree of prior preparation in situations in which the target is known in advance (e.g., Klapp, 1977). Moreover, accuracy differences between the lights off upon movement initiation conditions and the lights off with tone conditions indicate that a great deal of information about the movement environment decays quite rapidly, since most reaction times in this experiment were between 250 and 350 ms.

In summary then our studies to date indicate that a visual representation (or representations) of the movement environment may in some situations provide an adequate substitute for direct visual contact with the environment. While the majority of information useful for motor control decays very rapidly, there appears to be further loss of spatial information for up to about 10 s. Thus while iconic memory may have some role to play in guiding movement when vision is occluded for very brief periods of time, a less precise, more slowly decaying memory representation may have a role to play as well.

Continuous Limb Control

To this point all the work that I have discussed has been concerned with the role of vision in determining movement outcome, specifically movement error. In our more recent work, we (Elliott, Carson, Goodman, & Chua, in press) have also collected kinematic data in order to gain more insight into how error might be minimized.

In view of our work on intermittent visual pickup, it is interesting that many models of limb control have included some sort of intermittent error detection and correction mechanism. For example, multiple correction models of limb control (e.g., Crossman & Goodeve, 1963/1983; Keele, 1968) posit that target-aiming movements are made up of a series of submovements. The function of each submovement is to reduce error in the previous submovement. Alternatively, it has been proposed that only a single correction occurs (Beggs & Howarth, 1970; Carlton, 1979; Howarth, Beggs, & Bowden, 1971) near the end of the movement. In either case, discrete adjustments to the movement trajectory are based on the comparison of visual information about limb and target position.

Although a number of studies have demonstrated that discrete changes in target-aiming trajectories do occur (Carlton, 1979; Jagacinski, Repperger, Moran, Ward, & Glass, 1980; Langolf, Chaffin, & Foulke, 1976), very little work has been done to examine influence of vision on the kinematics of the movement. Consequently, we (Elliott, Carson, Goodman, & Chua, in press) decided to essentially repeat the Elliott and Madalena (1987) experiments, this time collecting both performance and kinematic data.

In an initial experiment, subjects aimed at a target 40 cm in front of them under both "fast" and "accurate" instructional sets. On some trial blocks, subjects performed with full vision, while on other blocks the lights were extinguished upon movement initiation. As well, we included a 2 s no vision delay condition in which subjects were required to sit in the dark prior to aiming. We used a 3-dimensional digitizing system (WATSMART), so that following a trial we were able to reconstruct the actual movement trajectory. While we examined a number of kinematic variables, of primary interest was the continuity of the movement trajectory. Specifically, we examined discontinuities in the acceleration profile or "zero-crossings" (see Brooks, 1974 and Crossman & Goodeve, 1963/1983). These zero-crossings are subsequent accelerations after primary acceleration and deceleration has already occurred, and are thought to reflect discrete adjustments to the movement (see Elliott et al., in press).

As we expected our error results were similar to those found in the Elliott and Madalena study (Experiment 1). Specifically, subjects performed with less error in the full vision condition than in the lights off upon movement initiation condition, particularly when the instructions stressed accuracy (i.e., long movement times). As well, there was further increase in error if subjects were required to sit in the dark 2 s before moving. Once again, we have suggested that this is because a representation of the movement environment has had time to at least partly decay.

Of primary interest in this study was the relationship between the accuracy and the number of discrete corrections or "zero-crossings". Following the predictions of traditional closed-loop error reduction models of limb control, we expected to find more zero-crossings in the full vision condition than in the two no vision conditions, especially when accuracy was stressed (cf. Jeannerod, 1984). Moreover, we expected to find more zero-crossings in the lights off upon movement initiation condition than in the 2 s delay condition. Our notion was that in the former situation error could be reduced by comparing feedforward and kinesthetic information about limb position with a visual representation of target position. In the 2 s delay condition, however this opportunity would be diminished because of the decay of the representation.

Our predictions in this study were not realized. For the fast instruction set, zero-crossings were in evidence on less than 12% of the trials and there were no differences between the three vision conditions. When accuracy was stressed, while discrete adjustments were made about four times as frequently, more zero-crossing actually occurred in the 2 s delay condition than in the full vision and lights off upon movement initiation conditions. Moreover, within-subject correlational analyses demonstrated that presence of zero-crossings did not predict accuracy in any of the vision or instructional conditions. Thus, while vision, and perhaps a visual representation of target position, have a large impact on movement error, increased accuracy does not appear to be achieved by making discrete adjustments to the movement trajectory.[7]

Because movement accuracy is influenced by the presence of visual information, but appears to be independent of the number or frequency of discrete adjustments to the movement trajectory, we (Elliott et al., in press)

---

[7]In a recent study in which subjects move a cursor to a target on a computer monitor by rotating their wrists, Meyer, Abrams, Kornblum, Wright, and Smith (1988) found no difference in the frequency of submovements (i.e., secondary changes acceleration) between conditions in which the cursor was visible and occluded. Incorporating this finding into a probabilistic account of error generation in manual aiming, they suggested that discrete adjustments are simply more accurate when vision is available. There are however a number of specifics of their model that are at odds with findings from more traditional target aiming experiments (e.g., Elliott, Carson, Goodman, & Chua, in press; Jeannerod, 1988).

have suggested that in some situations the visual control of aiming may be continuous, or at least pseudo-continuous. Thus rather than providing the basis for a corrective impulse, vision may tune the primary acceleration and deceleration impulse, by adjusting the gain on muscle activity. Alternately, vision may provide the basis for many overlapping discrete adjustments to the movement, thus giving the trajectory the appearance of continuity (i.e., pseudo-continuous). In either case modification of the trajectory proceeds extremely rapidly (see Carlton this volume). Perhaps this is accomplished by preparing peripheral levels of the nervous system in advance for environmental or feedback-based events which in turn control the intensity of muscle activity. This peripheral tuning process may even allow for relatively continuous control when target positions are perturbed (Goodale, Pélisson, & Prablanc, 1986; Prablanc, Pélisson, & Goodale, 1986). Certainly a system with these characteristics is quite different from one in which error inherent in one impulse is calibrated during an initial movement and corrected with second or third impulse (e.g., Meyer et al., 1988).

In proposing a more continuous mode of visual motor control, we are not denying that discrete adjustments also occur. Presumably the system can only be tuned in advance to operate within certain boundaries. Errors in preparation that take the movement outside of those boundaries, or the occurrence of unanticipated environmental events may require relatively large, discrete changes to the trajectory.[8] Preparation of these corrective impulses will be more time consuming (e.g., Keele, 1968) than the type of continuous control we have proposed (Elliott et al., in press).

Intermittent Visual Pickup and Continuous Motor Control
In the first section of this chapter, I made the case that in some conditions, intermittent visual pickup may be sufficient for reasonable movement accuracy. This situation exists because a brief representation of the movement environment serves as a temporary substitute for continuous visual contact with the environment. Presumably this representation can be used in conjunction with kinesthetic and feedforward information about limb position to provide aiming movements with continuity even when vision is occluded for a brief period of time. It was this issue of continuity that was the focus of the second part of the chapter.

Taken together our findings suggest relatively continuous motor control may proceed with intermittent visual pickup. The frequency of visual

---

[8]The majority of studies reviewed in this chapter have examined relatively simple aiming movements requiring only two-dimensional accuracy. With more spatially complex movements, secondary changes in acceleration (i.e., zero crossings) seem to be more important in determining end-point accuracy (Carson, Chua, Goodman, & Elliott, 1991).

sampling must be such that the internalized representation of one sample has not had a chance to decay when the next sample is taken. Interestingly enough, the only kinematic difference that we (Elliott et al., in press) have found between lights on and lights off upon movement initiation conditions has been the time spent following peak velocity. Specifically, when the instructional set dictated a long movement time, subjects spent less time following peak velocity in the lights off condition. It is our contention that in this situation subjects may be trying to complete the movement before the representation of the target position has had an opportunity to decay.

Although our theoretical framework has the potential to help us understand the performance of many day to day activities that require intermittent visual sampling, to date our experiments have been confined to rather impoverished laboratory aiming tasks. Moreover, our manipulation of vision has been rather crude. Perhaps recent technical advances (e.g., Milgram, 1987) will allow us to overcome these two problems.

Certainly another major shortcoming of our work has been our failure to consider how eye movements might mediate or contribute to our intermittent vision effects (see Abrams this volume and Carnahan this volume). To date, our failure to measure or manipulate eye movements has been strictly pragmatic, and we are attempting to remedy this shortcoming of our work. In a current experiment for example, we are trying to determine whether our proposed representation is coded retinotopically or spatiotopically. Our strategy is to present a target, induce an eye movement and then interfere with either the retinal or spatial location of the target. The results should give us some indication of what sources of information are important in movement situations in which saccades are made in order to sample different areas of the movement environment. This type of experiment will help us understand how we perform those important day to day tasks discussed earlier (i.e., getting the right tape into the tapedeck, finding that bottle of beer on the bar).

References

Beggs, W.D.A., & Howarth, C.I. (1970). Movement control in a repetitive task. *Nature, 225*, 752-753.

Brooks, V.B. (1974). Some examples of programmed limb movements. *Brain Research, 71*, 299-308.

Carlton, L.G. (1979). Control processes in production of discrete aiming responses. *Journal of Human Movement Studies, 5*, 115-124.

Carlton, L.G. (1981). Visual information: The control of aiming movements. *Quarterly Journal of Experimental Psychology, 33A*, 87-93.

Carson, R.G., Chua, R., Goodman, D., & Elliott, D. (1991). [A kinematic analysis of hand asymmetries in a visual aiming task]. Unpublished raw data.

Coltheart, M. (1980). Iconic memory and visual persistence. *Perception & Psychophysics, 27*, 183-228.

Crossman, E.R.F.W., & Goodeve, P.J. (1983). Feedback control of hand-movement and Fitts' Law. Paper presented at the meeting of the Experimental Psychology Society, Oxford, July 1963. Published in *Quarterly Journal of Experimental Psychology, 35A*, 251-278.

Elliott, D. (1986). Continuous visual information may be important after all: A failure to replicate Thomson (1983). *Journal of Experimental Psychology: Human Perception and Performance, 12*, 388-391.

Elliott, D. (1987). The influence of walking speed and prior practice on locomotor distance estimation. *Journal of Motor Behavior, 19*, 476-485.

Elliott, D. (1988). The influence of visual target and limb information on manual aiming. *Canadian Journal of Psychology, 42*, 57-68.

Elliott, D. (1990). Intermittent visual pickup and goal directed movement: A review. *Human Movement Science, 9*, 531-548.

Elliott, D., & Allard, F. (1985). The utilization of visual feedback information during rapid pointing movements. *Quarterly Journal of Experimental Psychology, 37A*, 407-425.

Elliott, D., & Calvert, R. (1990). The influence of uncertainty and premovement visual information on manual aiming. *Canadian Journal of Psychology, 44*, 501-511.

Elliott, D., Calvert, R. Jaeger, M., & Jones, R. (1990). A visual representation and the control of manual aiming movements. *Journal of Motor Behavior, 22*, 327-346.

Elliott, D., Carson, R.G., Goodman, D., & Chua, R. (in press). Discrete vs. continuous visual control of manual aiming. *Human Movement Science.*

Elliott, D., Jones, R., & Gray, S. (1990). Short-term memory for spatial location in goal-directed locomotion. *Bulletin of the Psychonomic Society, 28*, 158-160.

Elliott, D., & Jaeger, M. (1988). Practice and the visual control of manual aiming movements. *Journal of Human Movement Studies, 14*, 279-291.

Elliott, D., & Madalena, J. (1987). The influence of premovement visual information on manual aiming. *Quarterly Journal of Experimental Psychology, 39A*, 541-559.

Goodale, M.A., Pélisson, D., & Prablanc, C. (1986). Large adjustments in visually guided reaching do not depend on vision of the hand or perception of target displacement. *Nature, 320*, 748-750.

Holding, D.H. (1968). Accuracy of delayed aiming responses. *Psychonomic Science, 12*, 125-126.

Howarth, C.E., Beggs, W.D.A., & Bowden, J.M. (1971). The relationship between speed and accuracy of movements aimed at a target. *Acta Psychologica, 35*, 207-218.

Jagacinski, R.J., Repperger, D.W., Moran, M.S., Ward, S.L., & Glass, D. (1980). Fitts' law and the microstructure of rapid discrete movements. *Journal of Experimental Psychology: Human Perception and Performance, 6*, 309-320.

Jeannerod, M. (1984). The timing of natural prehension movements. *Journal of Motor Behavior, 16*, 235-254.

Jeannerod, M. (1988). *The neural and behavioural organization of goal-directed movements.* Oxford: Clarendon Press.

Klapp, S.T. (1977). Reaction time analysis of programmed control. *Exercise and Sport Sciences Reviews, 5*, 231-253.

Keele, S.W. (1968). Movement control in skilled motor performance. *Psychological Bulletin, 70*, 387-403.

Keele, S.W., & Posner, M.I. (1968). Processing of visual feedback in rapid movement. *Journal of Experimental Psychology, 77*, 155-158.

Langolf, G.D., Chaffin, D.B., & Foulke, J.A. (1976). An investigation of Fitts' law using a wide range of movement amplitudes. *Journal of Motor Behavior, 8*, 113-128.

Meyer, D.E., Abrams, R.A., Kornblum, S., Wright, C.E., & Smith, J.E.K. (1988). Optimality in human motor performance: Ideal control of rapid aimed movements. *Psychological Review, 95*, 340-370.

Milgram, P. (1987). A spectacle-mounted liquid-crystal tachistoscope. *Behavior Research Methods, Instruments and Computers, 19*, 449-456.

Neisser, U. (1967). *Cognitive psychology.* New York: Appleton-Century-Crofts.

Posner, M.I., Nissen, M.J., & Klein, R.M. (1976). Visual dominance: An information processing account of its origins and significance. *Psychological Review, 83*, 157-171.

Prablanc, C., Pélisson, D., & Goodale, M.A. (1986). Visual control of reaching movements without vision of the limb: I. Role of retinal feedback of target position in guiding the hand. *Experimental Brain Research, 62*, 293-302.

Proteau, L., & Cournoyer, J. (1990). Vision of the stylus in a manual aiming task: The effects of practice. *Quarterly Journal of Experimental Psychology, 42A*, 811-828.

Roy, E.A., & Elliott, D. (1989). Manual asymmetries in aimed movements. *Quarterly Journal of Experimental Psychology, 41A*, 501-516.

Smith, W.M., & Bowen, K.F. (1980). The effects of delayed and displaced visual feedback on motor control. *Journal of Motor Behavior, 12*, 91-101.

Sperling, G. (1960). The information available in brief visual presentations. *Psychological Monographs, 74*, (Whole No. 11).

Stubbs, D.F. (1976). What the eye tells the hand. *Journal of Motor Behavior, 8*, 43-48.

Thomson, J.A. (1983). Is continuous visual monitoring necessary in visually guided locomotion? *Journal of Experimental Psychology: Human Perception and Performance, 9*, 427-443.

Vince, M.A. (1948). Corrective movements in a pursuit task. *Quarterly Journal of Experimental Psychology, 1*, 85-103.

Woodworth, R.S. (1899). The accuracy of voluntary movement. *Psychological Review, 3*, (Monograph Supplement), 1-119.

Zelaznik, H.N., Hawkins, B., & Kisselburgh, L. (1983). Rapid visual feedback processing in single-aiming movements. *Journal of Motor Behavior, 15*, 217-236.

Acknowledgment

The majority of experiments reviewed in this chapter were supported by the Natural Sciences and Engineering Research Council of Canada.

VISION AND MOTOR CONTROL
L. Proteau and D. Elliott (Editors)

CHAPTER 3

VISUAL FEEDBACK PROCESSING AND MANUAL ASYMMETRIES: AN EVOLVING PERSPECTIVE

RICHARD G. CARSON

School of Kinesiology, Simon Fraser University, Burnaby, British Columbia, V5A 1S6

In spite of the considerable research effort which has been expended in the last decade, the phenomenon of human handedness remains puzzling and poorly explained. Nonetheless, the clarification of performance differences between the hands offers the prospect of insight into more general issues pertaining to the nature of movement regulation.

Visually guided reaching and aiming movements, are an important part of our repertoire of volitional movements. These tasks have also provided the focus for a number of accounts of manual asymmetries. Most studies examining hand differences[1] in these tasks have reported superior right hand accuracy or consistency (Elliott, in press). It has been customary to attribute at least some of this superiority to the processing characteristics of the contralateral (left) cerebral hemisphere. In continuing attempts to account for the right hand advantage, two hypotheses have been dominant. These suggest, respectively, that the left hemisphere/right hand system mediates more precise parameterization of force (e.g., Roy & Elliott, 1986, 1989), or more efficient execution of "error corrections" utilizing sensory information (e.g., Flowers, 1975; Todor & Doane, 1978; Todor & Cisneros, 1985). This chapter provides an account of the genesis and evolution of the latter hypothesis. The empirical work predicated on the "feedback processing" hypothesis is examined and critically discussed. I believe this process is illuminating, not only for the

---

[1]In this paper, any reference to hand advantage is with respect to right handers.

insight it provides into the nature of manual asymmetries but also as it allows us to trace the historical development of the associated explanatory constructs. Although the study of manual asymmetries has the potential to inform us as to the general nature of movement regulation, I will argue that this prospect has yet to be realised. Rather, the reverse has occurred. To date, all accounts of manual asymmetries have been prestructured by extant general theories of motor control.

Simply stated, the "feedback processing hypothesis" proposes that manual asymmetries are a function of the differential efficiencies with which sensory feedback is processed by the hand-hemisphere systems. In particular, the preferred hand is thought to be associated with a neural substrate which may more effectively use sensory information to effect "on-line" modifications required for accurate aiming responses (e.g., Doane & Todor, 1978; Flowers, 1975; Todor & Doane, 1978). This account of manual asymmetries has appeared parsimonious to many theorists, in that, it encapsulates widely accepted information processing characteristics of the left cerebral hemisphere. In terms of one favoured dichotomy, the left hemisphere is allegedly superior for the processing of information in a sequential manner, while the right hemisphere can deal more efficiently with the parallel processing of material (Cohen, 1973).

Preliminary Studies

The view that the functional characteristics of the visual guidance mechanisms may be directly related to asymmetries in the regulation of movement is not without precedent. Trevarthen (1974) observed that in "split brain" patients the perturbation of reaching movements upon the removal of peripheral vision was differentially expressed, depending upon the visual field through which the limb was moving. Trevarthen was led to conclude that "peripheral vision of the movement of either arm is governed, at least in the commissurotomy subject, more by the right hemisphere, while coincidence of eye and hand in fixation of the point target is governed more from the left hemisphere" (p. 253).

However, it is the work of Flowers (1975) which has been seminal in instigating the current program of research in normals. Flowers (1975) hypothesized that there existed during aimed movements a "corrective mode of control", and that it was differences in the efficiency with which this control regime was implemented which accounted for performance differences between the hands. In Flowers' original study, a Fitts' reciprocal tapping task was used as a means of manipulating the required level of control. It was postulated that if between-hand differences were observed to vary as a function of these control demands, it could be concluded that: "the essential dexterity difference between the preferred and non-preferred hands is in the sensory or feedback control of movements rather than in motor function per se" (p. 39). It is, of course, a central assumption of the Fitts' paradigm that the requirement for the use of feedback may be manipulated through the modification of very specific task parameters. Fitts (1954) suggested that the speed with which movements were made was governed by the capacity limits of an individual's motor system and by the "information" required to make any one movement. There is an implicit assumption that for cases in which the amplitude between, and size of targets is fixed, the resultant speed of movement may be taken as a measure of the capacity of the "hand system" to make the response. Flowers (1975) proposed that at low index of difficulty values, individuals are responding with "ballistic" movements, by which Flowers was presumably implying "unmediated by sensory feedback". Whereas, for more "difficult" combinations (high index of difficulty values) it was supposedly necessary for subjects to adopt some form of corrective procedure.

Flowers observed that performance differences between the hands, in terms of movement duration and proportion of target misses, were enhanced at index of difficulty values 4 bits and above, for subjects classified as strongly lateralized. It was therefore concluded that the preferred hand advantage for "non-ballistic" movements was due to the "lower rate of information transmission" for the non-preferred hand. As these experiments failed to elicit indications of asymmetries in movements having index of difficulty values smaller than 4 bits, and for simple tapping movements, it was concluded that the hands exhibited an equivalence for ballistic movements during which transmission of sensory information was considered to be of little consequence.

In a partial replication of the Flowers' study, Todor and Doane (1978) examined performance in conditions in which the required precision was manipulated while the index of difficulty was held constant. In also subscribing to Cohen's (1973) demarcation of the cerebral hemispheres, Todor and Doane (1978) supposed that "the parallel or nonserial processing mode of the right hemisphere may be associated with nonadaptive (i.e., limited feedback usage) movements... usually referred to as being preprogrammed" (p. 295). It was further presumed that the relative contribution of the ballistic or preprogrammed phase of a movement varied inversely with the demands for precision. As such, it was postulated that the right hand would exhibit a superiority in circumstances in which the demands for precision, and thus for feedback processing, were emphasised. Whereas, the left hand would be superior in conditions in which the demands for precision were low. Analysis of performance, assessed in terms of number of "hits" per ten second interval, partially supported these hypotheses. The performance of the preferred right hand was not distinguished by accuracy requirements. The left hand was, however, superior in circumstances requiring less precision (relative to the high precision condition rather than to the right hand).

Thus far, the most thorough examination of manual asymmetries on a Fitts' task has been the study conducted by Todor and Cisneros (1985). In their experiments discrete aiming movements were employed rather than reciprocal tapping (cf. Flowers, 1975; Todor & Doane, 1978). A stylus mounted accelerometer was used to derive movement kinematics. Subsequent examination of acceleration profiles, and the identification of "critical" transitions therein, led the investigators to partition the movement into two distinct stages. These were a ballistic or propulsive distance covering phase, and a "homing in" or "error correction" phase. In holding error rates relatively constant, Todor and Cisneros ensured that the effects of manipulating accuracy demands would be expressed largely in terms of temporal measures.

Their results indicated an overall increase in movement times for the left hand relative to the right, and that this difference was accentuated in response to increasing demands for accuracy. The majority of the accommodation to increased accuracy demands appeared to be accounted for by changes in the duration of the period from peak deceleration to target contact. The magnitude of this portion of the movement was greater for the left hand than for the right, a difference which was again extended for movements requiring greater precision. As such, Todor and Cisneros (1985) concluded that the differing efficiency with which accommodation was made to demands for precision "implicates hand differences in error corrective ability and/or the need for error correction in the terminal phase of the movement" (p. 366).

It is a central tenet of these approaches that feedback is evaluated in terms of some superordinate goal. Similarly, it is implicit that "corrective" procedures are mediated by visual feedback as it is vision which most obviously provides information pertaining to the relative positions of the limb and the target. However, the conclusions which have been drawn regarding the bases of manual asymmetries have been constrained by varying explications of what data obtained from the Fitts' task reveal of the underlying control processes. Indeed, the interpretation that the preferred hand advantage results from an enhanced efficiency in the use of feedback, rests upon the assumption that Fitts' Law may be accounted for in terms of feedback processes in general, and of visual feedback in particular (cf. Abrams, this volume; Abrams, Meyer, & Kornblum, 1990; Meyer, Abrams, Kornblum, Wright, & Smith, 1988; Schmidt, Zelaznik, Hawkins, Frank, & Quinn, 1979).

Direct Manipulations of Visual Feedback

The proposal that the hands differ in the efficiency with which feedback is processed (Flowers, 1975) has recently been explicitly interpreted as a suggestion that the right hand/left hemisphere is superior in the processing of visual information (e.g., Roy & Elliott, 1986). This position has the virtue of being clearly defined and thus amenable to testing.

In attempting to evaluate this hypothesis, Roy and Elliott (1986) employed a discrete aiming task, in which subjects made movements to a single target, in a range of movement times. The rationale was that with reduced movement time there would be less visual information available upon which to base corrections. There is a problem with this reasoning. It has been proposed that impulses of greater magnitude are typified by greater variability of both the duration and the magnitude of the impulse. As variability in the muscular impulses are thought to be a major determinant of accuracy, larger impulses are predicted to lead to reduced terminal accuracy (Abrams, this volume; Abrams et al., 1990; Meyer et al., 1988; Schmidt et al., 1979). Theorists have also suggested that the preferred hand exhibits reduced variability of motor output (Annett, Annett, Hudson, & Turner, 1979) and greater precision of force modulation (Peters, 1980). Roy and Elliott predicted that examination of the speed-accuracy trade-off function should reveal a steeper slope for the left hand as a result of its reduced efficiency of processing visual information. However, any such difference in the slope of the function could equally well reflect a difference in the elevation of impulse variability as movement speed increases. Fortunately, a second level of manipulation was employed. Ambient lighting was removed upon presentation of the stimulus. This was clearly a more direct means of manipulating the visual information available during movement. The authors again predicted that removal of visual information should affect right hand performance to a greater degree than left hand performance which, in line with Todor and Doane (1978), they suppose is dependent on a preprogrammed mode of control.

Roy and Elliott (1986) found that the right hand was more accurate than the left, that accuracy increased from the shortest to the longest movement times and that performances when ambient lighting was present were superior to those when it was removed. As movement speed increased, there was a greater deterioration in accuracy for the non-preferred (left) hand. In terms of the speed/accuracy function, the left hand had a "steeper negative slope". However, the difference in slope between the two hands did not vary between illumination conditions. This was interpreted as an indication that the hand differences were not due to variations in the efficiency with which visual information was processed.

In a second experiment, subjects were required to conduct movements in less than 200 ms. It was reasoned that, during movements of this duration, there would be insufficient time to effect modifications based upon visual information. However, there are indications that, visual feedback "loops" may operate over latencies much shorter than 200 ms (see Carlton, this volume). Therefore the imposition of this movement time constraint is unlikely to have precluded the use of vision as Roy and Elliott suggest. Indeed their analysis of movement times indicated a main effect for hand and a hand by feedback interaction. The right hand advantage for movement time was greatest in the illuminated condition. These data support the view that the right hand/left hemisphere system enjoys a superiority, at least in terms of the *efficiency*, with which visual information is utilised.

In the Roy and Elliott study, the use of the same target and starting location on every trial could have constrained the utility of the visual information present during the movement. Therefore, in a further investigation (Roy & Elliott, 1989), two spatially distinct starting locations were used. However, the same target position was again used on every trial. Three visual conditions were utilized. These were a no-vision condition in which ambient lighting was removed upon movement initiation, a condition in which ambient lighting was removed 10 s prior to movement initiation, and a full vision condition in which ambient illumination was present throughout the movement. The 10 s delay condition will not concern us here (but see Elliott, this volume, for more detail).

Movement time measures failed to differentiate either the hands or the vision conditions. This was also the case for constant error. The hands were distinguished only by variable error, with the left hand exhibiting greater movement endpoint variability than the right. There were no indications of higher order interactions involving hand and vision condition. Indeed only the measure of variable error revealed small differences between the conditions in which vision was present throughout the movement and that in which it was removed at movement initiation. The results of a second experiment, in which subjects were constrained to make movements in less than 200 ms, were also inconclusive. There were again no indications of hand by vision condition interactions.

Did these experiments provide a stringent test of Flowers' (1975) position? In the Roy and Elliott studies, the utility of visual feedback was limited by the use of a single target position. This was exemplified by a failure to demonstrate that vision during the movement was a significant determinant of performance. As such, we might wish to be remain cautious of drawing the conclusion that the hand-hemispheres systems are not differentiated by the processing of visual feedback.

Carson (1989) has postulated that a right hand advantage for the utilization of feedback information during the execution phase of a movement could coexist with a right hand advantage arising from a superiority in movement programming. As such, in circumstances/tasks in which the utility of visual information is reduced, right hand advantages may be attributable to advantages in "movement programming". As a consequence, differences between the hand-hemisphere systems in the processing of sensory feedback may be less prominent. In this situation the task context dictates not so much the relative contribution of the right and left hemispheres as much as it influences the manner in which the contribution of the left hemisphere is expressed. For a task in which the regulatory burden falls largely upon the "programmed" phase of the movement, asymmetries in the execution phase are likely to be less prominent.

In a recent experiment (Carson, Chua, Elliott, & Goodman, 1990) we attempted to provide a further test of the hypothesis that hand advantages are due to differences in the efficiency of utilizing visual feedback. We felt that it was necessary to utilize a task for which we could demonstrate that visual information contributes significantly to the successful completion of the task. Therefore, we employed a reaching task which required that subjects reach into space to acquire one of eight possible target positions presented on a screen. Four visual conditions were employed. In the Full-Vision condition subjects were afforded vision of both the hand and the target throughout the course of the movement. In the Ambient-illumination-Off condition, the room lights were extinguished at movement initiation, thus preventing vision of the moving limb. The target remained illuminated. In the Target-Off condition, the target was extinguished upon initiation of the movement. Ambient illumination and thus vision of the hand remained present. Finally there was a No-Vision condition in which ambient illumination was removed and the target was extinguished upon initiation of the response movement.

The manipulation of vision was found to be a potent determinant of the terminal accuracy and strongly influenced movement time. There was a consistent tendency for movements made under conditions of full vision to exhibit less error than than those made when the target was extinguished upon movement initiation. These movements were themselves more accurate those made when ambient illumination, and thus vision of the limb, was removed. Movements made with ambient illumination removed were, in turn, more accurate than those made without vision.

Although there was a trend toward greater accuracy for movements made with the right hand, there was no indication of an interaction involving hand and visual condition. This would appear to indicate that the hand/hemisphere systems did not differ in the extent to which visual information was utilised. As has been stressed, however, interactions involving movement time measures assume at least equal importance in this context. We are interested in the efficiency in addition to the absolute level of performance of the respective hand/hemisphere systems. The trends for movement time across vision conditions, in essence, paralleled those for radial error. Movements were of shorter duration when made without vision, while, the slowest movements were those made when both vision of the hand and of the target was available (cf. Meyer et al., 1988). Although movements were of greater duration when full vision was available, the hands did not differ in the extent to which this characteristic was expressed. In keeping with the results obtained for radial error, a tendency for movements made by the right hand to be more rapid than those made by the left hand, was not statistically reliable.

Although visual information contributed significantly to the successful completion of the task, effects attributable to hand were less clearly expressed. There were no consistently demonstrable advantages for either hand in the execution of the movement, although there were trends toward a right hand advantage for both movement time and radial error. The failure to elicit the anticipated asymmetries may, in part, have been due to the instructions administered to subjects. They were required to prepare and complete their reaching movements both as fast, and as accurately, as possible (cf. Elliott, Carson, Goodman, & Chua, in press). Given that these requirements are to some extent incompatible, it is likely that different strategies were adopted by individual subjects.

It should also be noted that Flowers' original (1975) hypothesis may be strictly interpreted as a proposal that the right hand system is more efficient in *modifying* movement trajectories on the basis of feedback regardless of its source, rather than that there is a differential use of visual information *per se*. In this regard, it is the discrete, modificational role of feedback information which is crucial. There exists some controversy as to the fashion in which visual information is used in the regulation of aiming movements (Elliott et al., in press; Meyer et al., 1988). It has been shown (Carson, Goodman, & Elliott, 1991) that vision of the limb only may be utilized in a continuous manner leading to improvements in terminal accuracy. Pélisson, Prablanc, Goodale, and Jeannerod (1986) have shown that vision of the target only may have a similar regulatory function. Recently, Elliott et al. (in press) have extended these findings, observing that, in a manual aiming task in which movements were made to targets coplanar with the starting position, full vision and no vision conditions were not distinguished by the number of discrete modifications of the movement trajectory. These conditions were, however, clearly different in terms of accuracy (cf. Meyer et al., 1988).

Although the pattern of movement times across vision conditions observed by Carson et al. (1990) was suggestive of changes in the "mode of control", it was not possible to determine whether visual information formed the substrate for discrete modifications of the movement trajectory or was used in a continuous fashion. The use of kinematic recording techniques allows one to distinguish between these possibilities. For example, the measure of zero crossings of the acceleration profile provides a formal distinction between what are superficially similar movements by diagnosing the relative continuity of trajectories (Brooks, 1974; Elliott, this volume).

It was with a view to clarifying these issues that we recently conducted a more comprehensive examination of the feedback processing hypothesis (Carson, Elliott, Goodman, & Chua, 1991). In addition to obtaining performance measures, kinematic indices were derived using the WATSMART motion analysis system. The relative continuities of movement trajectories were assessed by enumerating zero crossings of the acceleration profile in the period from peak velocity to the termination of the movement. In an attempt to ensure that subjects were consistent in their approach to the task, two instructional conditions were employed. In the first, subjects were required to move as fast as possible, whereas, in the other condition they were to complete the aiming movement as accurately as possible.

Our results indicated that the manipulation of instructional condition produced the required dissociation of response strategies. Accuracy was greatest for conditions in which subjects were instructed to emphasise this dimension. Radial error was again shown to be strongly influenced by visual condition. There was also an interaction of instructional condition and vision condition, suggesting that visual information was of greater utility when subjects were required to move as accurately as possible. In contrast to our previous experiment, a clear right hand advantage for terminal accuracy was observed. The extent of this advantage did not vary as a function of the visual information which was available.

Analysis of movement times revealed that the instructional condition was a strong determinant of the time required to complete the aiming movement. In agreement with expectations, movement times were greatest when accuracy was emphasised. Movements made under conditions of full vision were more time consuming than those made under no vision conditions. This trend was more clearly expressed when individuals were required to move accurately. There was no evidence to suggest that the hands differed in terms of the efficiency with which visual information was utilized, as there was no hand by vision condition interaction for movement time.

Measures of radial error and movement time, although suggestive, are not sufficient basis upon which to distinguish whether movements made under different visual conditions are differentiated by the mode of control which is operative in each case. More specific information is provided by analysis of the kinematics. Our data indicated that the number of zero crossings was the kinematics. Our data indicated that the number of zero crossings was substantially elevated when subjects were moving either under conditions of full vision, or when provided with vision of the limb but not of the target, relative to both conditions in which vision of the limb was not available.[2] Movements made by the left and right hands did not differ in terms of the number of zero crossings (reaccelerations) which were observed.

It was a fundamental tenet of Flowers' (1975) feedback processing hypothesis that during aimed movements, a corrective mode of control was operative. We needed to establish that a corrective mode of control did indeed prevail when sensory feedback was made available. Our data indicated that when individuals were afforded vision of the moving limb, discrete modifications of the movement trajectory were more prevalent. However, discrete modifications also occurred under conditions in which vision was not available. Therefore, we were required to consider whether the differences in the number of modifications represented a fundamental and qualitative change in mode of control or a quantitative change in a preexisting control function. Did modifications imply the occurrence of "corrections"? Elliott et al. (in press) reported that the number of modifications exhibited during the terminal phase of an aiming movement is poorly correlated with terminal accuracy. One reason this pattern of results could occur would be if only those movements which are initially inaccurate are corrected. In those instances, "on-line" correction may not be sufficient to compensate for inaccuracy of the initial impulse.

Calculation of correlations for individual trials for the Carson et al. (1991) data indicated that the overall correlation between terminal accuracy, as expressed by radial error, and discrete modifications, as expressed by the number of zero crossings, did not differ from zero ($r = -.01$), and did not vary as a function of the visual information which was available or of the instructional condition. Therefore, although the frequency with which discrete modifications occurred was influenced by the presence of visual information,

---

[2]Some studies have suggested that the relative frequency of secondary submovements or discrete modifications does not change in the presence of visual feedback (e.g. Elliott et al., in press; Meyer et al., 1988). However, the tasks employed by these authors were constrained in such a way that vision may simply have subserved variations in the "gain" of muscle activity responsible for moving the limb along the primary axis of the movement. For example, the task employed by Meyer et al. (1988) involved the imposition of a zero order control function relating the angular displacement of the wrist to the position of a cursor displayed on a screen. Given what was necessarily a fixed control/display gain ratio, subjects were required to estimate the gain they produced such that the combined machine/operator gain matched a step displacement to a target. As the system was not free to vary along any additional spatial dimensions, vision of the cursor (full vision) was merely likely to have permitted pseudocontinuous adjustments of the operator gain.

the modifications were not functional in any obvious sense. The presence of vision did not appear to promote the implementation of a corrective mode of control which was not otherwise present. Nonetheless, dynamic information pertaining to the position of the limb mediated some additional modificational procedure. However, the hands clearly did not differ with respect to the implementation of these modificational processes. Converging evidence has also recently been reported by Roy, Elliott, and Rohr (1990).

A Reassessment

In view of the failure to demonstrate differences between the hand-hemisphere systems in the processing of visual information, we might do well to critically reexamine the assumptions which gave rise to the feedback processing hypothesis. I will argue that the hypothesis in its varied expressions is largely an artefact of paradigms used to elicit more general principles of motor control.

Flowers (1975) suggested that the essential difference between the hands was in the feedback control of movements. His reasoning was based on the observation that differences between the hands were accentuated when movements of greater precision were required. The interpretation of feedback in general as visual feedback in particular was implicit in this account. However, Prablanc, Echallier, Komilis, and Jeannerod (1979), and Wallace and Newell (1983) have demonstrated that the relationship between movement time and the "index of difficulty" (the ratio of amplitude and target size described by Fitts' Law) is preserved when the movements are conducted without vision. Therefore the interpretation that the hand-hemisphere systems differ in the processing of visual feedback is not necessarily consistent with the observation that differences between the hands increase with index of difficulty. This anomaly illustrates a more general problem. That which has been conceived of as a means to differentiate the hands has often been dependent upon what has been regarded as a plausible basis for Fitts' Law (cf. Annett et al., 1979).

In more recent expressions of the feedback processing hypothesis, theorists have been concerned with whether the differential use of vision may contribute to the emergence of manual asymmetries. Nonetheless, we have witnessed how factors which are perceived of as having some centrality in accounting for visually based movement regulation are identified as those factors which underlie performance asymmetries. Clearly the contribution of vision is multidimensional, yet it is this very multidimensionality which has been seized upon and used as a means of resuscitating the feedback processing hypothesis. The reasoning has been roughly as follows. If visual information derived from a variety of sources, contributes through a variety of mechanisms to regulation, could it not be that the hand/hemisphere systems differ in the efficiency with which these procedures are implemented (cf. Carson et al.,

1990; Roy & Elliott, 1989)? This presents a difficulty primarily because the constructs which have been employed to account for differing levels of performance *within* the hands have been adopted as representing what must "obviously" account for the differences *between* the hands. The failure to generate independent explanatory constructs is a symptom of a more widespread malaise. Those variables we can easily conceive of or manipulate may not be those which are central to the regulation of movement. Similarly, they may not constitute the essence of what distinguishes the performance of the hand-hemisphere systems. Indeed we should have no expectation that they would.[3]

As we have witnessed, a series of attempts have been made to map what are presumed to be the information processing characteristics of the cerebral hemispheres onto variables which ostensibly capture essential aspects of movement regulation. There are both general and specific problems associated with this approach. For example, Cohen's (1973) serial/parallel dichotomy, which has been favoured by a number of theorists of motor control (e.g., Nachson & Carmon, 1975; Todor & Doane, 1978), has become established in the literature in a form which bears only a tenuous link to Cohen's original formulation. In its received guise, the left hemisphere processes all information in a serial fashion, while the right operates upon stimuli in a parallel manner. Cohen (1973) was more circumspect, limiting the claim to alphanumeric stimuli. However this "weak" version is in itself difficult to sustain. Discrimination between serial and parallel systems requires more complete and precise information than is conventionally obtained from "psychological experimentation" (Townsend, 1972).

Our failure to appreciate the essential nature of the physiological substrate, other than in the most trivial terms, is most obviously illustrated by the number of dichotomies which have been used to characterise the information processing propensities of the cerebral hemispheres. In addition to Cohen's (1973) separation of serial processing for the left hemisphere, parallel for the right, it has been proposed that the left cerebral cortex is specialized for analytic as opposed to holistic, global, synthetic or gestaltic apprehension (Nebes, 1978), verbal versus visuospatial activities (Kimura, 1961), focal rather than diffuse processing (Semmes, 1968), name matching in contrast to physical matching (Geffen, Bradshaw, & Nettleton, 1972), and in

---

[3]Attempts to account for performance asymmetries in terms of differences of force variability have been equally unsatisfying. There appears little evidence to suggest that the left hand exhibits proportionately greater variability as impulses are increased in magnitude (Carson, Thyer, Elliott, Goodman, & Roy, 1991; Roy & Elliott, 1989).

terms of the traditional verbal/nonverbal distinction (cf. Bradshaw & Nettleton, 1981). This list is by no means exhaustive.

While it has conventionally been maintained that strict dichotomies do not prevail (e.g., Bradshaw & Nettleton, 1981), that the capacities of each hemisphere lie at points on a continuum, and that differences are of degree rather than of kind (e.g., Corballis, 1981; Milner, 1971; Zangwill, 1960), the boundaries are imposed by the limits of our conceptualization rather than by the nature of the physiology they presume to describe. It is by no means evident that the brain "divides up its functions into categories that correspond to our concepts or vocabulary" (Bullock, 1965, p. 473). Although it has been suggested that the various labels are merely reflections of an underlying mechanism (Allen, 1983), it seems that "cerebral specialization is not likely to be less complex or more 'captured' by labelling one of its attributes" (McKeever, 1981, p. 74). This conflict indicates that the form of analysis has been inappropriate.

The critique is no less appropriate when we consider the variables which are supposedly "controlled" by the nervous system during limb movements (Stein, 1982). A persistent problem associated with the study of motor control processes has been that descriptive variables which appear to capture some aspects of system function have been seen as internalized prescriptions for regulation and control (e.g., Gottlieb, Corcos, & Agarwal, 1989). The problem is of course that descriptive rules cannot become prescriptive and there is no evidence to suggest that the nervous system is rule guided (cf. Dreyfus, 1987), although the consequences of its functioning may be well described by various rules. This failing applies particularly to accounts of manual asymmetries in that if there are variables which are "controlled" by the nervous system, it is also those variables which must be differentially controlled in giving rise to asymmetries. When schemes relating *control* of descriptive variables, derived from contemporary formulations of motor control, are combined with arbitrarily assigned characteristics of cerebral functioning, extracted from current cognitive/neuropsychology, the consequences are accounts of manual asymmetries which possess negligible generality.

Although the simplest description of any act is the act itself (cf. Maxwell, 1877/1952), we usually employ a scientific strategy which involves generating an abstraction of the act or system of interest (cf. Turvey, 1988). In attempting to derive the principles of organization which are presumed to underlie, for example, visually guided behaviour, or the abstraction of this behaviour, there is seldom recognition that the form of these principles is highly constrained not only by the the nature of that being described, but also by the nature of conceptual tools which are brought to bear. In most cases, the principles which are derived are unlikely to be those the system has evolved to embody. The continued failure to elucidate processes underlying manual asymmetries indicates that the "explanatory" principles which have been generated have not yet captured the essential attributes of the system's behaviour.

References

Abrams, R.A., Meyer, D.E., & Kornblum, S. (1990). Eye-hand coordination: Oculomotor control in rapid aimed limb movements. *Journal of Experimental Psychology: Human Perception and Performance, 16*, 248-267.

Allen, M. (1983). Models of hemispheric specialization. *Psychological Bulletin, 93*, 73-104.

Annett, J., Annett, M., Hudson, P.T.W., & Turner, A. (1979). The control of movement in the preferred and non-preferred hands. *Quarterly Journal of Experimental Psychology, 31*, 641-652.

Bradshaw, J.L., & Nettleton, N.C. (1981). The nature of hemispheric specialization in man. *The Behavioral and Brain Sciences, 4*, 51-91.

Brooks, V.B. (1974). Some examples of programmed limb movements. *Brain Research, 71*, 299-308.

Bullock, T.H. (1965). Physiological bases of behavior. In J.A. Moore (Ed.), *Ideas in modern biology* (pp. 451-482). New York: Natural History Press.

Carson, R.G. (1989). Manual asymmetries: Feedback processing, output variability and spatial complexity. Resolving some inconsistencies. *Journal of Motor Behavior, 21*, 38-47.

Carson, R.G., Chua, R., Elliott, D., & Goodman, D. (1990). The contribution of vision to asymmetries in manual aiming. *Neuropsychologia, 28*, 1215-1220.

Carson, R.G., Goodman, D., & Elliott, D. (1991). *Asymmetries in the discrete and pseudocontinuous regulation of visually guided reaching.* Manuscript submitted for publication.

Carson, R.G., Elliott, D., Goodman, D., & Chua, R. (1991). [Asymmetries in the regulation of visually guided aiming]. Unpublished raw data.

Carson, R.G., Thyer, L., Elliott, D., Goodman, D., & Roy, E.A. (1991). *The role of impulse and impulse-variability in manual aiming asymmetries.* Manuscript submitted for publication.

Cohen, G. (1973). Hemispheric differences in serial versus parallel processing. *Journal of Experimental Psychology, 97*, 349-356.

Corballis, M.C. (1981). Toward an evolutionary perspective on hemispheric specialization. *The Behavioral and Brain Sciences, 4*, 69-70.

Doane, T., & Todor, J.I. (1978). Motor ability as a function of handedness. In D.M. Landers & R.W. Christina (Eds.), *Psychology of motor behavior and sport* (pp. 264-271). Champaign, IL: Human Kinetics.

Dreyfus, H.L. (1987). Misrepresenting human intelligence. In R. Born (Ed.) *Artificial intelligence: The case against* (pp. 41-54). Croom Helm: London.

Elliott, D. (in press). Human handedness reconsidered. *The Behavioral and Brain Sciences.*

Elliott, D., Carson, R.G., Goodman, D., & Chua, R. (in press). Discrete versus continuous control of manual aiming. *Human Movement Science.*

Fitts, P. M. (1954). The information capacity of the human motor system controlling the amplitude of movements. *Journal of Experimental Psychology, 47*, 381-391.

Flowers, K. (1975). Handedness and controlled movement. *British Journal of Psychology, 66*, 39-52.

Geffen, G., Bradshaw, J.L., & Nettleton, N.C. (1972). Hemispheric asymmetry: Verbal and spatial encoding of visual stimuli. *Journal of Experimental Psychology, 95*, 25-31.

Gottlieb, G.L., Corcos, D.M., & Agarwal, G.C. (1989). Strategies for the control of voluntary movements with one mechanical degree of freedom. *The Behavioral and Brain Sciences, 12*, 189-250.

Kimura, D. (1961). Cerebral dominance and the perception of verbal stimuli. *Canadian Journal of Psychology, 15*, 166-171.

Maxwell, J.C. (1952). *Matter and motion.* New York: Dover. (Original work published 1877).

McKeever, W.F. (1981). On laterality research and dichotomania. *The Behavioral and Brain Sciences, 4*, 73-74.

Meyer, D.E., Abrams, R.A., Kornblum, S., Wright, C.E., & Smith, J.E.K. (1988). Optimality in human motor performance: Ideal control of rapid aimed movements. *Psychological Review, 95*, 340-370.

Milner, B. (1971). Interhemispheric differences in the localization of psychological processes in man. *British Medical Bulletin, 27*, 272-277.

Nachson, I., & Carmon, A. (1975). Hand preference in sequential and spatial discrimination. *Cortex, 11*, 123-131.

Nebes, R.D. (1978). Direct examination of cognitive function in the right and left hemispheres. In M. Kinsbourne (Ed.), *Asymmetrical function of the brain* (pp. 99-137). Cambridge: Cambridge University Press.

Pélisson, D., Prablanc, C., Goodale, M.A., & Jeannerod, M. (1986). Visual control of reaching movements without vision of the limb. II. Evidence of fast unconscious processes correcting the trajectory of the hand to the final position of a double step stimulus. *Experimental Brain Research, 62*, 303-311.

Peters, M. (1980). Why the preferred hand taps more quickly than the non-preferred hand: Three experiments on handedness. *Canadian Journal of Psychology, 34*, 62-71.

Prablanc, C., Echallier, J.F., Komilis, E., & Jeannerod, M. (1979). Optimal response of eye and hand motor systems in pointing. I. Spatio temporal characteristics of eye and hand movements and their relationships when varying the amount of visual information. *Biological Cybernetics, 35*, 113-124.

Roy, E.A., & Elliott, D. (1986). Manual asymmetries in visually directed aiming. *Canadian Journal of Psychology, 40*, 109-121.

Roy, E.A., & Elliott, D. (1989). Manual asymmetries in aimed movements. *Quarterly Journal of Experimental Psychology, 41A*, 501-516.

Roy, E.A., Elliott, D., & Rohr, L. (1990, October). *Manual asymmetries in visually directed aiming.* Paper presented to the Canadian Society for Psychomotor Learning and Sport Psychology, Windsor, Ontario.

Schmidt, R.A., Zelaznik, H.N., Hawkins, B., Frank, J.S., & Quinn, J.T. (1979). Motor output variability: A theory for the accuracy of rapid motor acts. *Psychological Review, 86*, 415-451.

Semmes, J. (1968). Hemispheric specialization: A possible clue to mechanism. *Neuropsychologia, 6*, 11-26.

Stein, R.B. (1982). What muscle variable(s) does the nervous system control in limb movements? *The Behavioral and Brain Sciences, 5*, 535-577.

Todor, J.I., & Cisneros, J. (1985). Accommodation to increased accuracy demands by the right and left hands. *Journal of Motor Behavior, 17*, 355-372.

Todor, J.I., & Doane, T. (1978). Handedness and hemispheric asymmetry in the control of movements. *Journal of Motor Behavior, 10*, 295-300.

Trevarthen, C. (1974). Analysis of cerebral activities that generate and regulate consciousness in commissurotomy patients. In S.J. Dimond & J.G. Beaumont (Eds.), *Hemispheric function in the human brain* (pp. 235-263). New York: John Wiley & Sons.

Townsend, J.T. (1972). Some results concerning the identifiability of parallel and serial processes. *British Journal of Mathematical and Statistical Psychology, 25*, 168-199.

Turvey, M.T. (1988). Simplicity from complexity: Archetypal action regimes and smart perceptual instruments as execution driven phenomenon. In J.A.S. Kelso, A.J. Mandell, & M.F. Shlesinger (Eds.) *Dynamic patterns in complex systems* (pp. 327-347). Singapore: World Scientific.

Wallace, S.A., & Newell, K.M. (1983). Visual control of discrete aiming movements. *Quarterly Journal of Experimental Psychology, 35A*, 311-321.

Zangwill, D.L. (1960). *Cerebral dominance and its relation to psychological function.* Edinburgh: Oliver and Boyd.

VISION AND MOTOR CONTROL
L. Proteau and D. Elliott (Editors)

CHAPTER 4

ON THE SPECIFICITY OF LEARNING AND THE ROLE OF VISUAL INFORMATION FOR MOVEMENT CONTROL

LUC PROTEAU

Département d'éducation physique, Université de Montréal,
P.O. Box Office 6128, Station A, Montréal, Québec, H3C 3J7

The nature of the changes that occur through learning and how an individual can control the wide variety of movements he/she masters are important questions to answer if we want to understand human motor performance. Over the years, numerous models have been proposed to help answer these questions. For example, Adams (1971) advocated that the processing of afferent information plays a primary role both in the learning and control of human movement. Alternatively, Keele (1968) proposed that afference was useful in learning a movement but that it was not used for its minute-to-minute control, while Schmidt (1975) softened that position by proposing that its importance for the control of an ongoing movement decreased as training at the task increased. Other hybrid control models have also proposed that motor control is achieved by an interplay between central planning and processing of afferent information (Abbs, Gracco, & Cole, 1984; Prablanc, Echallier, Jeannerod, & Komilis, 1979a; Van der Meulen, Gooskens, Denier Van der Gon, Gielen, & Wilhelm, 1990).[1] However, in the latter models the role played by afferent sources of information is considered to be as important for the learning of the task as for its regulation after modest or extensive practice. More precisely, it has been proposed that learning results

---

[1]Schmidt (1988) recently modified his position and proposed that movement related feedback can be used to detect and correct execution errors: "Once the program has been initiated, the pattern of action is carried out for at least one RT even if the environmental information indicates that an error in selection has been made. Yet, during the program's execution, countless corrections for minor errors can be executed that serve to ensure that the movement is carried out faithfully." (p. 237).

in the development of a sensorimotor store which, once the feedback is recognized, allows anticipation of the movement consequences associated with it. This predictive process makes these theoretical propositions much different from that of Adams (1971) and provides the performer with predictions about potential errors, thereby allowing corrective actions to be carried out before deleterious effects of the movement becomes manifest.

In terms of the latter type of control model, one can ask what the sensory bases of these corrections are, as well as their mode of intervention at the beginning and late in the learning of movements. This chapter is focused primarily on the potential role played by visual information regarding a self-performed ongoing movement under normal afference. Thus, I exclude the studies dealing with intra- and cross-modal judgements of kinesthetic and visual information (but see Jones, 1982) and also the studies using a prism adaptation paradigm (but see Redding & Wallace, this volume).

I begin this chapter by first critically reviewing the major experimental evidence that has been used to suggest that learning a motor skill can be equated with either a reduction of the need for sensory information or a decrease in the importance of visual afference in favor of kinesthetic feedback. I will then turn the discussion towards the central goal of this chapter, which is to shed some new light on the role played by visual information for movement control as a particular individual's expertise at the task increases. To reach that goal I first selectively review some studies in which the availability of visual information for the control of various types of movement has been manipulated. In so doing, I will report some of the results obtained in various types of aiming tasks, as well as in the "ball catching" task (see Bootsma, this volume, for more information on the latter type of task). From that analysis it will be clear that the normally available visual information is a major source of afference for movement control. I will nonetheless question the appropriateness of the experimental design used in these studies to answer the central question of this chapter. Thirdly, I will focus on the results obtained by researchers who have used a transfer paradigm to assess the effects of different sources of afference on movement learning and control. In so doing, some emphasis will be placed on research conducted in my own laboratory. Finally, I will propose a theoretical interpretation of the presented results.

The "Experts Do Not Need To Wear Their Glasses" Hypotheses Going From a Closed-Loop To an Open-Loop Process

The main line of support used by the advocates of the position that learning results in a decrease in the role of sensory information for movement control comes from a frequently-cited article authored by Pew (1966). The subjects' task was to align a dot shown on a cathode ray tube with a predetermined target by way of successive key presses performed by the index of each hand. The activation of the left key caused the displacement of the dot to the left, while the activation of the right key caused the displacement of the dot to the right. The subjects practiced the task for 16 1-hour sessions. The results clearly indicate that early in practice the subjects were producing discrete presses, that is, they were waiting for the result of one response before initiating the next one. This is illustrated by a mean interresponse time of 458 ms for the first three experimental sessions. However, a very different pattern of key presses emerged after extensive practice where short interresponse delays were observed. More specifically, for the thirteenth and fifteenth experimental sessions a mean interresponse time of 292 ms was observed. This result is generally interpreted as indicative of a reduction in the utilization of closed-loop control in favor of "gradually increasing reliance on higher-order strategy" (p. 771) and led Pew to conclude that "The underlying theme of these proposals is the hierarchical nature of the control of skilled acts which develop with practice beginning with strict closed-loop control and reaching levels of highly automated action with occasional 'executive' monitoring." (p. 771).

When discussing these results in his very important 1974 chapter, Pew wrote that "I infer from these subjects' performance, and I believe the result to be general, that they were not operating completely open-loop; they were not ignoring feedback in order to impose a structure on their skill but rather were using feedback to monitor and control their performance at a level removed from the representation of individual key strokes." (p. 34). It is thus clear that Pew himself does not believe that movement control evolves from a closed-loop process to an entirely open-loop one. More importantly, one should also consider the results obtained in a control condition included in Pew's study (1966). By the end of the sixteenth day of practice the subjects were asked to produce sequences of key presses as rapidly as possible with no requirement for movement control. It can thus be argued that a fully open-loop mode of control was used in that condition. The mean interresponse time observed was of 125 ms compared to a mean interresponse time of 292 ms for the thirteenth and fifteenth experimental sessions, when the subjects had to control the displacement of the dot on the oscilloscope. The 167 ms difference observed between these two conditions seems to have been long enough to have permitted utilization of visual feedback for the control of the dot displacement (see Carlton, this volume). In my view it is thus far from being clear whether Pew's (1966) results can be used to propose that the role of afferent information for movement control diminishes as training at the task increases.

The second main line of evidence comes from results reported by Schmidt and McCabe (1976). They used a coincident timing task in which the subjects had to hit a barrier by the end of a 2,000 ms delay. The barrier was 10 cm square and located 61 cm from the subject's starting point (index of difficulty of 3.49 bits; Fitts, 1954); the movement time required to perform the task was of approximately 750 ms. The subject thus had to initiate his/her response after having waited for some time (on the average: 2,000 ms - 750 ms = 1,250 ms). The subjects performed 50 practice trials followed by 200 acquisition trials realized on each of five consecutive days. The main dependent variable used was the index of preprogramming (Schmidt, 1972; Schmidt & Russell, 1972) which consisted of the within-subject correlation between starting time and the constant error of coincidence timing observed on each trial. The rationale for the utilization of the index of preprogramming was that, if the subjects were not using response-produced feedback to detect and correct errors, then there should be a high and positive correlation between the time at which the response was initiated and the constant error in timing. Furthermore, if practice results in a decrease in the utilization of afferent information for error detection and correction then the value of the index of preprogramming should increase as a function of practice. The results obtained for a restricted distribution of the performed trials indicated a marginally significant ($p = 0.06$) increase in the index of preprogramming going from 0.43 on day 1 to 0.68 on day 5.

The first thing that one should note about this study is that, as will be discussed in another section of this chapter, the index of difficulty of the task used might not have been difficult enough to encourage subjects to monitor their ongoing movement. In fact, from the task description provided by the authors, it might very well be that the task was such that the subjects were certain to hit the barrier on every single trial. Secondly, because of the low requirements for spatial accuracy, the task was very much like a slow tracking task in which the subject's movement had to "track" the sweeping hand of a clock. With practice it is thus likely that the subjects learned to choose the appropriate starting point (i.e., when the clock's sweeping hand reached a particular value on the clock) and then synchronized their response with the sweeping hand of the clock. With practice they would learn what was the most appropriate starting point and show less and less variability in their starting time. Further, if as I postulated above, there was no need to monitor whether or not the barrier would be hit on a particular trial and because the subject's movement was "linked" to that of the clock's sweeping hand, then a too late start would result in a positive constant error and a too early start would result in a negative constant error because the subject was tracking the clock. Thus contrary to Schmidt and McCabe's interpretation, the results might indicate that the subjects used a closed-loop mode of control and became more efficient at it as their expertise at the task increased!

Based on the two studies reported above it is thus very difficult to argue in favor of a mode of control which early in learning is based on a closed-loop process and then develops into an open-loop process as learning and/or expertise at the task increases.

Neglecting Visual Afference in Favor of Kinesthetic Afference

An alternative view to that of motor control becoming progressively open-loop is that practicing a motor skill results in a decrease of the role played by visual information in favor of the utilization of kinesthetic afference for movement control. Support for that proposition comes from a well-cited study by Fleishman and Rich (1963). They used a two-hand coordination task and observed that their subjects switched from a dependence on exteroceptive (visual-spatial) feedback early in learning to proprioceptive feedback later in learning. It should however be noted that Cox and Walkuski (1988) recently failed to replicate Fleishman and Rich's (1963) results. They showed that the correlations between kinesthesis and both pursuit rotor and ball tossing tasks were generally low, and this irrespective of the subjects levels of training at the tasks.[2]

Furthermore, just by looking at the content of the present volume, it is quite easy to conclude that vision is a main ingredient for the efficient control of posture, locomotion, prehension, catching and/or hitting a ball, aiming and also writing (Burton, Pick, & Holmes, 1990); all tasks that an adult subject normally masters. All these results can be interpreted by proposing that, although proprioception provides information relative to both the position and the displacement of the limbs (McCloskey, 1978; Kandel & Schwartz, 1985), this kinesthetic information is not sufficiently accurate to replace visually-based motor control. Three reasons can be invoked to explain the advantage of visual (or visual + kinesthetic) information over kinesthetic information alone. First, Gibson (1969) proposed that the quality of the sensorimotor store which is thought to develop during the learning of a task is mediated by the discovery of multimodal invariants which are required for skillfully directed action. From this, it can be suggested that the error correction mechanisms developed through learning are more effective when the number of pertinent afferent sources increases. A second proposition has been that, in the absence of visual information, one is uncertain as to where a particular limb is located in the environment (Howarth & Beggs, 1981; see also Meyer, Abrams, Kornblum, Wright, & Smith, 1988). Thus, it may be that visual information is mandatory to calibrate the easily decalibrated proprioceptive system (Held & Bauer, 1967; Shimojo, 1987; but see also Cordo & Flanders, 1990; Flanders & Cordo, 1989). Finally, Soechting and Flanders (1989a, b) suggest that a pointing

---

[2]Cox and Walkuski (1988) only used 30 trials in that experiment. However, near asymptotic performances were observed and the case can be made that learning was almost completed at the end of practice.

movement involves the transformation of the position of the target in space from an extrinsic frame of reference to an intrinsic one. The latter would define the position that the limb must adopt for the target to be reached. They propose that errors in pointing to virtual targets in the dark occur because the individual must implement a linear approximation of the transformation from intrinsic to extrinsic coordinates. Considering the fact that this transformation may be one of the steps that leads from the detection of a visually presented target to the organization and implementation of the proper response, its approximative nature can explain why proprioception alone does not lead to the same level of response accuracy as when visual information is also present.

In this last section I did not report much evidence opposing the view that, as learning increases, kinesthetic afference becomes more important to motor control than visual afference. I rather chose to review theoretical porpositions suggesting why such should not be the case. However, numerous evidence to support the proposition that visual afferences remain important for motor control even after extensive practice are reported in the remainder of this chapter.

The "Now You See It, Now You Don't" Paradigm[3]

Over the years, the role played by visual information for the control of an ongoing movement has received a lot of attention. Researchers interested in this problem have attempted to determine the minimum time required to effectively process this source of information and the nature of the processed information, as well as task constraints. As concluded by Carlton (this volume) for the minimal processing dilemma, it is very difficult to unequivocally answer these questions because it might very well be that the answer is task specific. However, one of the most striking facts to emerge from these experiments is that the number of trials performed by the subjects does not seem to be an influential factor.

The data reported in Table 4.1 are a representative sample of the data available in the literature. Furthermore, they also provide a good idea of the most frequently-used experimental designs and the wide spectrum of experimental tasks. In all but one of the comparisons (see Table, note "g") the subjects were found to be more accurate when vision of the ongoing limb (or displaced cursor) was visually available than when it was not. This was the case both when the subjects did not have much task experience or, on the contrary, had very extensively practiced the task. Thus it appears that visual

---

[3]I borrowed this title from a symposium presented at the 1983 North American Society for the Psychology of Sport and Physical Activity meeting, held at Michigan State University. The participants were Brian Hawkins, Karl M. Newell, Steven A. Wallace and Howard N. Zelaznik.

information is important even after extensive practice. Here one may wish to argue that these data were averaged over the number of trials performed by the subjects and therefore do not give a good idea of the level of accuracy reached in each condition at the end of practice. This is indeed a possibility. However, one must also recognize that the visual feedback manipulations were very often introduced as within-subject factors. In my view, this is likely to reduce the observed difference between the vision and no-vision conditions because of the well-known carry-over effect which is always a threat to within-subject designs (Poulton, 1981; but see Elliott and Madalena, 1987). In short, although these studies might not have been optimally designed to answer our specific question (which was not their purpose) they nonetheless give strong indications that for optimal aiming accuracy, extensive training cannot be substituted for visual information.

Before closing this section, I would like to draw the attention of the reader to the results obtained by Wallace and Newell (1983). These authors had their subjects realize a discrete tapping task under different levels of index of difficulty (Fitts, 1954). As can be seen in Table 4.1, the utilization of low indices of difficulty did not result in a better performance under a full vision than a no-vision condition whereas more difficult tasks were shown to be more accurately performed under the full vision condition. This important finding is paralleled in the so-called "ball catching" literature.

For instance, Smyth and Murray (1982) were interested in determining the role played by seeing one's hand in order to catch a ball. They asked their subjects to catch tennis balls that were delivered by a ball projection apparatus. In a first condition, the subjects simply had to catch the ball under normal visual information. For the other two conditions, the subject's sight was obstructed by either a transparent screen or an opaque one. The opaque screen blocked the subject's view of his own arm as well as the last 150-200 ms of the ball trajectory. The results indicated that the subjects caught significantly less balls under the opaque screen condition than the other two conditions, which led to authors to conclude that "It may be the ball requires most visual attention in catching, but accurate and effective use of the hand is increased when the hand can also be seen" (p. 151). Since then, it has been shown a number of times that seeing one's hand is important for optimal catching accuracy, even when the period of ball occlusion is reduced to 100 ms or below (Davids & Stratford, 1989; Diggles, Grabiner, & Garhammer, 1987; Fischman & Schneider, 1985; Smyth & Turton, 1983, as reported by Smyth, 1986; see also Smyth, 1982). Diverging results were, however, reported by Davids, Palmer, and Savelsbergh (1989). These authors used the above described screen paradigm to evaluate the importance of seeing one's limb in a

Table 4.1
The effect of visual information in different vision-no vision manipulations

Study	Experimental task	Within-subject factors	Number of trials	Performance Vision	No-vision
Abrams, Meyer, & Kornblum, 1990 experiment 2	Wrist rotation which caused the displacement of a dot showed on an oscilloscope	2 visual feedbacks x 2 eye movements x 2 target distances / Visual feedback is varied accross trials / Choice reaction time procedure	432	85.5%	32.0%[a]
experiment 3	Same as above	2 visual feedbacks x 3 eye movement-target combinations x 2 target distances	432 (?)	70.5%	30.0%[a]
Bard, Hay, & Fleury, 1985	Aiming without contact	4 visual feedbacks x 2 movement times x 4 target eccentricities / MT : < 130 ms and 250-300 ms	160	1.77°	2.9°[b]
Beaubaton & Hay, 1986	Aiming with finger	5 visual feedbacks x 4 target locations x 4 movement times / MT = from 110 ms to 270 ms	3840	7.0 mm	13.5 mm[c]
Elliott & Allard, 1985 experiment 1	Aiming with a stylus	2 visual feedbacks x 2 order of presentations x 4 target locations / Choice reaction time procedure	128	7.2 mm	8.9 mm[d]
Prablanc, Echallier, Komilis, & Jeannerod, 1979	Aiming with finger	2 visual feedbacks x 3 eye movement-targets combinations x 8 targets / Choice reaction time procedure	960	4.0 mm / 4.0 mm	22.0 mm[e] / 37.0 mm[f]
Wallace & Newell, 1983	Discrete tapping task	2 visual feedbacks x 3 target widths x 4 target distances / Choice reaction time procedure	480	5% / 9%	7%[g] / 25%[h]

Table 4.1 (continued)
The effect of visual information in different vision-no vision manipulations

Study	Experimental task	Within-subject factors	Number of trials	Performance Vision	No-vision
Zelaznik, Hawkins, & Kisselburgh, 1983 experiment 1	Aiming with a stylus	2 visual feedbacks x 8 movement times MT : from 100 ms to 240 ms	800	3.7 mm	4.1 mm[i]
experiment 2	Same as above	4 movement times MT : from 120 ms to 300 ms	480	4.0 mm	5.0 mm[i]
experiment 3	Same as above	2 visual feedbacks x 8 movement times MT : from 70 ms to 650 ms	1440	2.2 mm	3.6 mm[j]

a Proportions of "hits" for the saccade condition collapsed over the 2 movement distances. Computed from Table 4 (experiment 2) and Table 5 (experiment 3), respectively.

b Absolute directional error collapsed over the 2 movement times and 4 target eccentricities. Estimated from Figure 3.

c Radial error collapsed over the 4 movement times. The results are those obtained under the complete feedback and without feedback conditions. Estimated from Figure 2.

d Root mean square error obtained under the blocked condition. Reported from Table 1

e Absolute error for the 10 cm displacement on the right hand side under the foveal vision condition. Estimated from Figure 3.

f Absolute error for the 40 cm displacement on the right hand side under the foveal vision condition. Estimated from Figure 3.

g Proportions of "hits" collapsed for the 1.58, 2.58 and 3.58 index of difficulty tasks. This difference is not significant. Estimated from Figure 2b.

h Proportions of "hits" collapsed for the 4.58, 5.58 and 6.58 index of difficulty tasks. This difference is significant. Estimated from Figure 2b.

i Root mean square error collapsed for the distance and directional movement components and different movement times. For experiment 1, the data were computed from Table 1 and Table 2. For experiment 2, the data were estimated from Figure 1.

j Root mean square error collapsed for the distance and directional movement components and different movement times. Only the data for the 70 ms MT has been omitted. Estimated from Figure 2.

tennis volleying task. The subjects were thus required to perform the volleying task under conditions of full and occluded effector visual feedback. The results clearly indicated that the subjects performed equally well under these two conditions. Taken together, the results obtained under the ball catching paradigm and those reported by Wallace and Newell (1983) suggest that the tolerable margin of error can dictate whether or not visual information regarding the performing limb is used by the subject.

Overview

First, from the above-mentioned studies it is hardly arguable that the role played by the visual information available in a normal visual environment is useful for movement control. Secondly, and not unexpectedly, it also appears that the importance of visual information is task specific. If the task is such that the "permitted error" is very large then vision does not improve performance over a no-vision condition. Finally, when the accuracy requirements of a task are very stringent it appears that the role of visual information does not necessarily diminish as expertise at the task increases, because a full vision condition led most of the time to more accurate responses than a no-vision condition. In short, the above-reported studies provide good support for the proposition that visual information is important for movement control, even after extensive practice. Another way to test the utility of visual information for movement control is to use a transfer paradigm, in which subjects are trained at a particular task with different amounts of practice under normal visual conditions, and then asked to perform the same task under a no-vision condition. I now turn the discussion to those experiments in which such a transfer paradigm was used.

The Practice Makes Perfect (Or Does It?) Paradigm

The utilization of a transfer paradigm to evaluate the role played by different sources of afference, and more specifically visual afference, for the control of an ongoing movement is based on a two-part rationale. Firstly, if in an acquisition (or learning) phase in which vision of the ongoing limb is either permitted or not (between subject comparison), being able to see one's ongoing movement leads to a better aiming accuracy than when the task is performed under a no-vision condition, then one has to conclude that some visual information was used to control the ongoing movement. Secondly, if, in a transfer phase in which all subjects must perform the movement they had just learned under a no-vision condition, withdrawing the visual information available in the acquisition phase results in an increase in aiming error, then one must conclude that the visual information that had been withdrawn was indeed important for movement control.[4] On the other hand, if the withdrawal

---

[4]In my view it is very important that both condition be met before one can accept such a conclusion. For instance, let us suppose that two groups of subjects trained in an aiming task

of that visual information does not cause any increase in aiming error, then one must conclude that the withdrawn information was not used for control purposes. Therefore, by comparing the efficiency of an individual performing different types of tasks under a variety of feedback conditions, during both acquisition and transfer, one should be able to determine the sources of afferent information used to control the ongoing movement. Further, by comparing the results obtained using such a design after different amounts of practice, one should be able to assess the importance of the available afference as the expertise of the subject at the task increases. For example, if vision of the ongoing limb in an aiming task is very important early in learning, but its importance decreases as a function of practice, then having visual afference withdrawn in transfer should have a deleterious effect on movement accuracy early in practice, whereas its late withdrawal should lead to a very small decrease in accuracy. Alternatively, if it was shown that its late withdrawal leads to a larger increase in error than that found after its early withdrawal, then the only viable conclusion would be that visual information was indeed used, and even more so than in early acquisition.

Over the years the transfer paradigm described above has been used to assess the role played by different sources of afference for the learning of a motor task. Numerous types of tasks were used. They can, however, be subdivided into three general classes: oscilloscope "aiming", linear positioning, and manual aiming.

Oscilloscope Aiming

Annett (1959) used a task in which the subject had to apply pressure to a moveable bar in order for a point shown on an oscilloscope to move from a pre-determined starting position to a fixed target also shown on the oscilloscope (the movement of the luminous dot on the scope was larger than that of the cursor displaced by the subject). After 50 acquisition trials, the subjects were asked to perform the same task but this time without the benefit of seeing the dot displacement. The results indicated that the subjects grossly overestimated the pressure required to reach the intended target. From these results one can thus conclude that vision was an important source of afference

---

under either a full vision condition or a condition in which only the target is available. If the normal vision group is more accurate than the no-vision group after 20 trials, then one must conclude that seeing one's arm is important for movement control early in learning. Now if the two groups of subjects are kept prisoner in the laboratory until they have completed 20,000 trials and the results now indicate the same level of accuracy was reached for the two groups, can we conclude that seeing one's arm is no longer useful? Of course not. The only thing that can be concluded with some assurance is that extensive practice (probably with KR) under the no-vision condition can lead to the same level of accuracy as training under the normal vision condition.

in the acquisition phase of Annett's study (1959). However, considering the novelty of the task and the fact that only 50 acquisition trials were performed prior to the visual feedback withdrawal, these results might not come as much of a surprise. Furthermore, because a single level of practice was used, this study does not permit us to determine whether the role played by the visual information increased or decreased as a function of learning.

More recently, Smyth (1977) used a very similar type of task with a movement-to-display gain of 15:1. Given that the target shown on the oscilloscope was located 20 mm from the starting point, a 1.33 mm displacement of a bar, via finger pressure, was required to move the dot to the target. The task therefore involved a kinesthetic (pressure) -visual integration. Four experimental groups performed either 2, 5, 50 or 400 acquisition trials with visual feedback available prior to being transferred to a no-vision no-KR condition using the same task (10 trials). A control group performed 50 acquisition trials under a no-vision condition; they received verbal KR after each trial. Finally, the subjects who performed 50 and 400 acquisition trials "were warned before the end of practice that the light spot was about to be removed" (p. 278). The results obtained for the absolute error (AE) first indicated that the subjects who had either 2 or 5 acquisition trials were less accurate in the transfer task than the subjects who performed 50 or 400 acquisition trials (mean AE of 9.04 mm *vs.* 4.86 mm, respectively). Secondly, and very importantly, it was observed that the control group performed better in the transfer task than did any of the experimental groups (a mean AE of 2.63 mm). The better performance of the control group in transfer over that of the subjects who had visual information available during the acquision phase of the experiment was replicated using a movement-to-display gain of either 15:1, 7.5:1 (Smyth, 1977; experiment 2) or 1:1 (Smyth, 1978). Taken as a whole, these results thus indicate that visual information is important for optimal movement accuracy because its withdrawal leads to large errors. They further suggest that even though visual information was still used after moderate practice (400 trials), its importance decreases as a function of practice. However, before accepting this conclusion one has to remember that the subjects who trained for 50 and 400 trials were informed that the available visual information was about to be removed. Thus, it is possible that this warning caused them to ignore the visual information and to try to use some other available cue. In line with this interpretation is the fact that the transfer performance of the 50 and 400 trials groups was virtually identical. It thus appears very difficult to draw a strong conclusion on the role played by visual information as learning progresses on the basis of these results.

Linear Positioning

Adams, Goetz, and Marshall (1972) seem to have been the first to use a transfer paradigm to investigate the role played by different sources of afference after different amounts of practice. They had their subjects learn a linear positioning task (25.4 cm) for either 15 or 150 trials. Under each level of practice a group of subjects performed the task under either augmented or minimal afference. In the augmented condition, the subjects were able to see their ongoing movement, hear the slide as it moved along its supporting rod and feel the slide move against spring tension which was thought to provide augmented kinesthetic information. In the minimal afference condition, the subjects were blindfolded, prevented from hearing the slide move on the rod through the utilization of white noise, and moved an almost frictionless slide. Following the acquisition trials all subjects were submitted to a transfer task in which the afferent information available was either the same or different (in this section I only present the results of the subjects who went from the augmented to the minimal afference condition) as that available in acquisition; furthermore no verbal KR was given. The results obtained for AE at the end of acquisition indicated that the accuracy of the subjects' responses was positively related to the amount of feedback and number of acquisition trials (see also Adams & Goetz, 1973). In transfer, the subjects who only had KR withdrawn suffered an increase in error. This increase was of 2.7 mm for the subjects who had 15 trials of practice, while it was of 1.3 mm for the subjects who trained for 150 trials. It was, however, much less than that experienced by the subjects who went from the augmented to the minimal afference condition when going from acquisition to transfer (see also Johnson, 1980). More specifically, for the latter groups, the observed increase in error was of 33.1 mm and 38.8 mm for the subjects who trained for 15 and 150 trials, respectively. These results thus suggest that the sources of afference available during the acquisition phase of the experiment under the "augmented" condition, including the visual information, became more important for movement control as training at the task increased.

Adams, Gopher, and Lintern (1977) replicated the work of Adams et al. (1972). They used a movement of 20.3 cm and minimized auditory cues for all subjects through the utilization of white noise. However, instead of having all subjects experience the withdrawal of all the sources of afference available during acquisition when submitted to the transfer condition, different groups of subjects had either vision, proprioception, or both sources of afference withdrawn. The results obtained at the end of acquisition replicate quite well those reported above (Adams' et al., 1972). The same is also true for the subjects who trained under augmented vision and proprioception and were transferred to a minimal afference context. More specifically, after 15 acquisition trials, this transfer caused a net increase in error of approximately

28 mm, while the error was of approximately 39 mm for the subjects who had trained for 150 trials. However, the withdrawal of only visual afference in transfer caused a net increase in error of approximately 36 mm after 15 acquisition trials compared to approximately 25 mm after 150 acquisition trials. Again, this shows that the withdrawal of visual information has a detrimental effect on transfer performance.

The results presented in this section indicate that the visual information available in a linear positioning task when one is allowed to see his/her ongoing movement is an important piece of information for optimal accuracy. This was first demonstrated by the fact that this information leads to better performance in acquisition when it is available than when it is not available. It should be noted that this better performance cannot be accounted for by a visual *vs.* proprioceptive KR effect (Chew, 1976), although it is quite possible that some subtle cues inherent in the task helped the subjects to be more accurate when vision was permitted (Newell & Chew, 1975). The second line of support comes from the results observed for the transfer trials, where the withdrawal of visual information caused a significant increase in error. The third, and perhaps more interesting result lies in the fact that Adams et al. (1972, 1977) have shown that minimizing visual information in transfer causes a very large increase in error after moderate practice. In fact, it even appears that the detrimental effect of withdrawing visual information in transfer increases as a function of the number of trials performed in acquisition. From these results one can thus conclude that the role played by afference, and particularly visual afference, for the control of an ongoing movement *increases* as a function of learning. It is also suggested that what is learned is specific to the conditions under which learning occurred. Before trying to generalize these conclusions, it appears important to determine if they apply to situations in which the aimed target is directly visible to the subject and in which extensive practice is possible. These studies will be reviewed in the next section.

Manual Aiming

To help determine the role played by visual information regarding the ongoing limb, we (Proteau, Marteniuk, Girouard, & Dugas, 1987) decided to use a relatively long (90 cm) multi-degrees of freedom aiming movement. This movement was performed in the sagittal plane and involved rotation around the shoulder, elbow and wrist joints. It involved a vertical displacement of the stylus by approximately 35 cm from the starting position to the target (hereafter called main axis), while a lateral movement of approximately 5 cm was required laterally (hereafter called secondary axis). The subjects were further asked to execute the movement in 550 ms. We decided to use such a temporal requirement to avoid any possible trade-off between movement time (MT) and spatial accuracy. Four groups of subjects were used. Two groups practiced for 200 trials and two practiced for 2,000 trials. Knowledge of results regarding spatial and temporal accuracy was provided after each trial. Furthermore, for each level of practice there was a group of subjects who practiced with complete vision of both the performing limb and the target and a group in which only vision of the target was allowed. Finally, following the acquisition period, all subjects were transferred to a condition in which only the target was visually available and they did not receive KR.

As expected, the results indicated that the subjects' aiming error and temporal error (as evaluated by root mean square error) decreased as a function of practice. More interestingly, however, is the comparison of the results obtained late in acquisition and those obtained under the transfer condition. The results indicate that transfer caused a significant increase in aiming error regardless of the number of acquisition trials (i.e., 200 or 2,000). Secondly, for the subjects in the 2,000 trials condition, this increase in error was so large that they became less accurate than the subjects who had the same amount of practice, but under the target-only condition. These results suggest that the role played by visual information coming from the environment and the ongoing limb was still very important even after extensive practice at the task. Finally, the comparison of the results obtained in transfer for the subjects who trained under normal visual conditions indicated a significantly larger error after extensive rather than moderate training. These results are very similar to those reported by Adams et al. (1972, 1977) and suggest that the role played by the available visual information in acquisition was more important after extensive practice than it was after moderate practice.

This conclusion was strongly supported in a recent study (Proteau, 1991). Two groups of subjects practiced the same type of task as above but this time had to reach one of five randomly presented targets. The first group performed 1,000 acquisition trials under the target-only condition while the subjects of a second group performed 2,000 acquisition trials under a normal visual condition. Following each acquisition trial, the subjects received KR regarding the spatial accuracy of their movement and were informed of their MT when it fell outside a predetermined bandwidth (550 ms +/- 50 ms).[5] Transfer tests were similar to that defined above and occurred after 60, 160, 1,000 and 2,000 trials for the normal vision group, while they occurred after

---

[5]This small modification to the procedure was made to ensure that the MT requirement of the previous study did not obscure the results obtained for aiming accuracy because of the dual requirements of temporal and spatial accuracy. From the very beginning of the acquisition phase the subjects did not have any difficulty performing their movement within the prescribed bandwidth.

L. Proteau

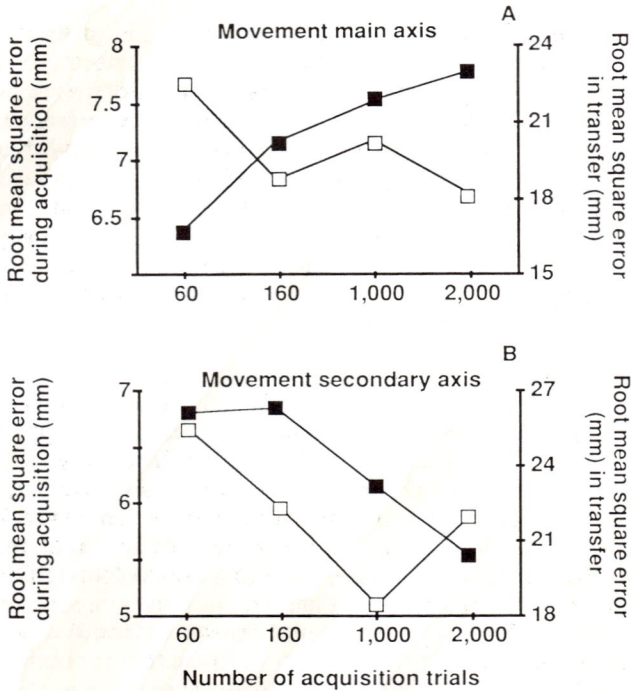

Figure 4.1. Root mean square error of aiming on the main (a) and secondary (b) axis of the movement. The open symbols indicate movement accuracy at the end of each acquisition period and are related to the left ordinate, whereas the filled symbols refer to the performance obtained during transfer and are related to the right ordinate. Note the change of scale on each ordinate.

60, 160 and 1,000 trials for the target-only group. The results first indicate that the subjects had no difficulty in acquisition or transfer to complete their movement within the prescribed time bandwidth. More importantly, and as depicted in Figure 4.1a, the more trials the subjects performed in acquisition, the more accurate they became on the main axis of the movement. However, a nearly perfectly reversed picture emerged from the results obtained in the transfer tests (also illustrated on Figure 4.1a). The subjects' responses became less and less accurate as a function of the number of trials performed in acquisition. Because no such decrease in accuracy was observed for the subjects who trained under the target-only condition, the above results cannot be explained by a KR withdrawal effect. The only conclusion that can be reached is that the visual information available during acquisition became more important for movement control as the number of practice trials increased. Regarding the secondary axis of the movement, the withdrawal of visual information in the transfer condition resulted in a severe increase in aiming error. However, as illustrated in Figure 4.1b, the results obtained on that particular axis were in some respects very different from those obtained on the movement's main axis. More specifically, a decrease in aiming error in transfer was observed as a function of the number of acquisition trials. It thus appears that what had been learned for the control of the movement on its secondary axis under normal visual condition was partially transferred in a situation in which this information was not available.

Taken together, the results illustrated in Figures 4.1a and 4.1b, might be viewed as supportive of Paillard's proposition of a dual visual system (1980; see also Paillard & Amblard, 1985) which was recently supported by Bard, Paillard, Fleury, Hay, and Larue (1990; see also Bard, Hay, & Fleury, 1985; Teasdale, Blouin, Bard, & Fleury, 1990). More specifically, Paillard and Amblard (1985) argued that a distinction has to be made between two independent visual channels which each processes different types of visual information. Both channels would operate in a closed-loop fashion and would provide either positional or directional error signals regarding an ongoing aiming movement. The first system would operate in foveal vision and account for the terminal correction of the trajectory during the homing-in phase of the aiming movement. The other channel would operate faster than the former and provide information regarding the directional error of the moving limb toward the target. Furthermore, this second channel would appear to be particularly effective in the first, usually faster, part of the movement. In light of the above study (Proteau, 1991), it might be suggested that the information which is processed in central vision is used to control movement even after extended practice, as evidenced by the results reported on the movement main axis. However, the information concerned with directional error would become slightly less important as training increases, which would support the results illustrated in Figure 4.1b. This interpretation is at the present speculative, and one should not forget that large errors were noted in transfer for both movement axes.[6]

---

[6]There is at least one alternative interpretation of the results just presented. On the one hand, one might argue that they illustrate an example of a spatio-spatial trade-off where the subjects devoted more attention to their performance on the accessory than the main axis of the movement in acquisition and/or transfer. If this was the case then it could explain why a learning effect was evidenced in transfer for the former axis and also why an inflated error was observed on the main axis of the movement. However, such an interpretation is fatally flawed because it implicitly recognizes that one has to devote attention to his/her movement in order for it to be accurate, and this even after extended practice. Otherwise, it would be impossible to explain the increase in error found in transfer on the main axis of the movement. Obviously,

Taken collectively, the results presented above (Proteau et al., 1987; Proteau, 1991) indicate that some aspects of the visual information available in a normal context are used to control an ongoing movement even after extensive practice.   However, which aspect(s) of the available information is it? Numerous candidates are possible.   First, it might be that only the dynamic information regarding the ongoing movement is still used after extended practice.   Secondly, it might also be that one or a combination of static cues such as the position of the hand before (Prablanc, Echallier, Komilis, & Jeannerod, 1979; Prablanc, Echallier, Jeannerod, & Komilis, 1979) or after (Beaubaton & Hay, 1986; Hay & Beaubaton, 1986) the aiming movement, or even the availability of a structured visual background (Conti & Beaubaton, 1976; Velay & Beaubaton, 1986) are required after extended practice to ensure optimal aiming accuracy.

Dynamic visual information.   Following the lead of Prablanc et al. (1979a, b) and Carlton (1981) it has generally been accepted that, contrary to Stubbs (1976) proposition, vision of the intended target coupled to kinesthetic information coming from the ongoing limb is not sufficient to ensure optimal aiming accuracy.   In fact, the former authors have demonstrated that vision of the ongoing stylus (or limb) was mandatory to ensure optimal accuracy. However, Elliott (1988) recently raised some doubts about this conclusion by showing that aiming accuracy was identical between a conventional no-vision condition (the target is however visible; experiment 2) and a situation in which both the ongoing stylus and the target to be reached were visually available. Faced with these conflicting results we reasoned that they might have been caused by the number of trials used by these researchers.   For instance, Prablanc et al. (1979a, b) used 960 trials, and Carlton had his subjects performed 40 trials under each of five visual feedback conditions, whereas Elliott (1988) had his subjects perform 7 trials a day for each of two consecutive days.   We (Proteau & Cournoyer, 1990) replicated three of the conditions used by both Carlton (1981) and Elliott (1988).   Specifically, full vision (similar to our above defined normal visual condition), vision of the stylus and target (a luminescent stylus was used) and a target-only condition. Three groups of subjects trained under one of these conditions for 15 trials (one group per condition), while three other groups performed 150 acquisition trials under the same conditions.   Following the acquisition trials all groups were submitted to a common target-only condition.   The results obtained on the movement's main axis are illustrated in Figure 4.2.

---

this interpretation runs directly against a theoretical position suggesting that the role of afference for movement control decreases as a function of practice. A second but minor weakness of such a proposition is that, at least to our knowledge, the concept of "spatial to spatial" trade-off has no roots in experimental psychology or neurophysiology.

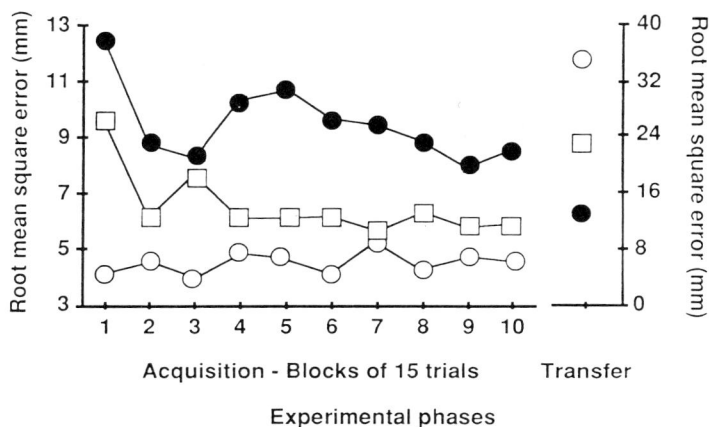

Figure 4.2. Root mean square error of aiming on the main axis of the movement. The filled circles indicate the performance observed for the subjects who learned the task under the no-vision condition. The open squares indicate the performance of the subjects who only had vision of the stylus permitted, while the open circles refer to the performance of the normal vision condition. The left ordinate refers to the performance observed during acquisition, whereas the right ordinate refers to the performance observed during transfer. Note the change of scale on each ordinate. (Adapted from Figures 3 and 4 from Proteau and Cournoyer [1990]).

The first observation that can be made is that the subjects learned to use the information provided by the visible stylus (see Figure 4.2). For instance, for the first block of 15 trials, it is quite clear that accuracy was similar under the vision of the stylus condition and the target-only condition, a result which replicates that of Elliott (1988). However as training increased, it is also clear that the subjects became more efficient in using the information provided by the luminescent stylus. This is indicated by the fact that they were nearly as accurate as the subjects who trained under the normal vision condition, a result which supports Carlton (1981) conclusion. Thus, the information provided by the stylus was (in fact, it became) more important for movement control after 150 trials than after 15 trials of practice. This fact is further substantiated by the results obtained in transfer. The more information available in acquisition (full vision > vision of the stylus > no vision) the less accurate the subjects became when this information was withdrawn. Finally, the comparison of the results obtained after 15 trials (not shown in Figure 4.2) and 150 trials of practice suggests that the more the subjects practiced with some sort of visual information available, the worse their performance became when that information was withdrawn.

The observation that one has to learn how to use the information provided by a luminous stylus before using that information optimally has some equivalence in the ball-catching literature. For instance, Rosengren, Pick, and Von Hofsten (1988) conducted four experiments in which they demonstrated that a ball was much more likely to be caught in a normal light situation than when: (a) only the luminescent ball was visible (experiment 1, 3 and 4), (b) the ball and the subject's hand were visible (experiment 3), (c) the ball and a spatial frame of reference or, the ball, a spatial frame of reference and the subject's hand were visible (experiment 3 and 4).[7] More importantly, they also found that the wearing of a glove with luminous referents did not permit the subjects to catch more balls than under the ball-only condition. These results thus apparently minimized the role played by visual information regarding the catching hand in such a task. However, opposite results were obtained by Savelsbergh and Whiting (1988) who showed that "good" and "poor" catchers caught less balls in a condition where they were not able to see their hand compared to a condition where the subjects wore a luminous glove. As in the Carlton (1981) and Elliott (1988) studies, the number of trials performed under each visual condition might be at the base of the conflicting results. In the Rosengren et al. (1988) experiment, the subjects performed only 24 trials (12 in each of the conditions ball + hand and, frame + ball + hand; experiment 3) with the "luminous glove" whereas in the Savelsbergh and Whiting (1988) study the subjects performed 40 trials (20 in each of the hand + ball and the hand + ball + string conditions). In the light of the Proteau and Cournoyer (1990) study, it might thus be suggested that the subjects of Rosengren et al. (1988) would have needed more experience with these visual cues before being able to use them optimally.

The results presented in this section suggest that the dynamic visual information available in an aiming task becomes more important for movement control as practice at the task increases. Further, it appears that what is learned during acquisition is specific to the sources of information available during that phase (Proteau & Cournoyer, 1990). These results cannot however be taken as evidence that the static visual cues available in a normal visual context are of no use. First, as previously reported, Prablanc et al. (1979a, b)

---

[7]The hand was made visible by having the subjects wear a glove on which dots of luminous paint had been painted at the wrist and finger joints.   A visual frame was provided by having luminous tape stripes (experiment 3) or light emitting diodes (experiment 4) activated on a wall located in front of the subject).

clearly showed that even after 960 trials of practice, vision of the surrounding environment prior to movement initiation helps the subjects to be more accurate in a finger-pointing task. In the next section I discuss the importance of this type of information as a function of practice.

Static visual information. The importance of static visual information for the control of an aiming movement as a function of practice was addressed in a recent study (Proteau & Marteniuk, 1991). In a first experiment we had eight groups of subjects practicing the task used by Proteau (1991). Four of these groups practiced the task for 15 trials whereas the four other groups practiced the task for 165 trials (modest and moderate practice, respectively). Under each level of training, the first two groups practiced the task under the normal visual information and target-only conditions. The subjects in the third group practiced the task under what we called the "prior environment" condition, whereas the subjects of the last group practiced under a visual KR condition. In the prior environment condition, the lights of the experimental room were off for the entire duration of the aiming movement, as well as during its return to the starting base. Vision of the surrounding environment was, however, permitted for a fixed period of time prior to movement initiation. For the visual KR condition, the lights of the experimental room were off for a fixed period of time prior to movement initiation and for the duration of the aiming movement. However, the lights went on as soon as the stylus touched the target area. Following the last practice trial, the subjects of all eight groups were submitted to a transfer condition using the target-only condition with no KR.

The results are illustrated in Figures 4.3a and 4.3b for modest and moderate practice, respectively. As illustrated in Figure 4.3a, the normal visual condition led to more accurate aiming movements than any of the three other conditions. Furthermore, both the prior environment and the visual KR conditions led to more accurate responses than the target-only condition. This suggests that the static cues available in the prior environment and visual KR conditions permitted the subjects to improve the accuracy of their movements. The second observation that can be made is that the passage to the transfer phase caused a significant increase in error only for those subjects who trained under normal visual condition. This replicates our previous observations.

Our main interest was in determining whether the visual information available in the prior environment and the visual KR conditions was still used after moderate practice. The results of interest are illustrated in Figure 4.3b. The first observation that can be made is that the 165 acquisition trials permitted the subjects who practiced under the target-only and the visual KR conditions to produce aiming errors which were approximately two thirds of those observed after 15 acquisition trials. No such decrease in error was found

L. Proteau

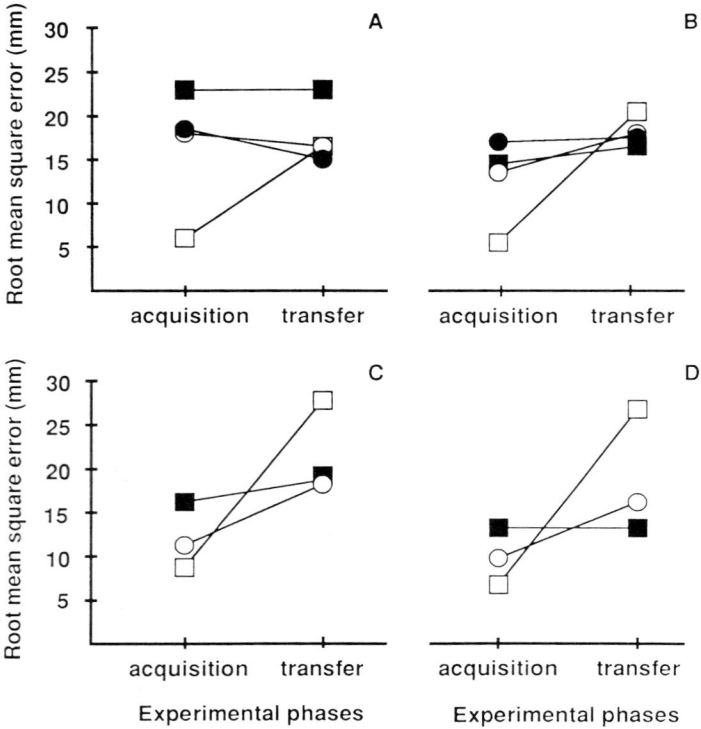

Figure 4.3. Root mean square error of aiming to one (c, d) or five different (a, b) targets. The results illustrate the performance observed at the end of practice and for immediate transfer after 15 (a), 165 (b), 200 (c) and 1,200 (d) trials of practice. The filled squares illustrate the performance under the no-vision condition, while the performance under the normal visual condition is illustrated by open squares. The performance under the prior environment and the visual KR conditions are illustrated by filled and open circles, respectively.

for the prior environment condition. More interestingly, however, is the fact that the subjects submitted to the normal vision and visual KR conditions suffered a larger increase in error when submitted to the transfer condition than those subjects who only performed 15 acquisition trials. Finally, and as was the case after modest practice, the passage to the transfer condition did not affect the response accuracy of the subjects who trained under the prior environment condition. Again, these results clearly support the idea that the more one practices a task under a normal visual condition, the more the available visual information becomes important for movement control. Moreover, it appears that one of the static parameters which is of particular importance is the opportunity to see the stylus just after it touched the target. Hay and Beaubaton (1986; see also Beaubaton & Hay, 1986) have suggested that this type of information is particularly useful as a feedforward process in that seeing one's error can be used to update one's movement representation on a following trial. In light of these results we decided to investigate the role played by the visual information in the visual KR condition after extensive practice.

In the second experiment six groups of subjects were used. Three of these groups practiced the task for 200 trials whereas the subjects in the other three groups practiced the task for 1,200 trials. Under each level of practice, there was a group of subjects who performed the task under the normal vision and the target-only condition, while the last group practiced under the visual KR condition. It should also be noted that a single target was used. The results obtained after 200 trials of acquisition replicate quite well those observed after 165 trials of practice in the preceding experiment (see Figure 4.3c). The only difference is that the accuracy of the subject's responses was slightly less than that illustrated in Figure 4.3b. On the one hand, it appears that having the opportunity to practice aiming movements to different targets helped the subjects to be more accurate than in a single-target condition. This aspect of the results support a variability of practice hypothesis (Schmidt, 1975; but see also Van Rossum, 1990).

The results obtained for the subjects who trained for 1,200 trials (see Figure 4.3d) are a close replication of those observed after only 200 trials of practice. In that regard the results obtained under the full vision condition are somewhat troubling, because they do not replicate other results from our laboratory. We have no ready explanation for this discrepancy. In terms of the results obtained for the visual KR condition, the joint results obtained for the last two experiments seem to indicate that seeing one's arm as it reaches the target is a valuable source of information to ensure optimal accuracy. It therefore appears that it indeed plays a role as a feedforward provider. More specifically, it can be proposed that this information is useful to update the motor program that is thought to control the first (so called "ballistic") part of an aiming movement (Abrams, this volume; see also Crossman & Goodeve, 1963/83; Woodworth, 1899; but see Teasdale et al., 1990 for a different interpretation). When this information is not present, as in the transfer phases of the previous experiments, this between-trial updating is no longer possible and the individual must rely solely on the movement representation developed through learning. This causes an almost constant decrease in accuracy, as we have shown. Finally, the results obtained for the prior environment condition indicate that the information available in such a context favors movement accuracy very early in learning. However, this beneficial effect is rather ephemeral. Such is the case because the subjects who practiced under the no-vision condition had not had the benefit of it, yet were still, after only moderate practice, as accurate as the subjects who had.

Overview

I see the results reported in this section as supportive of the idea that for movement information intrinsic to the task, learning does not result necessarily in a shift from one source of stimulus information to another (for example, from vision to proprioception), or in a gradual diminishing of the importance of stimulus information; rather, a major characteristic of motor learning seems to be its relative specificity to the feedback sources available when it occurred. Not surprisingly, it also appears that the visual dynamic cues available while one is performing an aiming or aiming-like movement are the main contributors to movement accuracy. The static cues available in the environment prior to movement initiation would only be beneficial very early in learning or would require that the individual see both his hand and the intended target immediately prior to movement initiation (Prablanc et al., 1979a, b). Finally, the static cues available immediately after movement completion would favor movement accuracy through a feedforward process (Hay & Beaubaton, 1986; Beaubaton & Hay, 1986).

Further Test of a Specificity of Learning Hypothesis

The literature reviewed above shows quite clearly that if the type or amount of feedback is changed, performance is negatively affected, and that this effect increases with amount of prior practice. Arising from the specificity hypothesis, the present theoretical position predicts that adding a relevant source of movement control information to a task previously learned without that information present will result in a performance decrement relative to an appropriate control condition, until the performer can integrate the new information into the existing reference store.

The first line of support for this proposition comes from a series of experiments performed by Held and his co-workers (Held, 1965; Held & Bauer, 1967; Held & Hein, 1963; Hein & Held, 1967). The basic idea behind these experiments was to rear young animals in situations that made it impossible for them to match proprioceptive and visual feedback while walking or reaching for an object. For example, in the Held and Bauer (1967) study, infant monkeys were reared for 34 days in an apparatus which permitted them to freely move their limbs but prevented them to see their moving limbs. During that period the monkeys were further trained to reach for a nursing bottle, while being prevented from seeing their reaching limb. After this acquisition or training period, one of the monkeys' forelimbs was exposed and a nursing bottle was presented to the monkey. The results from this transfer phase indicate that sight of the nursing bottle elicited a reaching movement toward the bottle. However, this reaching movement stopped as soon as the hand entered the monkey's view. Furthermore, subsequent reaching movements were inaccurate until what can be called a closed-loop reaching process (looking at the hand and the bottle alternatively) was implemented. These results thus clearly support a specificity of learning hypothesis. Further support for the above proposition comes from a variety of studies in which human adult subjects performed a linear positioning task or a manual aiming task. These studies will be briefly reviewed next.

Linear Positioning

Johnson (1980) used a linear positioning task where the subjects had to slide a cursor for 25.4 cm or 50.8 cm. Following 15 acquisition trials performed in a no-vision condition with KR available, the subjects were submitted to a transfer condition in which KR was withdrawn and where visual information regarding the ongoing movement was either made available, or not. The withdrawal of KR resulted in a 2.17 cm increase of AE, while KR withdrawal and the addition of visual information resulted in a 2.71 cm AE. Similar results were also obtained by Reeve and Cone (1980) and Reeve and Mainor (1983) using a similar task and also a modest number of acquisition trials. It thus appears that adding visual information resulted in an increase in error, which supports a specificity of learning hypothesis. What is not clear, however, is whether this specificity effect increases as the amount of practice at the task increases.

Adams et al. (1972) in a study I discussed in a previous section of this chapter reported some data that can be used to answer the previous question. As previously reported, they had eight groups of subjects learn a linear positioning for either 15 or 150 acquisition trials performed under either augmented or minimal afference conditions. I have already presented the results obtained in a transfer condition where the sources of afference went from augmented in acquisition to minimal in transfer. However, in other conditions, the subjects learned the task under the minimal afference conditions and, following 15 or 150 acquisition trials each followed with KR, were submitted to a transfer task in which the afferent information available remained the same as in acquisition, or went from minimal (no-vision, white noise to prevent hearing an almost frictionless slide move on its track) to augmented (the slide is moved against spring tension, and blindfold and white noise are withdrawn). No verbal KR was given in the transfer condition. In transfer, the subjects who only had KR withdrawn suffered an increase in error. This increase was of 9.2 mm for the subjects who had 15 trials of practice while it was of 7.0 mm for the subjects who trained for 150 trials. The effects of KR withdrawal *per se* thus appears to diminish as a function of practice. More importantly, it further was observed that the subjects who went from the minimal afference condition in acquisition to the augmented afference in transfer also suffered an increase in error. This increase in error was of 13.7 mm for the subjects who had trained for 15 trials while it was of 23.6 mm for those subjects who had trained for 150 trials! The results of the last two groups cannot be explained on the basis of a KR withdrawal effect because the transfer to the augmented condition caused an increase in error that was at least twice that observed when only KR had been withdrawn. Therefore, the increased error found for the subjects who went from the minimal to the augmented afference condition clearly indicated that adding a source of afference interfered with what had been learned. More importantly, the fact that this interference grew stronger as a function of the number of acquisition trials supports our contention of an increased specificity of the sources of afference used for motor control as practice at the task increases.

Manual Aiming

Elliott and Jaeger (1988) had three groups of subjects practice a manual aiming task, involving a 35 cm movement that had to be completed in an MT bandwidth ranging from 300 to 400 ms. The subjects of a first group performed the task in a full vision condition. The subjects of a second group performed the task under a target-only condition. Finally, for the third group, the lights of the experimental room were turned off two seconds prior to movement initiation (T + 2s). Following a 70 trials acquisition phase, all three groups performed under three transfer conditions that were identical to the three acquisition conditions described above. The subjects who trained in either the target-only or the T+2s conditions suffered an increase of error when transferred to the full vision condition. These results therefore replicated quite well the results reported above. That is, as predicted by a specificity of learning hypothesis, the addition of visual information regarding the ongoing movement caused a significant increase of error for those subjects who had practiced without it being available.[8]

We recently addressed the same question (Proteau, Marteniuk, & Lévesque, in press). Specifically, we wished to test the specificity of learning hypothesis by demonstrating that, for a movement aiming task that had been learned in the absence of an important source of sensory information (in this

---

[8]The results of the transfer tests indicated that the subjects who trained in the full vision condition suffered an increase of aiming error when performing under either the T or T+2s transfer conditions. These results thus replicated quite well those reported in the previous section.

case vision), the addition of this information would result in a movement decrement compared to a control condition. Our control condition was a baseline measure of performance established for the first 40 trials (pre-test) of an aiming task performed in a normal vision condition. For the experimental group, this baseline measure was compared to performance again obtained in normal vision condition, but this time measured in a transfer test following 200 (transfer 1) and 1,200 (transfer 2) practice trials in a target-only condition. Two control groups also participated in this study. The subjects of the first control group trained for 1,200 trials, but this time in the target-only condition. They were submitted to the same pre-test and transfer tests as our experimental group and thus, in transfer, were only submitted to the withdrawal of KR. Finally, the subjects of the second control group only participated in the pre-test and both transfer tests of the study.

Our hypothesis predicted that for the experimental group, transfer performance on the aiming task in a normal vision condition would be worse than the pre-test performance measured under exactly the same conditions, even though, in between these two tests, subjects had the benefit of either 200 or 1,200 trials of practice with knowledge of results and with vision of the target. No such effect should be seen for either control groups. Furthermore, we also wanted to determine if, as predicted by our theoretical position, the magnitude of the hypothesized increase in aiming error for the experimental group would be related to the amount of practice given under the no-vision condition. A larger aiming error for the second than the first transfer test would be supportive of the view that the sources of afference used to control an ongoing movement become more specific to the conditions prevailing during practice, and that therefore, the addition of a new source of afference interferes more with what has been learned.

Figure 4.4 illustrates the spatial accuracy results obtained for all three groups of subjects. As expected, the subjects of the second control group, who only participated in the pre-test and the transfer tests, were equally accurate in each of these three experimental phases. Also expected was the fact that the aiming error of the subjects of the first control group, who practiced in the normal vision condition, decreased from transfer 1 to transfer 2. This decrease of aiming error clearly reflects that the practice period helped the subjects to become more accurate. More interestingly, the specificity of learning hypothesis was well supported by at least two aspects of the results observed for the experimental group. First, when vision of the ongoing limb and surrounding environment were permitted after 1,200 trials of practice in the target-only condition, it caused a significant *increase* in error when compared to the aiming accuracy observed for the preceding 40 practice trials performed in a no-vision context (see Figure 4.4c). This result is clearly

consistent with the proposition that learning makes performance more specific
to the actual practice conditions. Secondly, the results showed an increase in
spatial error in the transfer tests when compared to the pre-test level (see
Figure 4.4a, b and c). These results indicate that 1,200 trials of practice
without vision of the arm does not transfer positively to a condition where
vision of the arm is available. Moreover, it appears that the use of vision
interferes with what has been learned, in that there was an increase in the
spatial error of aiming when vision of the performing limb and surrounding
environment were made available. Hence, withdrawing or adding a significant
source of information after a period of practice where it was respectively
present or absent results in a deterioration of performance. These results are
in direct contradiction with any proposition suggesting that learning involves
progressing from a closed-loop mode of control to an open-loop one, or
progressing from a visually to a kinesthetically guided movement.

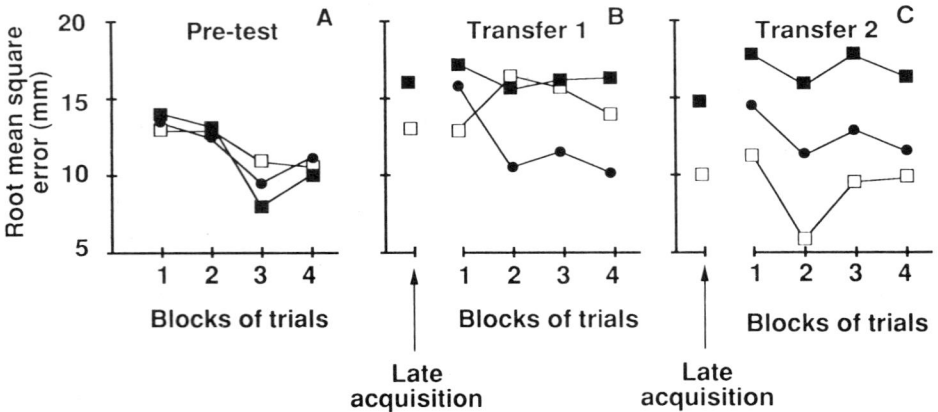

*Figure 4.4. Root mean square error of aiming in the pre-test, the end of an
acquisition phase and its associated transfer test. The transfer tests occurred
after 200 and 1,200 trials of practice, respectively. The filled circles indicate
the performance of the control group for which their was no practice. The
open and filled squares indicate the performance of the subjects who trained
under the normal vision and the no-vision conditions, respectively. Here it
should be noted that vision of the ongoing limb and surrounding environment
was present for all subjects in transfer. (Adapted with permission from
Proteau et al., in press).*

Our hypothesis of increased specificity as a function of practice was not directly supported because no significant increase in error took place for the subjects of the experimental group from transfer 1 to transfer 2. A possible explanation of this result might be that adding 1,000 trials of practice between transfer 1 and transfer 2 was not sufficient for the subjects to go from one level of stimulus discrimination to one of a higher order (Gibson, 1969) even if learning occurred between transfer 1 and transfer 2. Therefore, adding vision in transfer should lead to the same performance. It might also be that the fact the subjects were permitted to see where the stylus landed on the target area in the transfer experimental phase resulted in a ceiling effect that was already reached after 200 trials of practice. Both explanations are speculative and more work is clearly needed.

It should finally be noted that the increase in error found in the transfer conditions for the subjects who trained under the target-only protocol persisted for at least 40 trials. This aspect of the results might appear surprising because the subjects were able to see where the stylus hit the target area, and therefore may possibly have been able to evaluate the accuracy of their responses. However, through practice in the target-only condition the subjects had probably developed a close relation between kinesthetic feedback of the stylus touching the target area and the verbal KR provided by the experimenter. When, in transfer, the subjects were permitted to see where the stylus touched the target area, they benefited from a form of "visual KR" at the same time as they lost the "verbal KR" provided by the Experimenter in acquisition. The results indicated that the subjects were not able to use that "visual KR" effectively. This may mean that not only is movement control specific to the feedback conditions present during practice, but that the sensory channel(s) used for the evaluation of the movement outcome may also be specific to the practice conditions.

Conclusion

Numerous aspects of the results presented in this chapter argue against the view that the utility of sensory feedback information decreases (Schmidt, 1975) as a function of the amount of practice or expertise of the subjects, or that it is changed with practice (Fleishman & Rich, 1963). If the utility of sensory information was to decrease as the amount of practice at the task increased, one should be able to demonstrate that, for example, the passage to the transfer condition had less dramatic consequences on the subject's performance after modest than moderate practice. This would be the case because the subjects, after a larger amount of practice, would rely less and less on visual information to control their movements. Therefore, its withdrawal should have a smaller effect than after modest practice, where the subjects should have relied more heavily on it. On the contrary, the results presented in this chapter can be summarized by proposing that withdrawing or adding a significant source of information after a period of practice where it was respectively present or absent results in a deterioration of performance.

In line with the pioneering work of Held and his co-workers, who suggested that what is critical for the normal development of visuomotor coordination is the opportunity to correlate proprioceptive and visual feedback, I suggest that it is the visual dynamic cues available while one is performing an aiming or aiming-like movement that are the main contributors to movement accuracy. It further appears that the utilization of the visual information available when the hand and the target are in central vision can not be substituted for by extensive practice. Again this is in line with Held and Bauer's (1967) observations. Here I would like to recognize that this conclusion might very well be task specific. In line with Wallace and Newell (1983) and Davids et al. (1989) studies, it is very likely that the margin of tolerable error interacts strongly with the importance of visual information for the control of an ongoing movement. If the accuracy requirements of the task are very easy to meet, one can imagine that the first so-called ballistic part of an aiming movement can, with sufficient practice, be fine-tuned to the point that on-line control is not required for the attainment of the goal. In such a situation the withdrawal of visual information would not affect performance. In terms of the static cues available in the environment prior to movement initiation, it appears that they would only be beneficial very early in learning or else would require that the individual could see both his hand and the intended target immediately prior to movement initiation (Prablanc et al., 1979a, b). In the latter case, the static cues would thus play a major role for movement planning. Finally, the static cues available immediately after movement completion would favor movement accuracy through a feedforward process (Hay & Beaubaton, 1986, Beaubaton & Hay, 1986).

My collaborators and I have taken the results presented in this chapter as support for the notion that, at least for a task with very stringent accuracy requirements, learning results in a representation consisting of an integration of all relevant information about the movement task (here, task is defined as the interaction of the individual with the environment during the attempt to fulfill a movement goal) which becomes more tightly integrated with experience at the task. In more general terms, it could be proposed that early in the learning of a task, the different sources of available information that are used to help control the ongoing movement are compared *intramodally* (for example: visual afference to visual expected sensory consequences) to some internal representation of the movement. In fact, we agree with Schmidt's (1975, 1988) proposition that once the movement goal has been identified, the individual issues the appropriate motor commands and the expected sensory consequences associated with these motor commands. As the movement unfolds the actual sensory consequences are compared *intramodally* to the expected sensory consequences and, given enough time, appropriate corrections are issued to correct for the detected errors. This proposition is in agreement with the results reported by Adams et al. (1977), Elliott and Jaeger (1988) and Proteau (1991) who only showed a slight increase in spatial error when vision was withdrawn early in practice. However, as learning increases, it appears that the different sources of afference are used in a much different way. The results presented in this chapter led us to believe that the different sources of available sensory information are compared to an integrated or *intermodal* representation of the expected sensory consequences. We would like to argue that this integrated store provides more information than the mere addition of the different sources of afference and, therefore, leads to better performance as training at the task increases. However, the withdrawal of one source of information used to develop the intermodal store leaves the individual with an incomplete reference store and causes a decrease in performance.

In closing this chapter I would like to remind the reader that Gibson (1969) has argued in favor of a specificity hypothesis for perceptual learning. Her argument was that there are potential variables of stimuli which are not differentiated from each other early in learning, but which may be, given the proper conditions of exposure and practice. Learning would thus be the modulation of a particular response to a newly acquired level of stimulus discrimination. She went further and proposed that "It would be hard to overemphasize the importance for perceptual learning of the discovery of invariant properties which are in correspondence with physical variables" (p. 81), and that the discovery of multimodal invariants are required for skillfully directed action (p. 380). This suggests that higher order information might be inherent to multimodal stimulation or afference and that a reduction in either the utilization of afference or the number of different afferent sources is more likely to hamper learning than to favor it. Our interpretation of the results presented in this chapter is thus very close to that of Gibson (1969). It is, however, different in the sense that we do not limit our interpretation to the perceptual processes but extend it to the implementation of its associated motor responses. Finally, we suggest that once multimodal (or intermodal, in our terms) "perceptual-motor" invariants have been discovered by an individual through extended practice, he/she cannot instantly revert to a mode of control in which only intramodal information is used. This is the case because it was clearly shown that withdrawing a source of information after extended practice caused a larger error than when than information was withdrawn after modest or moderate practice.

References

Abbs, J.H., Gracco, V.L., & Cole, K.J. (1984). Control of multimovement coordination: Sensorimotor mechanisms in programming. *Journal of Motor Behavior, 16,* 195-231.

Abrams, R.A., Meyer, D.E., & Kornblum, S. (1990). Eye-hand coordination: Oculomotor control in rapid aimed limb movements. *Journal of Experimental Psychology: Human Perception and Performance, 15,* 248-267.

Adams, J.A. (1971). A closed-loop theory of motor learning. *Journal of Motor Behavior, 3,* 111-150.

Adams, J.A., & Goetz, E. T. (1973). Feedback and practice as variables in error detection and correction. *Journal of Motor Behavior, 5,* 217-224.

Adams, J.A., Goetz, E.T., & Marshall, P.H. (1972). Response feedback and motor learning. *Journal of Experimental Psychology, 92,* 391-397.

Adams, J.A., Gopher, D., & Lintern, G. (1977). Effects of visual and proprioceptive feedback on motor learning. *Journal of Motor Behavior, 9,* 11-22.

Annett, J. (1959). Learning a pressure under conditions of immediate and delayed knowledge of results. *Quarterly Journal of Experimental Psychology, 11,* 3-15.

Bard, C., Hay, L., & Fleury, M. (1985). Role of peripheral vision in the directional control of rapid aiming movements. *Canadian Journal of Psychology, 39,* 151-161.

Bard, C., Paillard, J., Fleury, M., Hay, L., & Larue, J. (1990). Positional versus directional control loops in visuomotor pointing. *European Bulletin of Cognitive Psychology, 10,* 145-156.

Beaubaton, D., & Hay, L. (1986). Contribution of visual information to feedforward and feedback processes in rapid pointing movements. *Human Movement Science, 5,* 19-34.

Burton, H.W., Pick, H.L., & Holmes, C.H. (1990). The independence of horizontal and vertical dimensions in handwriting with and without vision. *Acta Psychologica, 75,* 201-212.

Carlton, L.G. (1981). Visual information: The control of aiming movements. *Quarterly Journal of Experimental Psychology, 33A,* 87-93.

Chew, R.A. (1976). Verbal, visual and kinesthetic error feedback in the learning of a simple motor task. *Research Quarterly, 47,* 254-259.

Conti, P., & Beaubaton, D. (1976). Utilisation des informations visuelles dans le contrôle du mouvement: étude de la précision des pointages chez l'homme. [Utilization of visual information for the control of movement: Accuracy of pointing movements in human.]. *Le Travail Humain, 39,* 19-32.

Cordo, P.J., & Flanders, M. (1990). Time-dependent effects of kinesthetic input. *Journal of Motor Behavior, 22,* 45-65.

Cox, R.H., & Walkuski, J.J. (1988). Kinesthetic sensitivity motor learning. *Journal of Human Movement Studies, 14,*

Crossman, E.R.F.W., & Goodeve, P.J. (1983). Feedback cc movement and Fitts' Law. Paper presented at the n⸺ ⸺ ⸺ Experimental Psychology Society, Oxford, July 1963. Published in *Quarterly Journal of Experimental Psychology, 35A,* 251-278.

Davids, K.W., & Stratford, R. (1989). Peripheral vision and simple catching: The screen paradigm revisited. *Journal of Sports Sciences, 7,* 139-152.

Davids, K.W., Rex de Palmer, D., & Savelsbergh, G.J.P. (1989). Skill level, peripheral vision and tennis volleying performance. *Journal of Human Movement Studies, 16,* 191-202.

Diggles, V.A., Grabiner, M.D., & Garhammer, J. (1987). Skill level and efficacy of effector visual feedback in ball catching. *Perceptual and Motor Skills, 64,* 987-993.

Elliott, D. (1988). The influence of visual target and limb information on manual aiming. *Canadian Journal of Psychology, 41,* 57-68.

Elliott, D., & Allard, F. (1985). The utilization of visual feedback information during rapid pointing movements. *Quarterly Journal of Experimental Psychology, 37A,* 407-425.

Elliott, D., & Jaeger, M. (1988). Practice and the visual control of manual aiming movements. *Journal of Human Movement Studies, 14,* 279-291.

Elliott, D., & Madalena, J. (1987). The influence of premovement visual information on manual aiming. *Quarterly Journal of Experimental Psychology, 39A,* 541-559.

Fischman, M.G., & Schneider, T. (1985). Skill level, vision and proprioception in simple one-hand catching. *Journal of Motor Behavior, 17,* 219-229.

Fitts, P.M. (1954). The information capacity of the human motor system in controlling the amplitude of movement. *Journal of Experimental Psychology, 47,* 381-391.

Flanders, M., & Cordo, P.J. (1989). Kinesthetic and visual control of a bimanual task: Specification of direction and amplitude. *Journal of Neuroscience, 9,* 447-453.

Fleishman, E.A., & Rich, S. (1963). Role of kinesthetic and spatial-visual abilities in perceptual-motor learning. *Journal of Experimental Psychology, 66,* 6-11.

Gibson, E.J. (1969). *Principles of perceptual learning and development.* New York: Appleton-Century-Crofts.

Hay, L., & Beaubaton, D. (1986). Visual correction of rapid goal-directed response. *Perceptual and Motor Skills, 62,* 51-57.

Hein, A., & Held, R. (1967). Dissociation of the visual placing response into elicited and guided components. *Science, 158,* 390-391.

Held, R. (1965). Plasticity in sensory-motor systems. *Scientific American, 213*, 84-94.

Held, R., & Bauer, J.A. (1967). Visually guided reaching in infant monkeys after restricted rearing. *Science, 155*, 718-710.

Held, R., & Hein, A. (1963). Movement produced stimulation in the development of visually guided behavior. *Journal of Comparative Physiological Psychology, 53*, 236-241.

Howarth, C.I., & Beggs, W.D.A. (1981). Discrete movements. In D. Holding (Ed.), *Human skills* (pp. 91-117). New York: Wiley.

Johnson, P. (1980). The relative weightings of visual and nonvisual coding in a simple motor learning task. *Journal of Motor Behavior, 12*, 281-291.

Jones, B. (1982). The development of intermodal co-ordination and motor control. In J.A.S. Kelso & J.E. Clark (Eds.), *The development of movement control and co-ordination* (pp. 95-109). New York: Wiley.

Kandel, E.R., & Schwartz, J.H. (1985) (Eds.). *Principles of neural science, 2nd edition*. New York: Elsevier.

Keele, S.W. (1968). Movement control in skilled motor performance. *Psychological Bulletin, 70*, 387-403.

McCloskey, D.I. (1978). Kinesthetic sensibility. *Physiological Review, 58*, 763-820.

Meyer, D.E., Abrams, R.A., Kornblum, S., Wright, C.E., & Smith, J.E.K. (1988). Optimality in human motor performance: Ideal control of rapid aimed movements. *Psychological Review, 95*, 34-370.

Newell, K.M., & Chew, R.A. (1975). Visual feedback and positioning movements. *Journal of Motor Behavior, 7*, 143-158.

Paillard, J. (1980). The multichanneling of visual cues and the organization of a visually guided response. In G.E. Stelmach & J. Requin (Eds.), *Tutorials in motor behavior* (pp. 259-279). Amsterdam: North-Holland.

Paillard, J., & Amblard, B. (1985). Static versus kinetic visual cues for the processing of spatial relationships. In D.J. Ingle, M. Jeannerod, & D.N. Lee (Eds.), *Brain mechanisms of spatial vision* (pp. 367-385). La Haye: Martinus Nijhoff.

Pew, R.W. (1966). Acquisition of hierarchical control over the temporal organization of a skill. *Journal of Experimental Psychology, 71*, 764-771.

Pew, R.W. (1974). Human perceptual-motor performance. In B.H. Kantowitz (Ed.), *Human information processing: Tutorials in performance and cognition* (pp. 1-39). Hillsdale, NJ: Erlbaum.

Poulton, E.C. (1981). Human manual control. In V. Brooks (Ed.), *Handbook of physiology: Section I: The nervous system. Vol II. Motor control Part 2* (pp. 1337-1389). Baltimore: American Physiological Society.

Prablanc, C., Echallier, J.F., Komilis, E., & Jeannerod, M. (1979 a). Optimal response of eye and hand motor systems in pointing at a visual target: I.

Spatio-temporal characteristics of eye and hand movements and their relationships when varying the amount of visual information. *Biological Cybernetics, 35*, 113-124.

Prablanc, C., Echallier, J.E., Jeannerod, M., & Komilis, E. (1979 b). Optimal responses of eye and hand motor systems in pointing at a visual target: II. Static and dynamic visual cues in the control of hand movement. *Biological Cybernetics, 35, 183-187.*

Proteau, L. (1991). *Extensive practice in a manual aiming task : A test of the specificity of learning hypothesis.* Submitted for publication.

Proteau, L., & Cournoyer, J. (1990). Vision of the stylus in a manual aiming task: The effects of practice. *Quarterly Journal of Experimental Psychology, 42A*, 811-828.

Proteau, L., & Marteniuk, R.G. (1991). [On the role of static cues for the learning of a manual aiming task]. Unpublished raw data.

Proteau, L., Marteniuk, R.G., Girouard, Y., & Dugas, C. (1987). On the type of information used to control and learn an aiming movement after moderate and extensive training. *Human Movement Science, 6, 181-199.*

Proteau, L., Marteniuk, R. G., & Lévesque, L. (in press). A sensorimotor basis for motor learning: Evidence indicating specificity of practice. *Quarterly Journal of Experimental Psychology.*

Reeve, T.G., & Cone, S.L. (1980). Coding of learned kinesthetic location information. *Research Quarterly, 51,* 349-358.

Reeve, T.G., & Mainor, R., Jr. (1983). Effects of movement context on the encoding of kinesthetic spatial information. *Research Quarterly for Exercise and Sport, 54,* 352-363.

Rosengren, K.S., Pick, H.L., & Von Hofsten, C. (1988). Role of visual information in ball catching. *Journal of Motor Behavior, 20,* 150-164.

Savelsbergh, G.J.P., & Whiting, H.T.A. (1988). The effect of skill level, external frame of reference and environmental changes on one-handed catching. *Ergonomics, 31,* 1655-1663.

Schmidt, R.A. (1972). The index of preprogramming (IP): A statistical method for evaluating the role of feedback in simple movements. *Psychonomic Science, 27,* 83-85.

Schmidt, R.A. (1975). A schema theory of discrete motor skill learning. *Psychological Review, 82,* 225-260.

Schmidt, R.A. (1988). *Motor control and learning: A behavioral emphasis: 2nd edition.* Champaign, IL: Human Kinetics.

Schmidt, R.A., & McCabe, J.F. (1976). Motor program utilization over extended practice. *Journal of Human Movement Studies, 2,* 239-247.

Schmidt, R.A., & Russell, D.G. (1972). Movement velocity and movement time determiners of degree of preprogramming in simple movements. *Journal of Experimental Psychology, 96,* 315-320.

Shimojo, S. (1987). Attention-dependent visual capture in double vision. *Perception, 16,* 445-447.

Smyth, M.M. (1977). The effect of visual guidance on the acquisition of a simple motor task. *Journal of Motor Behavior, 9,* 275-284.

Smyth, M.M. (1978). Attention to visual feedback in motor learning. *Journal of Motor Behavior, 10,* 185-190.

Smyth, M.M. (1982). Sight of the hand in catching. *Journal of Motor Behavior, 14,* 255-256.

Smyth, M.M. (1986). A note: Is it a catch or a fumble? *Journal of Motor Behavior, 18,* 492-495.

Smyth, M.M., & Murray, M.A. (1982). Vision and proprioception in simple catching. *Journal of Motor Behavior, 14,* 143-152.

Smyth, M.M., & Turton, A. (1983). *Peripheral occlusion of the ball in one-handed catching.* Unpublished manuscript.

Soechting, J.F., & Flanders, M. (1989a). Sensorimotor representations for pointing to targets in three-dimensional space. *Journal of Neurophysiology, 62,* 582-594.

Soechting, J.F., & Flanders, M. (1989b). Errors in pointing are due to approximations in sensorimotor transformations. *Journal of Neurophysiology, 62,* 595-608.

Stubbs, D.F. (1976). What the eye tells the hand. *Journal of Motor Behavior, 8,* 43-58.

Teasdale, N., Blouin, J., Bard, C., & Fleury, M. (1991). *Visual guidance of pointing movements: Kinematic evidence for static and kinetic feedback channels.* Submitted for publication.

Van der Meulen, J.H.P., Gooskens, R.H.J.M., Denier Van der Gon, J.J., Gielen, C.C.A.M., & Wihelm, K. (1990). Mechanisms underlying accuracy in fast goal-directed arm movements in man. *Journal of Motor Behavior, 22,* 67-84.

Van Rossum, J.H.A. (1990). Schmidt's schema theory: The empirical base of the variability of practice hypothesis. A critical analysis. *Human Movement Science, 9,* 387-435.

Velay, J.-L., & Beaubaton, D. (1986). Influence of visual context on pointing movement accuracy. *European Bulletin of Cognitive Psychology, 6,* 447-456.

Wallace, S.A., & Newell, K.M. (1983). Visual control of aiming movements. *Quarterly Journal of Experimental Psychology, 35A,* 311-321.

Woodworth, R.S. (1899). The accuracy of voluntary movement. *Psychological Review, 3,* (Monograph Supplement), 1-119.

Zelaznik, H.N., Hawkins, B., & Kisselburgh, L. (1983). Rapid visual processing in single-aiming movements. *Journal of Motor Behavior, 15,* 217-236.

Acknowledgment

The research realized in my laboratory is supported by the Natural Sciences and Engineering Research Council of Canada and by the Fonds F.C.A.R., gouvernement du Québec. I would also like to thank Claude Alain, Tim Lee and Dan Weeks for their comments on a previous version of this chapter.

VISION AND MOTOR CONTROL
L. Proteau and D. Elliott (Editors)

CHAPTER 5

ADAPTIVE EYE-HAND COORDINATION: IMPLICATIONS OF PRISM ADAPTATION FOR PERCEPTUAL-MOTOR ORGANIZATION

GORDON M. REDDING* and BENJAMIN WALLACE**

*Department of Psychology, Illinois State University
Normal, IL, 61761

**Department of Psychology, Cleveland State University
Cleveland, OH, 44115

Perceptual-motor adaptability is certainly one of the more salient and important human capacities. Perhaps the most obvious example of such adaptability is skilled performance, ranging from primitive athletic and linguistic abilities to more modern skills such as machine control and performance in alien work environments. Less obvious, but in many ways more fundamental, is our ability to recover from systematic encoding dysfunctions arising from growth, pathology, or drift which misrepresent stimulus information and therefore specify inappropriate responses. The study of such perceptual adaptation is commonly called "prism adaptation".

Skill learning and perceptual adaptation, while both adaptive in the most general sense of the word, seem to be fundamentally different processes. Figure 5.1 illustrates a system analysis comparison of the two processes. The stimulus for skill learning is an objective change or temporal sequence of changes in the distal stimulus, while the stimulus for perceptual adaptation is a systematic distortion such that the relationship between distal and proximal stimuli is changed in some important respect.

Skill learning consists in determining the parameters of the changing stimulus (perceptual learning) and modifying motor programs (motor learning) such that the response becomes spatially congruent with the stimulus. For instance, the new visual and vestibular movement cues which occur when one first begins to ride a bicycle require development of new motor responses.

Once achieved, the skilled response can be immediately evoked whenever the situational cues are reinstituted, but a skill does not generalize well to situations very different from the conditions under which it was originally acquired.

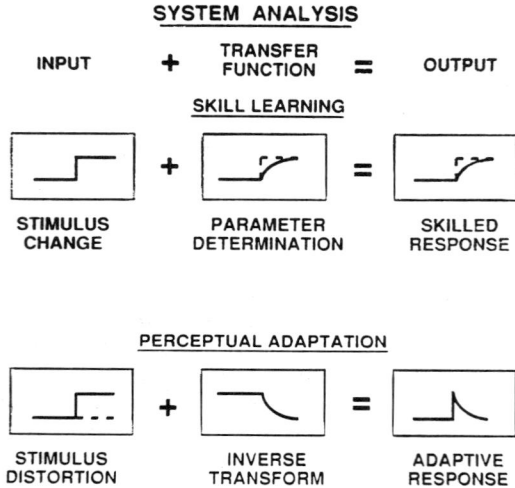

Figure 5.1. System analysis comparison of skill learning and perceptual adaptation. Differences are illustrated for input to the system, output from the system, and the mediating transfer function.

Perceptual adaptation consists in finding an inverse transform which will remove the stimulus distortion, thereby reestablishing veridical perceptual response. For instance, performance error arising from optical displacement of the visual field may be eliminated by retransformation of the proximal stimulus such that the distal stimulus is recovered and the original response is again appropriate. Once established, an inverse transform applies to all situations involving that particular perceptual modality, and any further stimulus distortion or removal of the distortion requires reinstitution of the adaptive process.

The prism-adaptation paradigm is the most direct method available for studying this kind of adaptive perceptual capacity. Prism adaptation tasks usually also involve skill learning, but the experimental paradigm includes controls which permit isolation of perceptual adaptation. It is this kind of adaptive capacity which is the focus of the present chapter. It is our belief, however, that prism adaptation cannot be understood outside the larger context of perceptual-motor organization.

In the following pages we selectively review the prism adaptation paradigm and research, suggesting some implications for perceptual-motor organization. Then, we sketch a model of adaptive eye-hand coordination which identifies prism adaptation as an experimental example of a more general mechanism which maintains spatial alignment among the various parts of the perceptual-motor system. Finally, we conclude with a summary of our attempts to test the model.

Prism Adaptation

Prism adaptation is a surprisingly enduring subject of research. Beginning with the work of Von Helmholtz (1911/1962) in Germany and Stratton (1896) in the United States, prism adaptation was a minor but persistent research area until the systematic work of Kohler (1951/1964) at Innsbruck and Held (1961) at MIT which produced an intense flurry of research in the '60s (e.g., Harris, 1965; Hay & Pick, 1966; Rock, 1966) and '70s (for a review see Welch, 1978). This momentum has abated somewhat, but still the prism adaptation paradigm is frequently used to investigate problems of perceptual and perceptual-motor plasticity (e.g., Bedford, 1989). This persistent popularity is due both to the methodological advantages of the paradigm and to the extensive perceptual-motor plasticity revealed by the research.

Experimental Paradigm

The prism adaptation paradigm consists in a comparison of performance on a criterion measure after prism exposure with performance measured before exposure. The criterion task is selected to measure that aspect of perceptual-motor behavior affected by the prismatic distortion. For example, laterally displacing prisms affect perceived radial location and the criterion task requires response to perceived location. The direction of change which would be adaptive can be specified beforehand and the difference between pretest and posttest provides a measure of the level of adaptation.

Nonperceptual controls. The paradigm includes features which control for some possible nonperceptual adaptive changes. For example, the fact that tests are performed without prisms makes it unlikely that behavior change arises from conscious rules acquired during exposure. Subjects have no consistent basis for rule-based responding after the prisms are removed. Thus, the measured adaptive changes occur at primitive levels of processing and are little influenced by higher-level cognitive processes. Also, the tests and exposure task can be so dissimilar as to preclude generalization of skill learning during exposure to the prisms. Thus, obtained adaptation can safely be attributed to changes in perceptual processes.

Locus-specific tests. Perceptual adaptation is assumed to be motivated by discordance between perceptual systems. For example, if the exposure tasks requires coordination of the eye and hand in the presence of lateral displacement, the discordance occurs between the visual eye-hand system and the proprioceptive hand-head system (see Figure 5.2). Adaptation consists in the removal of intersensory discordance such that the two systems signal the same position. Discordance reduction might be achieved by spatial recalibration of either or both of the conflicting systems.

Indeed, special tests which are sensitive to specific kinds of adaptation have shown that adaptation can occur in different parts of the total perceptual-motor system (e.g., Hay & Pick, 1966; Redding, 1978). For example, eye-head coordination tests that do not involve proprioceptive (hand-head) systems have shown visual adaptation. An instance of such a visual test is to have the subject verbally instruct the experimenter to position a visual target to look straight ahead of the nose. Pre-post differences provide a measure of visual shift (VS). Conversely, hand-head coordination tests which do not involve the visual system have shown proprioceptive adaptation. An instance of such a test is to have the subject point straight ahead of the nose with eyes closed. Pre-post differences measure proprioceptive shift (PS).

Theoretical Implications
Research with the prism adaptation paradigm suggests several conclusions about the nature of the adaptive process. These conclusions have theoretical implications for how the perceptual-motor system is organized.

Local adaptation and additivity. Perceptual adaptation consists largely of local recalibration of perceptual analysis (e.g., Howard, 1971, 1982), rather than adaptive changes at higher levels such as coordinative mappings between perceptual systems (cf. Hardt, Held, & Steinbach, 1971). This conclusion is based on repeated findings of additivity: the simple algebraic sum of adaptive changes in parts of a perceptual-motor coordination loop is usually equal to the total amount of adaptation. Adaptive changes at higher levels would predict greater total adaptation than could be accounted for by local adaptation. Figure 5.3 illustrates this generalization for the eye-hand coordination loop.

While evidence of adaptive recalibration at all of the local sites illustrated in Figure 5.3 is incomplete (e.g., Welch, 1978, pp. 55-59; Wallace & Garrett, 1975), there is ample justification of the generalization. For example, the sum of VS and PS usually equals the total adaptive shift (TS) measured by the common open-loop eye-hand coordination test which involves both visual and proprioceptive systems (e.g., Hay & Pick, 1966; Redding, 1978; Templeton, Howard, & Wilkinson, 1974; Wallace, 1977; Welch, 1974; Wilkinson, 1971). Exceptions to additivity can occur (e.g., Welch, Choe, & Heinrich, 1974) but they can usually be explained by nonperceptual contamination (e.g., Redding & Wallace, 1976, 1978; Wallace & Redding, 1979).

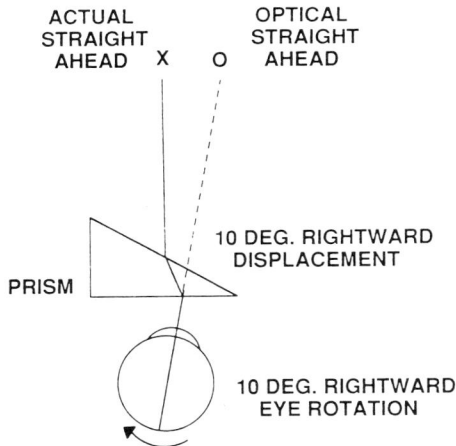

EXEMPLAR EXPOSURE TASK: COORDINATION OF EYE AND HAND

REQUIRED EYE POSITION (FIXATING TARGET): 10 DEG RIGHT
REQUIRED HAND POSITION (GRASPING TARGET): STRAIGHT AHEAD

Figure 5.2. Perceptual discordance. Prismatic displacement produces discordance between eye and hand positions required to "grasp" the same physical target.

Additive local adaptation suggests that the perceptual-motor system is organized around adaptive subsystems. The various subsystems (e.g., eye-head and hand-head) may be serially linked to perform more global perceptual-motor tasks (e.g., eye-hand coordination). In different situations adaptation will be localized in different subsystems.

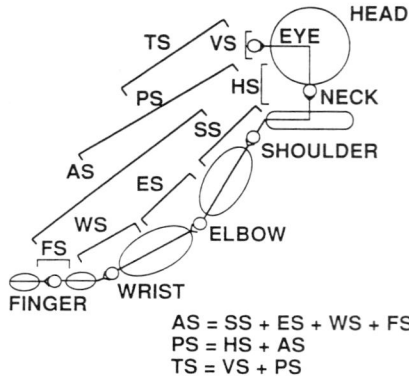

AS = SS + ES + WS + FS
PS = HS + AS
TS = VS + PS

Figure 5.3. Serial organization and component linkages of the eye-hand system. Total adaptive shift (TS) is equal to the sum of shifts in components. Visual shift (VS) refers to adaptive change in the eye-head subsystem while Proprioceptive shift (PS) refers to change(s) in position sense from head to hand. Head shift (HS) refers to adaptive shift in position sense of the head, while arm shift (AS) refers to adaptive shift in the position sense at the shoulder (SS), elbow (ES), wrist (WS), and/or finger (FS).

Variable locus and directional linkage. When discordant systems are linked to perform a coordination exposure task, adaptation is localized in the situationally subordinate system rather than the superordinate system (cf. Hamilton, 1964; Howard & Templeton, 1966). This conclusion is based upon interpretation of studies (e.g., Canon, 1970; Kelso, Cook, Olson, & Epstein, 1975) which have varied the locus (kind) of adaptation by manipulating the structure of the exposure situation (i.e., the coordination task).

For instance, Canon (1970) required subjects to track the location of a visible sounding target with their unseen right hand. In different conditions, subjects tracked either the optically displaced visual target or the pseudophonically displaced auditory target. Thus, either the eye or the ear was the superordinate system in the sense of specifying the target for the coordination response and the other systems (eye or ear and right hand) were subordinate in the sense of receiving rather than sending information. Since subjects were not permitted sight of the right hand and could never experience any discrepancy between seen and felt limb position, there was no discordance or basis for adaptation in this (subordinate) response system. Subjects could, however, experience the discordance between seen and heard target positions, and reduction of this discordance (i.e., adaptation) was largely localized in the situationally subordinate system (eye or ear). The direction of information flow (superordinate to subordinate) as specified by the structure of the exposure task determined the locus of adaptation. A similar conclusion for eye-hand coordination is suggested by the Kelso et al. (1975) study.

The implication is that the organizational linkages among subsystems of the perceptual-motor system are strategically flexible to reflect the demands of a particular task. The organization of subsystems to perform more global perceptual-motor task is achieved by directional linkages which establish a hierarchical coordination structure. Adaptation (i.e., discordance registration /reduction) occurs in the guided system(s).

Intention and coordinative linkage. Adaptation involves conscious intentional processes only indirectly in performance of the exposure (coordination) task, not directly in detection and reduction of the resultant discordance. Evidence for this conclusion comes from the repeated observation that adaptation can occur without any conscious awareness of the distortion or performance error (e.g., Howard, 1967; Howard, Anstis, & Lucia, 1974; Templeton, Howard, & Wilkinson, 1974; Uhlarik, 1973). Yet, adaptation may be depressed when subjects are required to perform a secondary cognitive task (e.g., mental arithmetic) simultaneously with the exposure task (e.g., Barr, Schultheis, & Robinson, 1976; Redding, Clark, & Wallace, 1985; Redding & Wallace, 1985) and error feedback enhances adaptation (e.g., Coren, 1966; Welch & Abel, 1970; Welch & Rhoades, 1969).

The suggestion here is that higher-level intentional processes are involved in establishing the coordinative linkages among subsystems required by the coordination (exposure) task. Awareness of performance error may increase the likelihood of coordinative linkage and distracting events may decrease the frequency of linkage, thereby indirectly affecting local registration/reduction of discordance.

Separable mechanisms and modularity. Perceptual adaptation is realized by a variety of separable but functionally similar mechanisms, rather than a single superordinate mechanism. The modular organization (cf. Fodor, 1983) of the perceptual-motor system is implicit in the observation that different kinds (loci) of adaptation occur, depending upon how the separate modules are coordinated. Local adaptive mechanisms in vision and proprioception are activated by the cooperative interaction of these systems. Further evidence comes from the impenetrability of adaptive processes by higher-level conscious processes. Higher-level cognitive processes are involved only in establishing cooperative linkage among modules. But, the strongest evidence of modularity comes from the observation that the adaptive process involves similar, but separable mechanism for different kinds of prismatic distortions. For instance, tilt and displacement adaptation are parametrically similar (Redding, 1973a, 1975a) even though the two kinds of adaptation do not interfere with each other (Redding, 1973b, 1975b).

Thus, the mind/brain seems to have solved in a similar fashion the ubiquitous problem of maintaining cross-calibration among its various parts. Each module includes the capacity to respond to remote calibration signals received from other modules.

Conclusions. The perceptual-motor system appears to be organized into distinct modules which are specialized in function. Cooperative interaction (coordination) requires a (sometimes intentional) configuration of coordinative linkages among the task-relevant subset of modules. Prism adaptation is an experimental example of the process which normally maintains spatial cross-calibration among the various modules in the face of naturally occurring misalignments arising from growth, pathology, or drift (cf. Held & Bossom, 1961; Robinson, 1976). Separate, but similar mechanisms reside in each module which serve to locally recalibrate the module in response to a remote, guiding signal from another module. The multiple directions of linkage required by the variety of everyday coordination assures that the various component of the perceptual-motor system are maintained in a veridical state of cross-calibration (cf. Craske, 1975).

Adaptive Eye-Hand Coordination

This general view of adaptive perceptual-motor organization has guided the formulation of the specific model of adaptive eye-hand coordination illustrated in Figure 5.4.

The model is certainly not the only way to conceptualize the organization of the eye-hand coordinative linkage, but it serves as a framework for research. The eye-hand coordination system is assumed to be organized around semi-autonomous sensory-motor systems, each of which includes a unique spatial map. For instance, a visual system may consist of sensory and motor capacities inclusive of the eye-to-head relationship, whereas a proprioceptive system could include these capacities between the hand and head. Each such sensory-motor system is capable of autonomous operation, but many tasks involve coordination of sensory-motor systems. Coordinative linkages between subsystems carry spatial information which can be used to guide one sensory-motor system to positions coded by another system. The particular linkage of subsystems depends upon the nature of the coordination task, but such linkages are assumed to be directional in the sense that the same subsystem cannot be guiding and guided at the same time. Moreover, coordination of subsystems requires translation of spatial coordinates of one system into those of another. Spatial translation is assumed to be mediated by a single spatial representation common to all subsystems, and (encoding/decoding) operators in each subsystem translate between locally unique spatial maps (M) and this noetic (N) space.

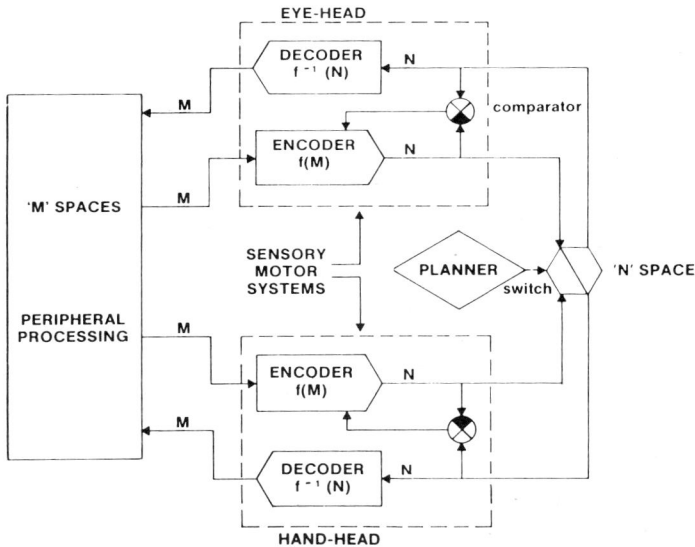

Figure 5.4. Adaptive eye-hand coordination. Coordination requires directional guidance linkage between semi-autonomous eye-head (visual) and hand-head (proprioceptive) sensory-motor system. Encoding operators (and inverse decoders) translate between a particular motor (M) space and noetic (N) space common to all sensory-motor systems. when misalignment occurs (e.g., in the visual system), discordance between the remote guidance signal and the local positional code prompts recalibration of the adaptive encoder in the guided system. The linkage direction shown is such as would produce proprioceptive adaptation. A switch setting connecting the other pair of input/output lines would produce visual adaptation. (Reprinted with permission from Redding & Wallace, 1988a).

Now, when misalignment occurs among subsystems, either for artificial reasons like prisms or natural causes, the resultant spatial discordance between subsystems is registered in the guided system whose local operators are then recalibrated in response to the remote, guiding signal from the other sensory-motor system. The comparator in the guided subsystem detects any spatial discrepancy (discordance) between the remote guidance signal and local position, and this discordance causes a gradual parametric change in encoding (and the inverse decoding operation) to bring the guided subsystem back into spatial alignment with the guiding subsystem. For example, if the coordination task requires eye-to-hand linkage, adaptive recalibration will occur in the proprioceptive (hand-head) system, but the reverse direction of guidance will produce visual (eye-head) adaptation. Although similar in function, the local adaptive mechanisms are unlikely to have the same time constant and adaptive components may be expected to develop at different rates.

Depending upon the particular mix of linkage directions required by a particular coordination task, the proportion of total perceptual adaptation which is localized in eye-head or hand-head subsystems will vary. Thus, the model incorporates the common observation of additivity of adaptive components. Indeed, within the framework of the model additivity has important methodological implications. Substantial deviations from additivity may mean that the component sensory-motor systems were not linked appropriately during exposure and the component tests were biased. Additivity, therefore, provides a convergent indicator that the necessary conditions for studying perceptual components of adaptation have been established.

Note carefully that adaptation is assumed to be the automatic response to discordance registration within a sensory-motor system (see Figure 5.4) and is not directly affected by performance error. The proximal cause of adaptation is a discrepancy between position codes, corresponding to a common spatial location, for two sensory-motor systems. This discordance is not available to higher-level processes to serve as an error signal. Performance error may, however, motivate coordinative linkage of subsystems, thereby facilitating adaptation. Subsystems must be coordinatively linked in order for discordance registration/reduction (i.e., recalibration) to occur, but variables which affect how and when subsystems are coordinated do not directly affect adaptation.

The planner element in Figure 5.4 responds strategically to characteristics of the coordination task to optimize performance. For example, in target pointing the planner sets the direction of coordinative linkage between subsystems to make optimal use of available feedback. When the limb becomes visible early in the pointing movement, the optimal linkage would seem to be eye-to-hand. Target position and hand position can be compared visually, errors detected, and corrective guidance signals sent to the hand-head system. When the limb becomes visible only late in the pointing movement, however, visual feedback is less effective for error correction. An alternative strategy may be to track the unseen limb with the eye utilizing proprioceptive feedback from the hand-head system. In this case, the direction of coordinative linkage between subsystems would be hand-to-eye. When the limb finally becomes visible late in the pointing movement, error information becomes available for correction of the next pointing movement. Thus, the direction of coordinative linkage between subsystems and consequently the locus of prism adaptation should be dependent upon the availability of visual feedback.

Consistent with motor control research (for reviews, see Klein, 1976, and Stelmach, 1982) planning operations are assumed to be subject to limitations on central processing capacity. A secondary cognitive task may interfere with performance of a primary eye-hand coordination task when the two tasks are performed simultaneously. Thus, central processing capacity may, in some cases, be required to establish and maintain coordinative linkages among subsystems. In general, routine coordination tasks place fewer demands on central processing capacity than unusual tasks which require deliberate monitoring to assure proper linkage of subsystems. We assume that, everything else being equal (e.g., feedback availability), a routine linkage will be selected when it is available in order to minimize demand on limited central processing capacity. For example, eye-to-hand linkage (visual guidance) may be elected over hand-to-eye linkage (proprioceptive guidance) because such visual guidance is the more common mode of eye-hand coordination and may be executed with less "effort". The distinction here corresponds to relatively slow controlled (attentional, effortful) processing which requires limited central processing capacity and faster automatic processing which has less or perhaps no capacity limitation (e.g., Hasher & Zacks, 1979; Posner & Snyder, 1975; Schneider & Shiffrin, 1977; see also Redding, Clark, & Wallace, 1985).

The model accounts for cognitive interference with adaptation by less frequent controlled linkage of discordant subsystems under cognitive load. However, automatic linkages may also be available which are less subject to interference from cognitive load. In selecting from among the available strategic linkages, differing in direction and automaticity, the planner is assumed to operate according to a kind of "minimal-effort" principle, optimizing performance only so far as is required to meet the performance level expected for the task. For example, if a high level of pointing accuracy is not required, visual guidance may occur even if performance could be improved by more deliberate proprioceptive guidance, because such a linkage places fewer demands on central processing capacity and still meets the required performance level for the task. Thus, cognitive interference with adaptation might fail to appear, not because planning operations do not normally draw on central processing capacity, but because a less effortful (more automatic) linkage is selected which produces "good enough" performance.

In summary, the model distinguishes between processes which maintain spatial alignment in the perceptual-motor system and other adaptive processes which mediate task-appropriate perceptual-motor coordination. Coordination activates automatic mechanisms within sensory-motor modules which test for spatial alignment and reduce any detected discordance (i.e., perceptual adaptation). Development of coordinative linkage (skill learning) and strategic selection from among available linkages (skilled performance) only indirectly affect perceptual adaptation by determining (1) the frequency of coordinative linkage and therefore the amount of discordance registration/reduction and (2) the direction of coordinative linkage and therefore the locus of discordance registration/reduction. Our empirical exploration of the model has, thus far, been largely restricted to investigation of conditions which affect strategic linkage of discordant subsystems.

Research Summary

Because the model assumes that coordinative linkage is strategically dependent upon the structure of the coordination task, it is important to specify, as exactly as present knowledge allows, the relevant characteristics of the experimental task. In our investigations so far we have employed a paced saggital pointing task. The limited visual field imposed by prism-bearing goggles prevents view of the limb in the starting position near the trunk. Lateral corrections after a movement is completed are discouraged and appear infrequently. Pointing movements are made in time to a metronome signal and are unconstrained by surface contacts. Subjects are instructed to initiate a movement on one beat, move the limb such that it is fully extended on the next beat, withdraw the limb such that the hand arrives back at the starting position on the third beat, and so on. In this manner, a smooth movement cycle is achieved for every three beats of the metronome with minimal hesitation on a beat and relatively constant velocity between beats. Typically, movements are relatively slow (e.g., with the metronome set to beat every 3 s) and few in number (e.g., 60).

In addition to measures of VS and PS, TS measures are routinely obtained so that additivity can be assessed (i.e., VS + PS = TS). Unless otherwise indicated, there were no substantial deviations from additivity in the following experiments and we can be reasonably sure that the proper conditions were present for studying perceptual adaptation. Consequently, only results for the component tests (VS and PS) are usually reported here.

Visual Feedback

One prediction is that the direction of coordinative linkage between subsystems (and consequently the locus of prism adaptation) should be dependent upon the availability of visual feedback. When the limb becomes visible early in a pointing movement, the available visual feedback should be used to guide the limb. Coordinative linkage should be largely eye-to-hand and adaptation should be primarily proprioceptive in nature, a change in the limb position sense. Conversely, when the limb becomes visible only late in a movement, proprioceptive feedback may be used to guide the eye in tracking the unseen limb. Linkage may be largely hand-to-eye and adaptation may be mostly visual, a change in the eye position sense.

This prediction has now been verified several times (e.g., Redding & Wallace, 1988a, 1988b, 1990a; Uhlarik & Canon, 1971). The point in the movement, early to late, where the limb first becomes visible (timing of visual feedback) seems to be more important in determining the locus of adaptation than the amount of the movement during which the limb is visible (duration of visual feedback). For instance, Figure 5.5 illustrates results when timing was manipulated with duration held constant (Redding & Wallace, 1990a, Experiment 2). All subjects saw 2.5 cm of their finger tip at the end of the 50 cm pointing movement and an additional 2.5 cm view of the moving limb was provided for different groups at 2.5 cm increments from the point in the movement where it was first possible to see the moving limb (i.e., the prism goggles restricted view for the first 25 cm of the movement). Visual adaptation increased smoothly with feedback delay, but corresponding decreases in proprioceptive adaptation underwent an additional sharp change when feedback was delayed until about three-fourths of the way to the terminal limb position. These results suggest that the limb may be ballistically released as it nears the terminal position, and, thereafter, any opportunity for visual guidance (and consequential proprioceptive adaptation) is not effective. On the other hand, since proprioceptive feedback is available throughout the movement, proprioceptive guidance (and visual adaptation) is a simple linear function of delay in visual feedback.

An obvious interpretation of these results is that the causal bases for visual and proprioceptive adaptation reside in the manner in which different parts of a movement are controlled. Visual guidance and PS may occur early in the movement while proprioceptive guidance and VS may occur late in the movement. An alternative hypothesis is that the direction of linkage responsible for a particular kind of adaptation can be set for an entire movement. As visual feedback is increasingly delayed, the proportion of visual guidance trials and PS decreases while proprioceptive guidance trials and VS

increase. Visual and proprioceptive guidance might, then, be more global
strategies.

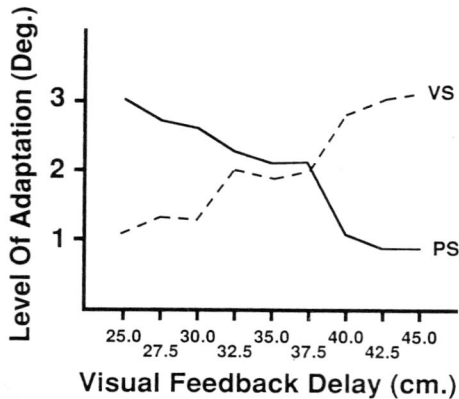

Figure 5.5. Mean level of visual shift (VS) and proprioceptive shift (PS) as a
function of the distance in the (50 cm) pointing movement where the limb first
became visible. The amount of the movement during which the limb was
visible was constant at 5 cm. (Reprinted with permission from Redding &
Wallace, 1990a, Experiment 2).

Figure 5.6 illustrates the results of one attempt to test between these
hypotheses (Redding & Wallace, 1991, Experiment 1). Adaptation was
measured after each of six blocks of 10 pointing trials for conditions where
visual feedback was delayed until the very end of the movement (Terminal
exposure) and where visual feedback was concurrently available with
proprioceptive feedback over most of the movement (Concurrent exposure).
The critical data are observations where one component increased while the
other did not change over a block of trials. This pattern of results could only
happen if linkage was exclusively in one direction over the entire block of
trials. Such evidence of trial-to-trial change in direction of coordinative
linkage was statistically present for Blocks 4 and 5 with Terminal exposure and
for all but Block 3 for Concurrent exposure. Results for the remaining blocks
(4 out of 10), where both components increased or neither changed, are
consistent with either hypothesis.

These results do not mean that the direction of coordinative linkage cannot be changed within a movement. They do, however, suggest that perceptual-motor coordination is strategically flexible, not exclusively set by the kinematics of the response system. Investigation using different pointing tasks may be expected to outline the limits of this strategic flexibility.

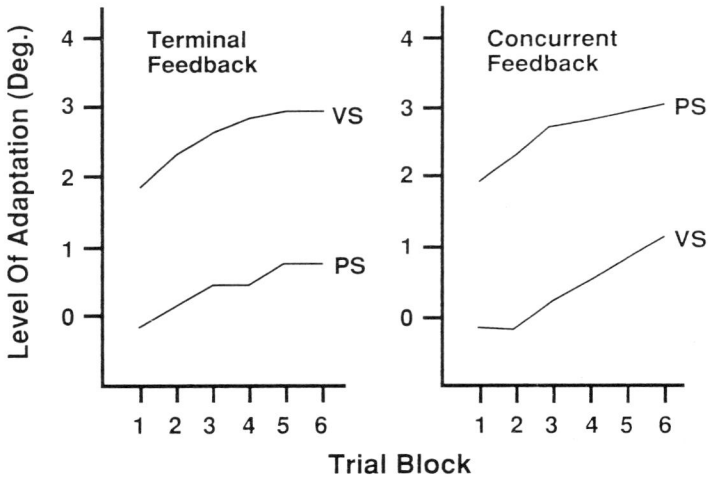

Figure 5.6. Mean level of visual shift (VS) and proprioceptive shift (PS) as a function of six blocks of 10 pointing trials when visual feedback was delayed until the end of the movement (terminal) or concurrently available with proprioceptive feedback during the movement (concurrent). (From Redding and Wallace, 1991)

Pointing Rate

Another implication of the model is that coordinative linkage requires time to activate. Some linkages may be more habitual (automatic) and can be activated more quickly than more controlled linkage. For instance, proprioceptive guidance of the eye, which is presumed to underlie visual adaptation, may be relatively slow to occur. Using proprioceptive feedback to form a visual image of the unseen hand seems to be an uncommon strategy which would require time to implement. Moreover, there may be limits on the speed with which the eye can track the unseen hand in this manner. In contrast, visual guidance of the hand may be a more common strategy and relatively fast. Visual guidance may also be faster because it involves the simpler and more direct comparison between visual positions of hand and target. The prediction, then, is that the availability of visual feedback should interact with pointing rate in affecting the direction of coordinative linkage and therefore the locus of adaptation.

This prediction is supported by the data illustrated in Table 5.1 (Redding & Wallace, 1990b). The usual differential effect of Terminal and Concurrent feedback on adaptive components occurred when subjects pointed slowly (once every 3 s), but with faster pointing (once every 2 s) proprioceptive adaptation was greater than visual adaptation for both feedback conditions. These results are consistent with the idea that visual guidance (with consequential proprioceptive adaptation) is a faster, more automatic process, while proprioceptive guidance (with consequential visual adaptation) is a slower, more controlled process.

Table 5.1
Mean level of adaptation in degrees and 95 percent confidence limits for visual shift (VS) and proprioceptive shift (PS) following fast (2 s) or slow (3 s) pointing at prismatically displaced targets when visual feedback was delayed until the end of the movement (terminal) or concurrently available with proprioceptive feedback during the movement (concurrent). (From Redding & Wallace, 1990b).

	Fast pointing			Slow pointing		
Feedback	VS	PS	MEAN	VS	PS	MEAN
Terminal	3.8	0.2	2.0	1.4	2.6	2.0
	(0.9)	(0.6)		(0.3)	(0.3)	
Concurrent	3.8	0.2	2.0	2.7	1.2	2.0
	(0.6)	(0.1)		(0.5)	(0.4)	
Mean	3.8	0.2	2.0	2.0	1.9	2.0

Another aspect of these data suggest a further effect of pointing rate beyond determining the effectiveness of visual feedback in setting the direction of coordinative linkage. Pointing rate affected the rate of growth of adaptation but availability of visual feedback did not. This effect is illustrated in Figure 5.7. More specifically, visual adaptation reached a limit much earlier in exposure with faster pointing, while proprioceptive adaptation tended to reach a limit earlier with slower pointing. Thus, development of visual adaptation is encouraged by a slow pointing rate while development of proprioceptive adaptation is encouraged by a fast pointing rate.

This differential effect of pointing rate on the development of adaptive components may reflect differences in the temporal characteristics for the adaptive encoders in eye-head and hand-head systems. Maximal rate of growth of an adaptive component may be dependent upon the degree to which pointing rate (and rate of discordant inputs) matches the optimal response rate of the adaptive encoder in the guided system (see Figure 5.4). Less than optimal input rates may allow decay of adaptive between successive inputs. Greater than optimal input rates may preclude complete processing of successive inputs. Either situation would produce a limit to the amount of adaptation that occurs. Further manipulation of pointing rate may enable estimation of the operational parameters of adaptive encoders.

Cognitive Load

The suggestion that visual guidance is largely automatic for the present pointing task has been investigated more directly by varying cognitive load. In one experiment (Redding, Rader, & Lucas, 1991, Experiment 2), subjects were required to perform mental arithmetic (e.g., "47 + 63 = ?") simultaneous with target pointing. To maximize the probability of visual guidance, a fast pointing rate (2 s) was employed with concurrent feedback.

The average VS was quite small (.4 deg), indicating that the attempt to elicit visual guidance was successful. Proprioceptive adaptation under cognitive load (2.4 deg) was slightly less (23 percent) than when no problems were given (3.1 deg), but this effect was not statistically reliable, even though subjects found the mental arithmetic difficult, successfully solving an average of only 75 percent of 20 problems attempted. On average subjects attempted a problem every 6 s (3 pointing movements) which is close to the time required to verbalize problem and answer. Further evidence that the cognitive load manipulation was successful is the fact that subjects were reliably less accurate (1.1 deg) at the pointing task when attempting to simultaneously solve problems (1.5 deg) compared to the no-problems condition (.4 deg).

These data suggest that adaptation is not substantially affected by a secondary cognitive task when the exposure (coordination) task can be performed in a repetitive, routine manner (cf. Redding, Clark, & Wallace, 1985; Redding & Wallace, 1985). Apparently, more controlled processing is necessary for perfectly accurate pointing, but automatic processing is sufficient for eye-to-hand coordinative linkage and discordance registration/reduction occurs without error-free performance. The cognitive load manipulation provides a means of assessing the degree of controlled/automatic linkage involved in a coordination task and of separating effects of coordinative linkage from those of discordance *per se*.

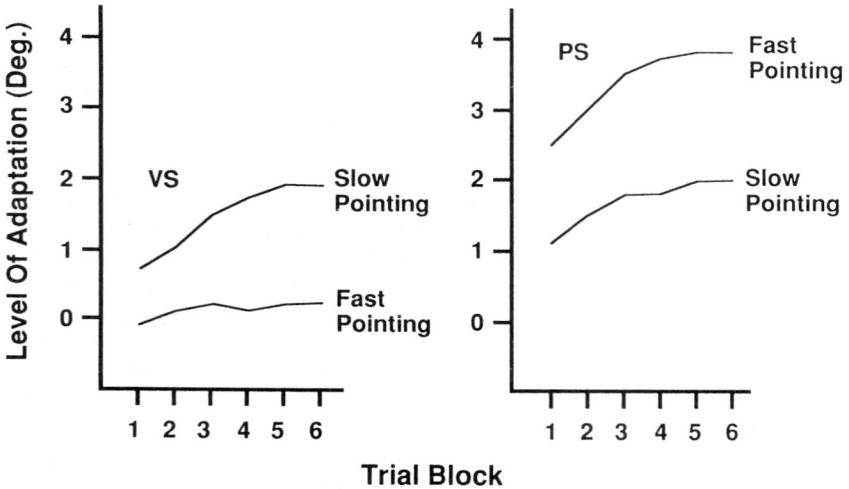

Figure 5.7. Mean level visual shift (VS) and proprioceptive shift (PS) as a function of six blocks of 10 fast (2 s) or slow (3 s) pointing trials when visual feedback was delayed until the end of the movement (terminal) or concurrently available with proprioceptive feedback during the movement (concurrent). (From Redding & Wallace, 1990b).

Target Uncertainty

The observation that adaptation is largely unhindered by the presence of significant pointing errors suggest that error feedback is not directly involved in discordance registration/reduction. This suggestion is consistent with the present model where error feedback may encourage controlled linkage to optimize performance, but is not itself the proximal cause for adaptation. So long as subsystems are coordinatively linked, discordance registration/reduction will occur independent of how well the coordination task is performed. Further support for this aspect of the model comes from manipulation of target uncertainty. Redding and Wallace (1988a, Experiment 1) found that neither VS or PS under Terminal and Concurrent feedback conditions were significantly affected by whether subjects pointed at explicit or implicit ("center-of-the-field") targets. If error information were directly involved in adaptation, it seems that more clearly defined (explicit) targets should enable better error discrimination and enhance adaptation. Explicit targets did produce greater TS, but this was arguably a nonperceptual effect since underadditivity (i.e., VS + PS < TS) also appeared in this condition (cf. Welch, 1974; Welch, Choe, & Heinrich, 1974). Such deviation from additivity is what would be expected if error information affects higher-level coordination rather than low-level spatial alignment. Error information can be expected to have substantial effects on adaptation only if the coordination task involves largely controlled processing.

Conclusions

The suggestion that spatial alignment and coordination are separable processes has implications for research and theory in perceptual-motor organization. If adaptive spatial alignment occurs within modules of the perceptual-motor system, then for most investigations of coordination under normal conditions these processes are dormant and can largely be ignored. In the absence of discordance-producing conditions, spatial alignment can be assumed and there is no clear theoretical imperative to explain translation among the various spatial representations differing, for example, in dimensionality (but see, Churchland, 1986; Grossberg & Kuperstein, 1989). By the same token, coordination under normal (non-discordant) conditions is not very informative of the spatial alignment problem (for a review see, Howard, 1982). The introduction of spatial discordance, however, implicates processes at different levels of the perceptual-motor system and incurs the methodological difficulty of separating the normal coordination processes which establish the occasion for spatial realignment from the realignment process itself. Understanding experimental instances of spatial realignment like prism adaptation also involves understanding of normal coordination processes and is, therefore, informative of how the total perceptual-motor system is organized.

References

Barr, C.C., Schultheis, L.W., & Robinson, D.A. (1976). Voluntary, non-visual control of the human vestibulo-ocular reflex. *Acta Otolaryngology, 81*, 365-375.

Bedford, F. (1989). Constraints on learning new mappings between perceptual dimensions. *Journal of Experimental Psychology: Human Perception and Performance, 15*, 232-248.

Canon, L.K. (1970). Intermodality inconsistency of input and directed attention as determinants of the nature of adaptation. *Journal of Experimental Psychology, 84*, 141-147.

Churchland, P.S. (1986). *Neurophilosophy: Toward a unified science of the mind/brain.* Cambridge, MA: MIT Press.

Coren, S. (1966). Adaptation to prismatic displacement as a function of the amount of available information. *Psychonomic Science, 4*, 407-408.

Craske, B. (1975). A current view of the processes and mechanisms of prism adaptation. *Les Colloques de l'Institut National de la Santé et de la Recherche Médicale, 43*, 125-138.

Fodor, J.A. (1983). *The modularity of mind.* Cambridge, MA: MIT Press.

Hamilton, C.R. (1964). Studies on adaptation to deflection of the visual field in split-brain monkeys and man. Unpublished doctoral dissertation, California Institute of Technology.

Grossberg, S., & Kuperstein, M. (1989). *Neural dynamics of adaptive sensory-motor control: Expanded Edition.* New York: Pergamon.

Hardt, M.E., Held, R., & Steinbach, M.J. (1971). Adaptation to displaced vision: A change in central control of sensorimotor coordination. *Journal of Experimental Psychology, 89*, 229-239.

Harris, C.S. (1965). Perceptual adaptation to inverted, reversed, and displaced vision. *Psychological Review, 72*, 419-444.

Hasher, L., & Zacks, R.T. (1979). Automatic and effortful processes in memory. *Journal of Experimental Psychology: General, 108*, 356-388.

Hay, J.C., & Pick, H.L. (1966). Visual and proprioceptive adaptation to optical displacement of the visual stimulus. *Journal of Experimental Psychology, 71*, 150-158.

Held, R. (1961). Exposure-history as a factor in maintaining stability of perception and co-ordination. *Journal of Nervous and Mental Disease, 132*, 26-32.

Held, R., & Bossom, J. (1961). Neonatal deprivation and adult rearrangement: Complementary techniques for analyzing plastic sensory-motor coordinations. *Journal of Comparative and Physiological Psychology, 54*, 33-37.

Howard, I.P. (1971). Perceptual learning and adaptation. *British Medical Bulletin, 27*, 248-252.

Howard, I.P. (1982). *Human visual orientation.* New York: Wiley.

Howard, I.P., Anstis, T., & Lucia, H.C. (1974). The relative lability of mobile and stationary components in a visual-motor adaptation task. *Quarterly Journal of Experimental Psychology, 26*, 293-300.

Howard, I.P., & Templeton, W.B. (1966). *Human spatial orientation.* New York: Wiley.

Kelso, J.A.S., Cook, E., Olson, M.E., & Epstein, W. (1975). Allocation of attention and the locus of adaptation to displaced vision. *Journal of Experimental Psychology: Human Perception and Performance, 1,* 237-245.

Kohler, I. (1964). The formation and transformation of the perceptual world (H. Fiss, Trans.). *Psychological Issues, 3,* 1-173. (Original work published in 1951).

Klein, R.M. (1976). Attention and movement. In G.E. Stelmach (Ed.), *Motor control: Issues and trends* (pp. 143-174). New York: Academic Press.

Posner, M.I., & Snyder, C.R.R. (1975). Attention and cognitive control. In R.L. Solso (Ed.), *Information processing and cognition* (pp. 55-86). Hillsdale, NJ: Erlbaum.

Redding, G.M. (1973a). Visual adaptation to tilt and displacement: Same or different processes? *Perception & Psychophysics, 14,* 193-200.

Redding, G.M. (1973b). Simultaneous visual adaptation to tilt and displacement: A test of independent processes. *Bulletin of the Psychonomic Society, 2,* 41-42.

Redding, G.M. (1975a). Decay of visual adaptation to tilt and displacement. *Perception & Psychophysics, 17,* 203-208.

Redding, G.M. (1975b). Simultaneous visuo-motor adaptation to optical tilt and displacement. *Perception & Psychophysics, 17,* 97-100.

Redding, G.M. (1978). Additivity in adaptation to optical tilt. *Journal of Experimental Psychology: Human Perception and Performance, 4,* 178-190.

Redding, G.M., & Clark, S.E., & Wallace, B. (1985). Attention and prism adaptation. *Cognitive Psychology, 17,* 1-25.

Redding G.M., Rader, S.J., & Lucas, D.R. (1991). [Cognitive load and prism adaptation]. Unpublished raw data.

Redding, G.M., & Wallace, B., (1976). Components of displacement adaptation in acquisition and decay as a function of hand and hall exposure. *Perception & Psychophysics, 20,* 453-459.

Redding, G.M., & Wallace, B. (1978). Sources of "overadditivity" in prism adaptation. *Perception & Psychophysics, 24,* 58-62.

Redding, G.M., & Wallace, B. (1985). Cognitive interference in prism adaptation. *Perception & Psychophysics, 37,* 225-230.

Redding, G.M., & Wallace, B. (1988a). Components of prism adaptation in terminal and concurrent exposure: Organization of the eye-hand coordination loop. *Perception & Psychophysics, 44,* 59-68.

Redding, G.M., & Wallace, B. (1988b). Adaptive mechanisms in perceptual-motor coordination: Components of prism adaptation. *Journal of Motor Behavior, 20,* 242-254.

Redding, G.M., & Wallace, B. (1990a). Effects on prism adaptation of duration and timing of visual feedback during pointing. *Journal of Motor Behavior, 22*, 209-224.

Redding, G.M., & Wallace, B. (1990b, November). Effects of pointing rate and availability of visual feedback on components of prism adaptation. Prepared for presentation at the Annual Meeting of the Psychonomic Society, New Orleans, LA.

Redding, G.M., & Wallace, B. (1991). [Acquisition of visual and proprioceptive adaptation to prismatic displacement during eye-hand coordination]. Unpublished raw data.

Robinson, D.A. (1976). Adaptive gain control of vestibulo-ocular reflex by the cerebelum. *Journal of Neuropsychology, 39*, 954-969.

Rock, I. (1966). *The nature of perceptual adaptation.* New York: Basic Books.

Schneider, W., & Shiffrin, R.M. (1977). Automatic and controlled information processing in vision. In D. LaBerge & S.J. Samuels (Eds.), *Basic processes in reading* (pp. 127-154). Hillsdale, NJ: Erlbaum.

Stelmach, G.E. (1982). Information-processing framework for understanding human motor behavior. In J.A.S. Kelso (Ed.), *Human motor behavior: An introduction* (pp. 63-92). Hillsdale, NJ: Erlbaum.

Stratton, G.M. (1896). Some preliminary experiments on vision without inversion of the retinal image. *Psychological Review, 3*, 611-617.

Templeton, W.B., Howard, I.P., & Wilkinson, D.A. (1974). Additivity of components of prismatic adaptation. *Perception & Psychophysics, 15*, 249-257.

Uhlarik, J.J. (1973). Role of cognitive factors on adaptation to prismatic displacement. *Journal of Experimental Psychology, 98*, 223-232.

Uhlarik, J.J., & Canon, L.K. (1971). Influence of concurrent and terminal exposure conditions on the nature of perceptual adaptation. *Journal of Experimental Psychology, 91*, 233-239.

Von Helmholtz, H. (1962). *Treatise on physiological optics.* (J.P.C. Southall, Trans.). New York: Dover. (Original work published in 1866).

Wallace, B. (1977). Stability of Wilkinson's linear model of prism adaptation over time for various targets. *Perception, 6*, 145-151.

Wallace, B., & Garrett, J.B. (1975). Perceptual adaptation with selective reductions of felt sensation. *Perception, 4*, 437-445.

Wallace, B., & Redding, G.M. (1979). Additivity in prism adaptation as manifested in intermanual and interocular transfer. *Perception & Psychophysics, 25*, 133-136.

Welch, R.B. (1974). Speculations on a model of prism adaptation. *Perception, 3*, 451-460.

Welch, R.B. (1978). *Perceptual modification.* New York: Academic Press.

Welch, R.B., & Abel, M.R. (1970). The generality of the "target-pointing effect" in prism adaptation. *Psychonomic Science, 20*, 226-227.

Welch, R.B., Choe, C.S., & Heinrich, D.R. (1974). Evidence for a three component model of prism adaptation. *Journal of Experimental Psychology, 103*, 700-705.

Welch, R.B., & Rhoades, R.W. (1969). The manipulation of informational feedback and its effects upon prism adaptation. *Canadian Journal of Psychology, 23*, 415-428.

Wilkinson, D.A. (1971). Visual-motor control loop: A linear system? *Journal of Experimental Psychology, 89*, 250-257.

Author's note

Correspondence should be addressed to Gordon M. Redding, Department of Psychology, Illinois State University, Normal, Illinois, 61761.

VISION AND MOTOR CONTROL
L. Proteau and D. Elliott (Editors)

CHAPTER 6

COORDINATION OF EYE AND HAND FOR AIMED LIMB MOVEMENTS

RICHARD A. ABRAMS

*Department of Psychology, Washington University,
St. Louis, MO, 63130*

It is obvious that the eyes provide important information that can be used in the control of limb movements. Many of the daily activities that we perform so effortlessly would be difficult or impossible to perform with the eyes closed. Since the time of Woodworth (1899) researchers have been interested in the nature of the information provided by the eyes for use in controlling movements of the limbs. A good deal has been learned about various sources of visual feedback information (Abrams, Meyer, & Kornblum, 1990; Carlton, 1981b; Elliott & Madalena, 1987), about the time needed to process such information (Carlton, 1981a; Keele & Posner, 1968; Vince, 1948; Zelaznik, Hawkins, & Kisselburgh, 1983), and about the relative timing and spatial accuracy of eye and hand movements (Abrams et al., 1990; Biguer, Jeannerod, & Prablanc, 1982; Prablanc, Echallier, Komilis, & Jeannerod, 1979b). Nevertheless, a number of issues regarding the exact nature of the coordination of eye and hand remain unresolved. The present chapter reports some studies that provide insight into those issues.

We first outline the nature of the information that the eyes can provide that might be used to guide limb movements. Next, we discuss some features of limb movements that might yield some insight into how the information provided by the eye may actually be used. Finally, we report the results of three experiments designed to address different questions about the relation between eye and hand. In addition to an interest in the role of various types of visual information provided by the eyes (e.g., vision of the moving limb or of the target for the movement), we are also interested in the role of eye movements in enhancing non-visual information that could be useful in guiding limb movements. Our approach differs somewhat from that taken by previous researchers because we directly manipulate the behavior of the eyes during limb movements, and assess the impact of these manipulations in order to make inferences about the role that the eye movements serve. The results of our research show that both visual and non-visual sources of information from the eyes can provide important contributions to the accuracy of aimed limb movements.

Retinal and Extraretinal Information

There are two distinct types of information that can be obtained from the eyes for use in controlling limb movements: retinal and extraretinal. Retinal information includes information obtained from the patterns of stimulation on the retina regarding the location and movement of the limb and of the target for a movement. Extraretinal information includes information about the movement and position of the eyes obtained from non-retinal sources (such sources could include the outflowing oculomotor commands used to produce eye movements, or inflowing proprioceptive information about eye position).

Retinal information may play several roles in guiding limb movements. Prior to a movement, information about the location of a peripheral target may be available only through retinal information. During a movement, retinal information can provide the visual feedback necessary to monitor the status of the ongoing movement. This feedback may include not only information about the moving limb (Carlton, 1981b), but also information about the location of the target and the relative positions of limb and target (Prablanc, Pélisson, & Goodale, 1986).

Extraretinal information may also play an important role in guiding the limb. Several researchers have shown that people can accurately guide a limb to a target location based only on extraretinal cues about eye position (Hansen & Skavenski, 1977, 1985; Hill, 1972; Morgan, 1978). Thus, extraretinal information might be used to localize the target for an aimed movement (Prablanc et al., 1979b).

Although a good deal is already known about retinal and extraretinal information individually, it is not clear precisely what role is played by these different sources of information during actual aimed limb movements. One of the major challenges in understanding coordinated visual-motor behavior is to learn when and how retinal and extraretinal information can contribute to the guidance of an aimed limb movement.

Phases of an Aimed Movement

As a first step toward this goal, we reasoned that the information that is important for movement control might be different during different component phases of a limb movement. Researchers have identified three distinct component phases of a rapid aimed movement during which the information requirements may differ considerably: movement preparation, initial impulse, and error correction (e.g. Carlton, 1981a; Kerr, 1978; Meyer, Abrams, Kornblum, Wright, & Smith, 1988; Meyer, Smith, Kornblum, Abrams, & Wright, 1990; Woodworth, 1899). Below we consider what kind of information might be provided by the eye during the different component phases.

Movement-preparation phase. The movement-preparation phase occurs immediately before the onset of overt movement. During this time the subject is believed to construct some sort of representation of the goal for the movement, and assemble the initial motor commands that are necessary to attain that goal. Retinal information about the target and limb (if they are visible) and their relative positions can be used to accomplish this (Prablanc, Echallier, Jeannerod, & Komilis, 1979a). Also, extraretinal information may help the subjects locate the target if the eyes are pointing at the target during movement preparation (Hill, 1972; Morgan, 1978).

Initial-impulse phase. The next phase of movement consists of the execution of the motor commands that begin to move the limb toward the target. Because the initial impulse is usually ballistic (Carlton, 1981a; Crossman & Goodeve, 1963/1983; Meyer et al., 1988; Woodworth, 1899), information obtained from the eyes during this phase may not be useful until the initial impulse has ended. Nevertheless, in at least some situations the initial impulse may be modifiable before it has ended (Pélisson, Prablanc, Goodale, & Jeannerod, 1986). As a result, the behavior of the eyes may provide information that would be of use in planning the modifications. Furthermore, the behavior of the eyes during the initial impulse may have important consequences for the next movement phase.

Error-correction phase. After the initial impulse, a limb movement may enter an error-correction phase in which adjustments are made in the position of the limb as it homes-in on the target. The error corrections are typified by discontinuities in the position, velocity, and acceleration of the limb near the end of the movement (Abrams et al., 1990; Carlton, 1981a; Meyer et al., 1988). The success of the error corrections depends to a great extent on the availability of information about the current state of the limb. If retinal information about the limb (visual feedback) is unavailable shortly before or during the error-correction phase, then the accuracy of the limb movement will suffer (Carlton, 1981a; Keele & Posner, 1968; Meyer et al., 1988; Prablanc et al., 1986; Wallace & Newell, 1983; Woodworth, 1899; Zelaznik et al., 1983). Retinal information about the limb would be most informative if the eyes were near the target during the error corrections, because the limb is usually near the target then also. Additionally, the error corrections may be enhanced by extraretinal information about the target position which would also be available during this movement phase if the eyes were pointing at the target (Prablanc et al., 1979b).

Behavior of the Eyes During Aimed Limb Movements

It should be clear from the preceding discussion that the nature of the information provided by the eyes that can be used for the control and guidance of a limb movement depends to a great extent on the behavior of the eyes before and during the movement. A considerable amount of work has already been done examining movements of the eyes and limbs to visual targets (e.g., Angell, Alston, & Garland, 1970; Biguer et al., 1982; Mather & Fisk, 1985; Megaw & Armstrong, 1973; Prablanc et al., 1979a; Prablanc et al., 1986). A common finding has been that the eyes usually move to the target for a limb movement shortly before the limb begins to move. Thus, extraretinal cues could provide information about the target for use in the preparation of the initial impulse and the guidance of error corrections. Additionally, having the eyes at the target before the error corrections begin would enhance the useful retinal information available about the limb (i.e. visual feedback) as it approaches the target near the end of the movement. Nevertheless, in each of the preceding studies, there was temporal or spatial uncertainty about the location of the movement target; subjects were instructed to move their eyes to the target; and the latencies of the eye or hand movements were to be minimized. Thus, it is not clear whether the observed eye movements were performed in order to improve the information available to guide the limb, or whether subjects produced eye movements simply to satisfy the demands of the tasks. Our first goal was to determine precisely what people do with their eyes during rapid aimed limb movements when there is no uncertainty about the location of the target, and no pressure to minimize eye movement or hand movement latencies (Abrams et al., 1990).

To examine this issue, we recorded eye movements while people performed a wrist-rotation task similar to one that has been used by a number of investigators (Crossman & Goodeve, 1963/1983; Meyer et al., 1988; Wright & Meyer, 1983). Clockwise and counterclockwise rotations of a light-weight handle produced rightward and leftward movements of a cursor on a video display. Subjects were required to rotate the handle in order to move the cursor from an initial starting point (the *home* position) into a target region. The movements were to be as brief as possible subject to the constraint that the cursor was to stop moving in the target on the majority of trials. Subjects were under no pressure to minimize the latency of the wrist-rotation movements. During the wrist rotation task we recorded the position of the subjects eyes. No constraints were imposed on the eye movements other than a requirement that the subject fixate on the home position prior to the beginning of each trial. Subjects understood that after successful fixation they could do whatever they wanted with their eyes. We didn't even require them to move their eyes at all.

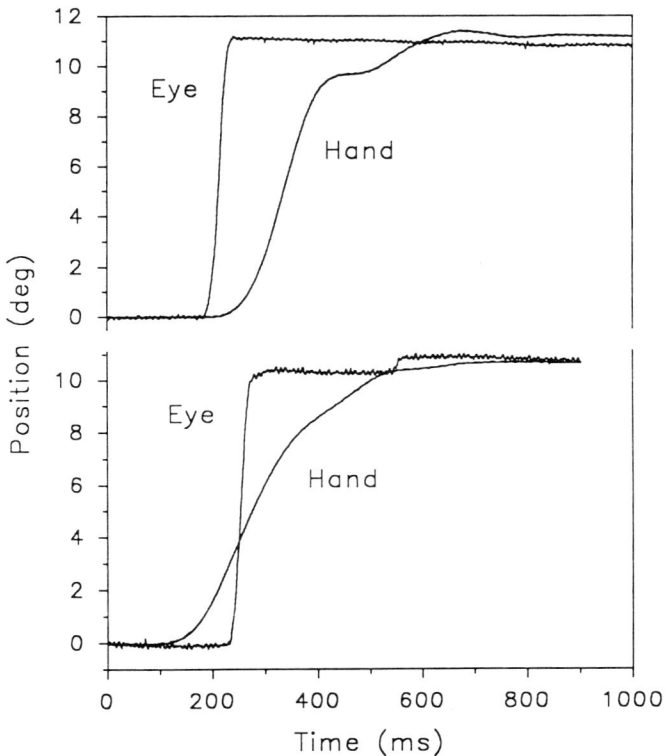

Figure 6.1. Eye movement and hand movement trajectories from two typical trials. Position is in degrees of visual angle, reflecting the distance between the home position and the target region on the video display. The movements shown in the figure required 39.5° of wrist-rotation in order to move the cursor to the center of the target. (Adapted with permission from Abrams et al., 1990).

We observed saccadic eye movements on 98% of the trials even though subjects were not instructed to make any eye movements at all. The mean eye movement amplitude was 96% of the distance between the home and the target. Because saccades often undershoot their goal slightly, these data show that subjects moved their eyes directly to the target region.

We also evaluated the times at which the eye movements occurred relative to the hand movements. Figure 6.1 shows examples of eye and wrist-rotation trajectories on two typical trials from the experiment. The upper panel shows data from a trial in which the eye moved to the target before the beginning of the wrist-rotation. The lower panel shows a trial on which the wrist-rotation movement began first. The former type of trial represented the majority. The eye began to move before the wrist rotation on 76% of the trials in the present experiment. Nevertheless, the particular order of eye and hand movements varied considerably from subject to subject. Some subjects began to move their eyes first on most of the trials, whereas others initiated the hand movement first. Despite these individual differences, the eye and hand almost always began to move at approximately the same time. Furthermore, because eye movement durations are much briefer than hand movement durations, the eyes almost always arrived at the target region well before the hand did (see Figure 6.1). The duration and accuracy of the wrist-rotation movements did not depend at all on the relative onset of the eye and hand movements.[1]

To further examine the relation between eye and hand movements we analyzed kinematic features of the wrist rotation movements in order to identify the beginnings and ends of the initial-impulse and error-correction phases of movement. This "parsing" of the movement trajectories involved identifying features of the movements such as velocity and acceleration zero-crossings and inflection points that were believed to be indicative of the transition from an early, ballistic initial-impulse phase, to a later, corrective error-correction phase. The criteria used for locating the transition from one movement phase to another were extensions of similar analytic techniques used by previous researchers (e.g., Carlton, 1980; Jagacinski, Repperger, Moran, Ward, & Glass, 1980; Langolf, Chaffin, & Foulke, 1976).[2] Figure 6.2 shows an example of a typical wrist-rotation movement divided into its component submovements.

---

[1] We evaluated the movement durations, the mean movement end location, and the standard deviation of the movement endpoints in space . None of these measures were affected by the relative onset of eye and hand movements, even though these measures were sensitive enough to reflect the effects of manipulations such as target distance and target width.

[2] Details of the movement parsing algorithm are provided in Abrams et al. (1990), and Meyer et al. (1988).

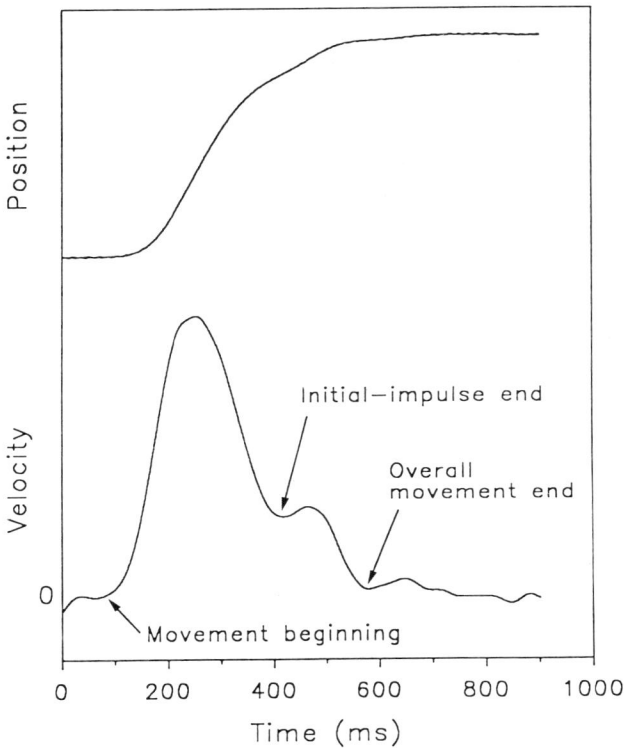

Figure 6.2. Example of position and velocity curves from a typical wrist-rotation movement. Indicated on the figure is the end of the initial impulse, and the beginning and end of the overall movement, as identified by the movement parsing algorithm.

Figure 6.3 shows a schematic representation of the relative timing of eye and hand movements from the experiment. Shown in the figure are the mean times at which the eye movements began and ended relative to: (a) the onset of the wrist-rotation movement (time zero in the figure); (b) the end of the initial-impulse phase of the wrist-rotations (and hence the beginning of the error-correction phase); and (c) the end of the overall wrist-rotation movement. Two aspects of this figure deserve attention. First, the eye movements were almost always initiated around the same time as the wrist-rotation movements. On the average, the eye began to move 57 ms before the onset of wrist-rotation.[3]

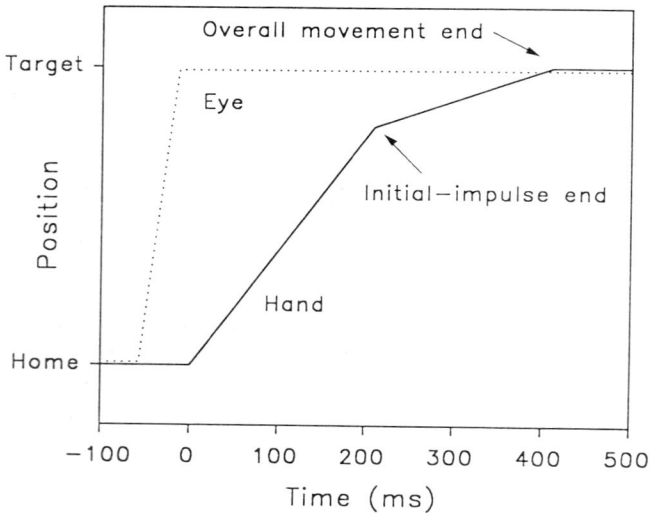

Figure 6.3. Schematic representation of the relative timing of the eye and hand movements from the first experiment. The times at the start and end of the eye and hand movements and at the end of the initial impulse are the means from the experiment. Only temporal aspects of the figure are meaningful.

The second interesting feature of the data in Figure 6.3 involves the times at which the eye movements ended relative to the wrist-rotations. On the average, the eye landed at the target 224 ms before the end of the wrist-rotation initial-impulse and the beginning of error-corrections. Furthermore, the error corrections themselves had a mean duration of 199 ms (including the small proportion of trials that had no error-correction phase). Thus, a

---

[3]We studied movements to several different combinations of target distance and target width. For some target conditions, the eye began to move as early as 74 ms before the hand, on the average. For other target conditions, the onset of eye movement preceded the hand movement by only 18 ms. Additionally, some subjects (in this and other experiments) consistently initiated the limb movements before the eye movements, although most subjects showed the opposite pattern.

considerable amount of time elapsed between the end of the saccades and the end of the wrist-rotation movements.

The most important finding of the present experiment is that subjects spontaneously produced saccadic eye movements to the target for the limb movement at around the same time that the limb began to move.[4] Some subjects initiated the eye movement first on most trials, others tended to initiate the wrist-rotation first. Nevertheless, because eye movements are usually much briefer than limb movements, the eyes almost always landed at the target well ahead of the limb. This pattern differs somewhat from results reported by others who have studied coordinated movements of the eyes and limbs. In those studies, the eye almost always completed its movement to the target before the limb movement ever began. As discussed earlier this may be due to temporal and spatial uncertainty about the target for the movement, or to pressure on the subjects to minimize their movement latencies.

Because subjects were not instructed to move their eyes at all in the present experiment it seems likely that the eye movements were produced because they served some useful function for the control and guidance of the limb movement. A number of possible benefits could arise from the eye movements that were observed. First, movement of the eye to the target could provide an extraretinal signal that could be used to better localize the target. Because the eye landed at the target well before the onset of the final error-corrections, this signal could have been used to control them. As a number of researchers have shown, people are able to localize targets based only on extraretinal eye-position information (Hansen & Skavenski, 1977, 1985; Hill, 1972; Morgan, 1978). Another possible benefit of producing saccades to the target would be an enhancement of the retinal visual feedback available about the moving limb as it approached the target. The eyes reached the target location 224 ms before the onset of the error-corrections. This is well within the time needed to process visual feedback information and incorporate that information into the error-corrections (Carlton, 1981a; Keele & Posner, 1968; Zelaznik et al., 1983). Finally, movements of the eyes to the target may lead to an improved sense of the target location through enhanced retinal information about the target. Prablanc et al. (1986) have shown that such information can improve the terminal accuracy of some movements. The results from the present experiment, however, do not allow us to distinguish between these various possibilities.

---

[4] Additional data come from a correlational analysis of the movements, performed separately for each subject and each of the several movement targets that we studied: The interval between the end of the eye movement and the end of the initial impulse was positively correlated with the duration of the initial impulse. This shows that the eye and hand movements were initiated together.

Role of Eye Movements to the Target

Given the different possible roles that could be served by moving the eyes to the target of an aimed limb movement, our next goal was to learn more about the precise nature of the benefit, if any, afforded to subjects by the eye movements. Two issues are of interest here. The first involves the extent to which the movement of the eye to the target plays a role in addition to any possible benefit of having the eye at the position of the target. For example, one way that information about movement could be important would involve the use of extraretinal cues arising from the oculomotor commands used to move the eye to the target location. Such information might be of use in guiding the limb movements. Alternatively, it might be that the entire benefit of moving the eye is due to having the eye pointing at the target position -- and thus information about the eye movement *per se* would not be useful. One of our objectives was to distinguish between these possibilities.

If eye position is important, then there is a second issue that we are interested in. This involves the extent to which the benefit of having the eyes at the target arises from retinal information about the limb (i.e., visual feedback), or to extraretinal information about the target. Both of these sources of information would be enhanced with the eyes at the target, and either might yield a benefit.[5]

In order to examine these possibilities we conducted an experiment in which we explicitly manipulated the eye movements that subjects produced during the wrist-rotation task (Abrams et al., 1990). The task in this experiment was similar to that in the previous experiment, with exceptions noted below. We studied three different eye-movement conditions. In the *saccade* condition, subjects were permitted to do whatever they wanted with their eyes, as they had in the first experiment. In the *fixate* condition, subjects were required to remain fixated on the home position throughout the entire wrist-rotation movement. Finally, in the *pursuit* condition, subjects were required to follow (by eye) a small dot that moved smoothly from the home position into the target region prior to the limb movement, thus producing a smooth-pursuit eye movement to the target. They were then required to maintain fixation on the target region until after the end of the wrist rotation.

Any differences in the wrist-rotation movements in the different eye movement conditions will be informative about the extent to which the movement of the eyes or eye position is important. For example, in both the saccade and pursuit conditions, subjects performed the wrist-rotation movements while fixating on the target region. Thus, if the only benefit of eye

---

[5]Having the eye at the target could also provide improved *retinal* information about the target (Paillard, 1980). This possibility is addressed in the next experiment we present here.

movements to the target is due to having the eye in the correct position, performance in these two conditions should be equal, and better than performance in the fixate condition. Alternatively, if there is some benefit gained by having produced a saccadic eye movement to the target in addition to the improvement gained by the new eye position, then wrist-rotation performance in the saccade and pursuit conditions should differ. This is because extraretinal information about eye movements is believed to differ after saccadic and pursuit eye movements (Festinger & Canon, 1965).[6]

We also manipulated the visibility of the cursor in order to learn more about the nature of the information that may be provided by having the eye at the target position. On one half of the trials in each of the three eye-movement conditions the cursor disappeared from the video display at the onset of the wrist-rotation movement. Because subjects were unable to see their hands, they were performing these movement without any retinal visual feedback information. If movements of the eye enhance the wrist rotation solely through improved retinal information about the moving limb, then the eye movement condition should not affect the wrist-rotations at all when the cursor is invisible. This is because there is no retinal information available about the cursor when it is invisible -- regardless of what the eyes do. However, if at least some of the benefit of moving the eyes to the target is due to an improved sense of the target's location, then there should still be a benefit of having the eyes at the target even when the cursor is invisible.

Some results from this experiment are shown in Figure 6.4. Plotted in the figure are constant errors for the wrist-rotation movements in the various conditions.[7] The constant error represents the mean deviation between the center of the target region and the endpoints of the wrist-rotation movements. As can be seen, constant errors increased (i.e., the movements travelled farther) when the cursor was invisible as opposed to when it was visible, regardless of the eye movement condition. This shows that retinal information about the moving limb plays an important role in the accuracy of aimed movements, consistent with previous findings (e.g., Carlton, 1981b).[8] Next,

---

[6]Saccadic movements are actively produced by the subject whereas pursuit movements are more passive, being driven by the motion of the pursued stimulus.

[7]We studied two different combinations of target distance and target width. Targets were centered either 10° (degrees of wrist-rotation) or 39.5° from the home position, and they were 2.5° wide. Effects of eye movement condition and target visibility were similar for movements to each of the targets, and all results are shown averaged over the two targets.

[8]Other aspects of the wrist rotations also suggest that cursor visibility is important. For example, the standard deviations of the spatial endpoints of the wrist rotations were much

the constant errors also differed in the different eye-movement conditions. Constant errors were equal in the saccade and pursuit conditions, but were considerably greater in the fixate condition. This pattern suggests that merely having the eye at the target location can improve the accuracy of the limb movement -- and it apparently does not matter how the eye arrived at the target (i.e., whether by a saccadic or a smooth-pursuit movement). Additionally, the differences among eye-movement conditions did not depend on cursor visibility, suggesting that the benefit of having the eye at the target location was not limited to providing improved visual feedback about the cursor. Rather, having the eye at the target improved the sense of target location -- perhaps through some extraretinal information about eye position.

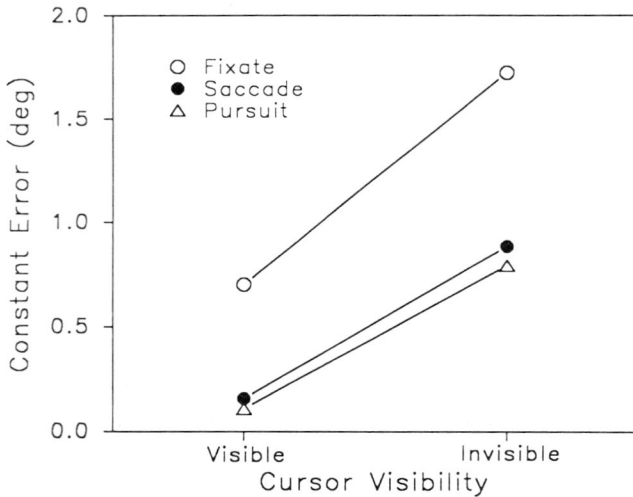

Figure 6.4. Constant errors of the wrist rotations, from the second experiment. The constant error is the mean deviation between the end of the movements and the center of the target region, in degrees of wrist-rotation.

We also evaluated features of the initial-impulse and error-correction phases of the wrist-rotations. If the information requirements for aimed limb movements differ during the different component submovements, then our manipulations of cursor visibility and eye movements might have distinct

greater when the cursor was invisible than when it was visible, and movements were much more likely to end in the target region with a visible cursor.

effects on the different submovements. Although the analysis of constant errors revealed no differences in the wrist rotations under the saccade and pursuit conditions, there were some differences in the wrist rotations that did depend on the how the eye reached the target location.

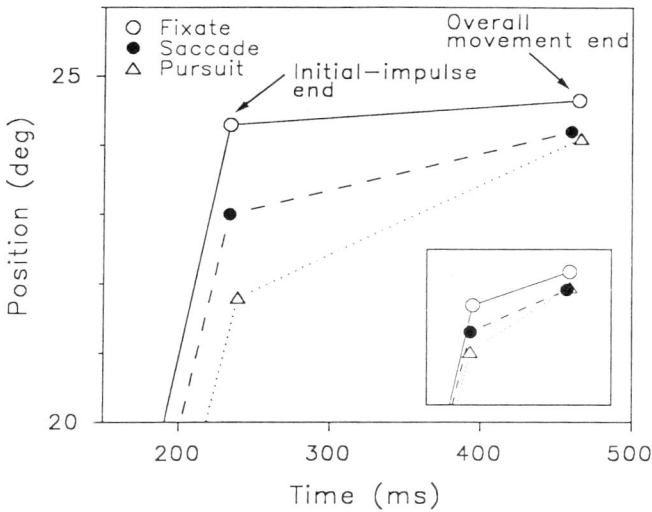

Figure 6.5. Schematic representation of initial-impulse and error-correction phase of the wrist-rotations under the visible-cursor condition in the second experiment. The plotted points indicate the durations of the initial impulses and the overall movements (the overall duration is the sum of the initial impulse and the error correction durations), as well as the distances traversed during the initial-impulse and error-correction phases of movement. Distances are in degrees of wrist-rotation. All movements began at time = 0 and position = 0. The inset shows data from the invisible-cursor condition.

Figure 6.5 shows the wrist rotation movements under the different eye movement conditions divided into initial-impulse and error-correction phases. Shown in the figure are schematic representations of the position of the wrist as a function of time during the wrist rotations. The plotted points indicate the durations of the initial impulses and the overall movements (the overall movement duration is the sum of the initial impulse and the error correction durations), as well as the distances traversed during the initial-impulse and error-correction phases of movement. Several important results are shown in the figure. First, movements in the fixate condition travelled farther overall than movements in the saccade and pursuit conditions, with the latter two conditions yielding movements of the same amplitude. This is the same pattern that was seen in the constant errors, as discussed above. Next, there were no differences in the durations of the initial-impulse or error-correction phases as a function of eye movement condition (nor were there differences in the overall movement durations). Thus, temporal aspects of the movements were unaffected by the behavior of the eyes. Finally, the most important feature of the data shown in the figure is that initial impulses in the saccade condition travelled farther than those under the pursuit condition, but error corrections in the saccade condition travelled less far. As a result, wrist-rotations in the saccade and pursuit conditions ended equally accurately despite the differences in the initial impulses. Each of these observations applies to movements during which the cursor was visible, as well as to those with an invisible cursor (shown in the inset of Figure 6.5).

The results of the present experiment provide considerable insight into the nature of the information provided by eye movements for use in controlling limb movements. First, it appears that the benefit of the spontaneous eye movements that subjects make to the target region is not limited to improved retinal information about the limb. This is because the eye movements improved the wrist-rotations even when the cursor was invisible. Thus, the eye movements must have provided some information about the target -- either retinal information (because the target was always visible) or extraretinal information. Next, because there were differences in the wrist rotation movements between the saccade and pursuit conditions, we can conclude that some of the information provided by the eye arises from the movement of the eye *per se*. It is not sufficient that the eye was at the target before the onset of movement. Rather, the manner in which the eye reached the target location affected the limb movement.

Perhaps the most important result of the present experiment is the finding that different sources of information are used in the programming of the initial impulse and the error corrections. If the programming of each of these movement phases shared the same information, then any manipulation that affected the initial impulses should also have had the same effect on the error corrections. But the results show that the type of eye movement affected the end location of the initial impulses but not the end of the error corrections (i.e., the end of the overall movement). These differences are discussed more fully below.[9]

---

[9]It is also of interest to note the nature of the influence that the eye movement condition had on the initial impulses and error corrections. While the distances traversed during the two

Role of Target Visibility

Results from the previous experiment suggest that having the eye at the target provided an improved sense of the target's location. Regardless of how the eye reached the target, the wrist rotations were more accurate with the eye at the target (saccade and pursuit conditions) than when it remained at the starting position (fixate condition). We attributed the enhanced perception of the target when the eyes were gazing at it, at least in part, to extraretinal information about eye position. But it is also possible that some benefit was due to improved retinal information about the target (cf. Paillard, 1980). Because the target was always visible in the previous experiment, it is not possible to determine precisely what contribution was provided by retinal information about the target. Our next goal was to explore the nature of the contribution of a visible target.

Questions about target visibility have received some attention previously, but no consensus has yet been reached regarding the role of a visible target. For example, Prablanc et al. (1986) showed that vision of the target early in a limb movement can help improve the accuracy of the movement. Similarly, Elliott and Madalena (1987) reported that some useful information can be obtained from a target that is visible shortly before and during a movement, even if the limb will not be visible. Nevertheless, Carlton (1981b) found that the removal of the target at the onset of a limb movement had no effect on movement accuracy. Taken together, these results suggest that retinal information about the target can be beneficial during the movement-preparation phase, and perhaps also shortly after the movement has begun, but it may play only a minor role later in the movement.

One reason that a visible target may be beneficial primarily during movement preparation may be related to the eye movements that subjects produce. A visible target may allow subjects to program and produce an accurate saccade to the target. Once the eye movement has been completed (which occurs around the time of limb movement onset, as seen in the first experiment reported here), they may not need a visible target any longer. If these possibilities are true, then target visibility may be unimportant only in situations in which subjects are permitted to look at the target, but target visibility may indeed play a role when subjects are not allowed to produce eye movements to the target.

---

movement phases depended systematically on the eye movement condition, the durations of the component phases did not change at all. These results suggest that the changes observed in the wrist rotations were accomplished by a rescaling of the *force parameter* for the movements, but not the *time parameter* for them (Meyer, Smith, & Wright, 1982). Similar results have also been reported for modifications in saccadic eye movements (Abrams, Dobkin, & Scharf, 1990).

To examine this issue we had subjects produce rapid pointing movements to target regions in a manner similar to that used in the previous experiments. Here, instead of wrist-rotations, the subjects moved a handle that slid from side to side on a track mounted directly beneath and in front of a video display. A shield prevented the subjects from viewing their hand or the handle directly. A cursor moved across the display contingent on movements of the handle, thus providing complete visual feedback about the status of the movement. Also shown on the display was a target region.   As in the previous experiments, the subjects' task was to move the handle in order to move the cursor from a home position to the target region. Subjects were to minimize the movement durations but also maintain a high level of accuracy.

Two different eye movement conditions were studied in the present experiment. In the *saccade* condition, subjects were permitted to do whatever they wanted with their eyes. In the *fixate* condition, subjects were required to remain fixated on the home position throughout the limb movement. Additionally, on half of the trials in each of the eye movement conditions, the target disappeared at the onset of the limb movement (*invisible-target* condition); on the other half of the trials the target remained visible throughout the movement (*visible-target* condition). The cursor remained visible throughout the limb movement in all conditions.

If a visible target serves mainly to enhance the accuracy of eye movements during the movement preparation phase, then target visibility may not influence the limb movements in the saccade condition. This is because subjects would be able to produce saccades to the target before it disappears. However, if subjects are unable to produce saccades to the target, as is the case in the fixate condition, then target visibility may affect the limb movement. If it did, then that would suggest that retinal information about the target can be used to help guide the limb movements even after the onset of movement.

Some results are shown in Figure 6.6. Plotted in the figure are constant errors of the limb movements for each eye-movement and target-visibility condition.   Several interesting patterns are apparent in the figure.   First, consider the differences between the saccade and fixate conditions.   The requirement to remain fixated at the home position caused the limb movements to travel much farther than when subjects were allowed to produce a saccade to the target. This result is similar to that seen in the preceding experiment. Apparently, the inability to produce saccades to the target induces a change in the perceived location of the target (Hill, 1972; Morgan, 1978). Next, the effects of target visibility depended on the eye movement condition. Target visibility affected the movements only in the fixate condition, when subjects were not allowed to produce saccades. But target visibility did not matter at all in the saccade condition, when subjects could produce eye movements to the target.

These results provide further insight into the role of eye movements and target visibility in aimed movements. Apparently, the information provided by a visible target is redundant with the information available after making a saccadic eye movement to the target. There are two types of information that a visible target could provide that might be unnecessary after a subject has produced an eye movement to the target. First, retinal target information might enhance the sense of target location when the target is on the peripheral retina (i.e., when the subject is unable to look at the target). This information might not be needed if extraretinal information about either eye movement or eye position was available to signal the target's location, as would be the case after a saccade to the target. Second, retinal information about the relative positions of the limb and the target could be used to help guide the limb into the target region. This information may also not be necessary with the eye at the target. In that case, absolute retinal information about the limb might be sufficient to guide it accurately.

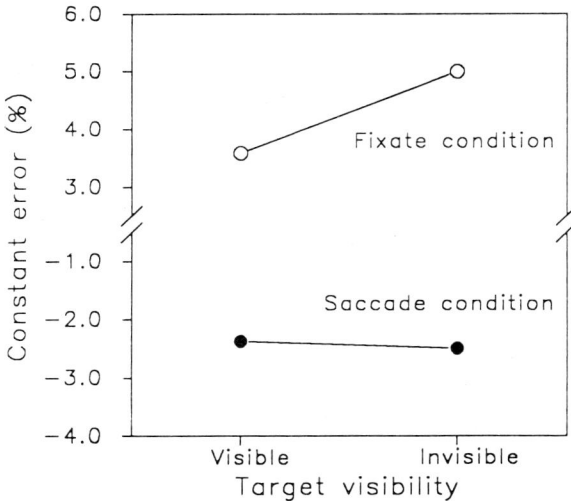

Figure 6.6. Constant errors for limb movements in the third experiment. Constant error is shown as a percentage of the total distance between the starting point and the center of the target region.

Because the cursor was always visible during the limb movements in the present experiment, it is not possible to determine the extent to which each of these two possibilities accounts for the results. Nevertheless, based on the results of the preceding experiment, it seems likely that extraretinal information about the target after an eye movement plays at least some role, in addition to whatever benefit may be due to improved retinal information about the cursor. Experiments are in progress in order to learn more about these issues.

Toward a Model of Eye-Hand Coordination
The results described above permit a number of conclusions about the role of eye movements in the coordination and control of aimed limb movements. These conclusions are consistent with a variety of observations in the literature on motor control and spatial localization, and they provide the basis for a tentative model that describes the processes underlying coordinated eye-hand behavior.

Two important findings described above must be accounted for in any model of eye-hand coordination. First, extraretinal information is available to the limb-movement control system from two distinct sources: (a) information about the position of the eye, and (b) information about movements of the eye. Second, the information that is used to guide the initial impulse is different from the information used to control the error corrections. The results of the present study suggest that information about movement of the eye is used for the programming and production of the initial impulse. This is because initial impulses differed in the saccade and pursuit conditions of the second experiment we reported. Because eye position during the limb movements was equivalent in those two conditions, the only difference between conditions was the nature of the eye movements that the subjects produced. Next, the results suggest that the error corrections rely on information about eye position and not eye movements. The reason for this conclusion is that the error corrections ended equally accurately in the saccade and pursuit conditions of the above mentioned experiment. Subjects had their eyes at the target location during the error corrections in both of these conditions. Thus, the initial impulses appear to be sensitive to the nature of the eye movements that subjects produce, whereas the error corrections depend only on the position of the eye.

What might be the nature of the difference in the information used to guide the initial impulses as opposed to that used for control of the error corrections? We hypothesize that the initial impulses are programmed on the basis of the perceived distance between the starting position and the target, whereas the error corrections are programmed to arrive at the perceived target location. If these ideas are correct, then perceived distance apparently depends on the nature of the eye movements that people produce, whereas the perceived location of the target depends only on eye position.

Support for these conclusions comes from past research on spatial localization, motor control, and perceptual-motor performance. Consider first the initial-impulse phase. Initial impulses in the pursuit condition of the second experiment travelled less far than those in the saccade condition. This is consistent with a number of results showing that people underestimate the extent of smooth pursuit eye movements compared to saccades of the same size (Abrams, Owens, Zeffren, & Drotman, 1990; Festinger & Canon, 1965; Honda, 1984, 1985, 1990; Mack & Herman, 1972; Miller, 1980).[10] If the initial impulses were programmed on the basis of the perceived distance to the target, then the differences we observed in our second experiment are exactly what would be expected.

Additional support for the programming of initial impulses on the basis of perceived distance comes from some models that have been proposed for the mechanisms that underlie rapid aimed limb movements. A number of such models characterize limb movements as arising from the application of a pulse of force to the limb with a specific magnitude and a specific duration (e.g., Meyer et al., 1988; Meyer, Smith, & Wright, 1982; Schmidt, Zelaznik, Hawkins, Frank, & Quinn, 1979). According to these models, the parameters of the force pulse may be programmed before movement onset to allow the limb to traverse a particular desired distance.

Next, consider the error corrections. According to our hypotheses, the error corrections are assumed to be programmed to arrive at a particular location in space, and not in terms of the distance needed to reach that location. As a result, the error corrections ended equally accurately in the saccade and pursuit conditions of our second experiment despite the fact that they began at different locations (i.e., the initial impulses had travelled different distances). Error corrections in the fixate condition however (in both the latter two experiments reported here), where subjects could not look at the target, were much less accurate than those in the saccade and pursuit conditions. These results suggest that the assessment of the location of the target depends in large part on extraretinal information about eye position.[11] Such information would

---

[10]Additionally, initial impulses in the fixate condition travelled farther than those in the other two conditions. This is also consistent with earlier work showing systematic differences in the localization of visible targets as a function of their retinal posit ion (Hill, 1972; Morgan, 1978).

[11]The results of our last experiment showed that the assessment of the target location may also depend on retinal information about the target -- but only when subjects are unable to move their eyes to it. The degradation in the movements attributable to target invisibility, however, was much smaller than that attributable to the inability to produce saccades (see Figure 7.6).

not be available when the eye was not fixating on the target. These conclusions are consistent with previous results showing that people are able to accurately localize targets after either saccadic or pursuit eye movements based only on extraretinal information about eye position (Hansen, 1979; Hansen & Skavenski, 1977, 1985).

As was true for the initial impulses, there is also some precedent for the kind of motor programming that we propose to be occurring during the error corrections. Some models of motor control characterize movements as arising from the adjustment of the resting lengths and tensions of opposing muscle pairs (Polit & Bizzi, 1979; Sakitt, 1980). By specifying the new desired resting lengths and tensions, the commands to the muscles could directly specify the final location for the limb. As a result, such movements would be *location-seeking*; able to arrive at a particular location in space regardless of the distance needed to reach that location, and regardless of any perturbations encountered along the way. The results of the second experiment reported above suggest that the error corrections may be programmed in just that way.

One implication of the mechanisms that we have proposed requires some additional discussion. If our hypotheses are correct, then the type of eye movements that subjects produce can influence the perceived distance to the target more than the perceived location of the target. This is because the type of eye movement produced in our second experiment affected the initial-impulse endpoints, but not the ends of the error corrections. These results imply that perceived distances and locations must be derived from different types of spatial representations. If perceived distances and locations were derived from a common spatial representation, then any manipulation that affected perceived distance should also have had the same effect on perceived location. Indeed, support does exist for a distinction between perceived distances and locations. Abrams and Landgraf (1990) had subjects provide judgments of either the distance moved, or the starting and ending locations of a visual target that underwent a combination of real and illusory movement. Judgments of distance were more affected by the illusory movement than were judgments of initial and final locations. These results suggest that perceptions of distance are at least partially independent of perceptions of locations, a conclusion that has also been made by others (Mack, Heuer, Villardi, & Chambers, 1985).

Summary
The results of the experiments described above show that people spontaneously produce saccades to the target for an aimed limb movement. The eye and limb begin to move at about the same time, but the eye arrives at the target well ahead of the limb. If subjects are unable to gaze at the target region during the movement, then the accuracy of the limb movement will

suffer, especially if the target is not visible. If subjects are able to gaze at the target, then the movements will end accurately, but the manner in which the eyes arrive at the target will affect kinematic features of the movements. The results support a model of rapid-movement control in which initial impulses are programmed to traverse a specific distance, and error corrections are programmed to arrive at a particular final location. The assessment of distance is affected by the nature of the eye movements that subjects make, whereas the perceived location of the target depends only on where the eyes are pointing.

Of course, many questions regarding the precise nature of the coordination of eye and hand remain unanswered. In particular, our experiments have focused somewhat separately on questions about retinal limb information, retinal target information and extraretinal information. But, a thorough understanding of coordinated eye-hand behavior must also consider interactive effects of the various types and sources of information. More work is needed to understand the extent to which each of these sources may work together to help guide movements. Our hope is that the initial framework that we have proposed will help to guide future inquiries into details of the mechanisms that underlie the control of aimed limb movements.

References

Abrams, R.A., Dobkin, R.S., & Scharf, M.K. (1990). *Adaptive modification of saccadic eye movements.* Manuscript submitted for publication.

Abrams, R.A. & Landgraf, J.Z. (1990). Differential use of distance and location information for spatial localization. *Perception & Psychophysics, 47,* 349-359.

Abrams, R.A., Meyer, D.E., & Kornblum, S. (1990). Eye-Hand Coordination: Oculomotor control in rapid aimed limb movements. *Journal of Experimental Psychology: Human Perception and Performance, 15,* 248-267.

Abrams, R.A., Owens, P., Zeffren, M., & Drotman, M. (1990, November). *Eye-Hand coordination: Spatial localization after saccadic and pursuit eye movements.* Paper presented at the meeting of the Psychonomic Society, New Orleans, LA.

Angell, R.W., Alston, W., & Garland, H. (1970). Functional relations between the manual and oculomotor control systems. *Experimental Neurology, 27,* 248-257.

Biguer, B., Jeannerod, M., & Prablanc, C. (1982). The coordination of eye, head, and arm movements during reaching at a single visual target. *Experimental Brain Research, 46,* 301-304.

Carlton, L.G. (1980). Movement control characteristics of aiming responses. *Ergonomics, 23,* 1019-1032.

Carlton, L.G. (1981a). Processing visual feedback information for movement control. *Journal of Experimental Psychology: Human Perception and Performance, 7*, 1019-1030.

Carlton, L.G. (1981b). Visual information: The control of aiming movements. *Quarterly Journal of Experimental Psychology, 33A*, 87- 93.

Crossman, E.R.F.W., & Goodeve, P.J. (1983). Feedback control of hand-movement and Fitts' Law. Paper presented at the meeting of the Experimental Psychology Society, Oxford, July 1963. Published in *Quarterly Journal of Experimental Psychology, 35A*, 251-278.

Elliott, D., & Madalena, J. (1987). The influence of premovement visual information on manual aiming. *Quarterly Journal of Experimental Psychology, 39A*, 541-559.

Festinger, L., & Canon, L.K. (1965). Information about spatial location based on knowledge about efference. *Psychological Review , 72*, 373-384.

Hansen, R.M. (1979). Spatial localization during pursuit eye movements. *Vision Research, 19*, 1213-1221.

Hansen, R.M., & Skavenski, A.A. (1977). Accuracy of eye position information for motor control. *Vision Research, 17,* 919-926.

Hansen, R.M., & Skavenski, A.A. (1985). Accuracy of spatial localizations near the time of saccadic eye movements. *Vision Research, 25*, 1077-1082.

Hill, A.L. (1972). Direction constancy. *Perception & Psychophysics, 11*, 175-178.

Honda, H. (1984). Functional between-hand differences and outflow eye-position information. *Quarterly Journal of Experimental Psychology, 36A*, 75-88.

Honda, H. (1985). Spatial localization in saccade and pursuit-eye-movement conditions: A comparison of perceptual and motor measures. *Perception & Psychophysics, 38*, 41-46.

Honda, H. (1990). The extraretinal signal from the pursuit-eye-movement system: Its role in the perceptual and egocentric localization systems. *Perception & Psychophysics, 48*, 509-515.

Jagacinski, R.J., Repperger, D.W., Moran, M.S., Ward, S.L., & Glass, B. (1980). Fitts' law and the microstructure of rapid discrete movements. *Journal of Experimental Psychology: Human Perception and Performance, 6*, 309-320.

Keele, S.W., & Posner, M.I. (1968). Processing of visual feedback in rapid movements. *Journal of Experimental Psychology, 77*, 155-158.

Kerr, B. (1978). Task factors that influence selection and preparation for voluntary movements. In G. E. Stelmach (Ed.), *Information processing in motor control and learning* (pp. 55-69). New York: Academic Press.

Langolf, G.D., Chaffin, D.B., & Foulke, J.A. (1976). An investigation of Fitts' law using a wide range of movement amplitudes. *Journal of Motor Behavior*, *8*, 113-128.

Mack, A., Heuer, F., Villardi, K., & Chambers, D. (1985). The dissociation of position and extent in Muller-Lyer figures. *Perception & Psychophysics*, *37*, 335-344.

Mack, A., & Herman, E. (1972). A new illusion: The underestimation of distance during pursuit eye movements. *Perception & Psychophysics*, *12*, 471-473.

Mather, J.A., & Fisk, J.D. (1985). Orienting to targets by looking and pointing: Parallels and interactions in ocular and manual performance. *Quarterly Journal of Experimental Psychology*, *37A*, 315-338.

Megaw, E.D., & Armstrong, W. (1973). Individual and simultaneous tracking of a step input by the horizontal saccadic eye movement and manual control systems. *Journal of Experimental Psychology*, *100*, 18-28.

Meyer, D.E., Abrams, R.A., Kornblum, S., Wright, C.E., & Smith, J.E.K. (1988). Optimality in human motor performance: Ideal control of rapid aimed movements. *Psychological Review*, *95*, 340-370.

Meyer, D.E., Smith, J.E.K., Kornblum, S., Abrams, R.A., & Wright, C.E. (1990). Speed-accuracy tradeoffs in aimed movements: Toward a theory of rapid voluntary action. In M. Jeannerod (Ed.), *Attention and performance XIII* (pp. 173-226). Hillsdale, NJ: Erlbaum.

Meyer, D.E., Smith, J.E.K., & Wright, C.E. (1982). Models for the speed and accuracy of aimed movements. *Psychological Review*, *89*, 449-482.

Morgan, C.L. (1978). Constancy of egocentric direction. *Perception & Psychophysics*, *23*, 61-68.

Miller, J.M. (1980). Information used by the perceptual and oculomotor systems regarding the amplitude of saccadic and pursuit eye movements. *Vision Research*, *20*, 59-68.

Paillard, J. (1980). The multichanneling of visual cues and the organization of a visually guided response. In G.E. Stelmach & J. Requin (Eds.), *Tutorials in motor behavior* (pp. 259-279). Amsterdam: North-Holland.

Pélisson, D., Prablanc, C., Goodale, M.A., & Jeannerod, M. (1986). Visual control of reaching movements without vision of the limb. II. Evidence of fast unconscious processes correcting the trajectory of the hand to the final position of a double-step stimulus. *Experimental Brain Research*, *62*, 303-311.

Polit, A., & Bizzi, E. (1979). Characteristics of motor programs underlying arm movements in monkeys. *Journal of Neurophysiology*, *42*, 183-194.

Prablanc, C., Echallier, J.E., Jeannerod, M., & Komilis, E. (1979a). Optimal response of eye and hand motor systems in pointing at a visual target: II. Static and dynamic visual cues in the control of hand movement. *Biological Cybernetics, 35*, 183-187.

Prablanc, C., Echallier, J.F., Komilis, E., & Jeannerod, M. (1979b). Optimal response of eye and hand motor systems in pointing at a visual target. I. Spatio-temporal characteristics of eye and hand movements and their relationships when varying the amount of visual information. *Biological Cybernetics, 35*, 113-124.

Prablanc, C., Pélisson, D., & Goodale, M.A. (1986). Visual control of reaching movements without vision of the limb. I. Role of retinal feedback of target position in guiding the hand. *Experimental Brain Research, 62*, 293-302.

Sakitt, B. (1980). A spring model and equivalent neural network for arm posture control. *Biological Cybernetics, 37*, 227-234.

Schmidt, R.A., Zelaznik, H.N., Hawkins, B., Frank, J. S., & Quinn, J. T., Jr. (1979). Motor-output variability: A theory for the accuracy of rapid motor acts. *Psychological Review, 86*, 415-451.

Vince, M.A. (1948). Corrective movements in a pursuit task. *Quarterly Journal of Experimental Psychology, 1*, 85-103.

Wallace, S.A., & Newell, K.M. (1983). Visual control of discrete aiming movements. *Quarterly Journal of Experimental Psychology, 35A*, 311-321.

Woodworth, R.S. (1899). The accuracy of voluntary movement. *Psychological Review, 3*, (Monograph Supplement), 1-119.

Wright, C.E., & Meyer, D.E. (1983). Conditions for a linear speed-accuracy trade-off in aimed movements. *Quarterly Journal of Experimental Psychology, 35A*, 279-296.

Zelaznik, H.N., Hawkins, B., & Kisselburgh, L. (1983). Rapid visual feedback processing in single-aiming movements. *Journal of Motor Behavior, 15*, 217-236.

Acknowledgment

Preparation of this chapter was supported by grant R29-MH45145 from the National Institutes of Health. Linda Van Dillen provided important contributions to the third experiment.

VISION AND MOTOR CONTROL
L. Proteau and D. Elliott (Editors)
1992 Elsevier Science Publishers B.V.

CHAPTER 7

VISUAL-MOTOR CONTROL
IN ALTERED GRAVITY

MALCOLM M. COHEN and ROBERT B. WELCH

NASA-Ames Research Center, Moffett Field, CA, 94035-1000

All life on Earth has evolved and developed under the influence of the terrestrial environment, which includes a unique mixture of atmospheric gases, periodic changes in radiant energy from the sun, a limited range of temperatures, and the presence of magnetic and gravitational fields. Gravity is a particularly important environmental variable because it imposes both time and velocity constraints on the behavioral capabilities of organisms; the action of gravity shapes anatomical structures, influences neuromuscular activities, and presents physical demands for the functioning of the nervous system.

With the advent of modern vehicles such as aircraft and spacecraft, human beings were subjected for the first time to substantial, and sometimes prolonged, alterations of normal gravitational-inertial forces (GIFs). These forces are exerted on the entire organism and stimulate several types of receptors, including those in tendons, muscles, and skin, as well as the otolith organs and the cardiovascular baroceptors (Gauer & Zuidema, 1961). This stimulation causes several dramatic modifications of both perception and perceptual-motor control (e.g., Glenn, 1962; Hernandez-Korwo, Kozlovskaya, Kreydich, Martinez-Fernandez, Rakhamanov, Fernandez-Pone, & Minenko, 1983). Spatial orientation is often disrupted (Gazenko, 1983; Gilson, Guedry, Hixson, & Niven, 1973; Young, Oman, Watt, Money, & Lichtenberg, 1984), and symptoms of motion sickness are sometimes elicited (Lackner & Graybiel, 1982; Nicogossian & Parker, 1982).

Many of the changes in perception and behavior occur not only during exposure to the altered GIFs, but afterwards as well. Elevated thresholds for linear accelerations (Parker & Reschke, 1983), altered gait and posture (Homick, Reschke, & Miller, 1977; Schmitt & Reid, 1985), and changed durations and amplitudes of various reflex responses (Baker, Nicogossian,

Hoffler, & Johnson, 1976; Reschke, Anderson, & Homick, 1984) have been observed in astronauts upon their return to Earth following exposure to the hypogravic conditions of space flight. Transient aftereffects of exposure to hypergravity have also been reported (Cohen, 1970a; Lackner & Graybiel, 1982).

This chapter examines changes in hand-eye coordination that result when GIFs are altered in magnitude alone. We have deliberately excluded the effects of exposure to situations in which the GIF vector is changed in both magnitude and direction; we have done so because such changes are generally transient, and even when prolonged, as in a fixed-cab centrifuge (Graybiel & Brown, 1951) or a "slow rotation room" (Graybiel, Clark, & Zarriello, 1960), they are not representative of the natural environment in which astronauts function. Further, because these more complex changes of both vector magnitude and direction, produced by rotation in a reference G-field, usually introduce several additional interacting variables, they can complicate our understanding of the effects of GIF magnitude.

After describing the perceptual and perceptual-motor effects of exposure to altered GIF magnitude, we present an historical review of experiments in which measures of adaptation to these altered conditions were obtained. This is followed by an analysis of various experimental factors that should be considered in studies of these effects. Next, we outline our general theoretical approach for systematically examining the effects of altered GIFs, and illustrate this approach with a description of our recent and ongoing research activities in this area. Finally, we conclude by proposing several potentially fruitful avenues of research for future investigation.

The Perceptual and Behavioral Effects of Altered GiFs

Under stationary terrestrial conditions, intersensory relationships are generally consistent, external mechanical and inertial forces are well defined and constant, and each physical action has predictable sensory consequences. Thus, except under very unusual circumstances, we each can readily distinguish the motions of our body from those of objects in the external world, and we seldom experience difficulty in moving about. In short, we function superbly as spatially coordinated organisms.

In contrast, under acceleration conditions such as those that can be produced by modern transportation vehicles, intersensory relationships may not be consistent, external mechanical and inertial forces may be variable, and each physical action may have unexpected sensory consequences. The combined effects of acceleration due to gravity and to changes in the velocity of the vehicle are such that the net force vector is frequently changed in magnitude, in direction, or in both magnitude and direction. As a result of any of these changes, the occupants of the vehicle may become disoriented and may experience severe disruptions of both perception and perceptual-motor coordination.

Muscle Loading/Unloading

When GiFs are increased or decreased in magnitude, the resulting mechanical loading or unloading of the limbs requires a change in the muscular forces necessary to make a particular movement. Under hypergravic conditions, the weight of the limbs is increased, and, as a result, the normal efferent signals are insufficient to move the limbs along a given path at a given velocity. If information regarding the changed environmental conditions is not provided, the increased weight of the limbs will result in an initial reach to a point below the desired location. With hypogravity, the reduced weight of the limbs makes these same efferent signals largely superfluous, resulting in initial over-reaching. Thus, on the basis of mechanical loading or unloading alone, one would expect initial attempts to reach for a visual target to be too low in hypergravity and too high in hypogravity (Cohen & Welch, 1988; Ross, 1991).

Elevator Illusion

This situation is complicated by another effect of exposure to altered GIF magnitude, known as the "elevator illusion" (Cohen, 1973a; Correia, Hixson, & Niven, 1968; Schöne, 1964; Whiteside, 1961). In this illusion, an observer exposed to hypergravity perceives a visual target in an otherwise empty field to move above its objective position; when the observer is exposed to hypogravity, the target appears to move below its objective position (Gerathewohl, 1952; Graybiel, Clark, & MacCorquodale, 1947). In the presence of a well-structured visual field, this illusion is diminished, or eliminated (Schöne, 1964).

It is well established that the localization of visual targets depends on both retinal and extra-retinal signals (Matin, 1976; Matin & Fox, 1989; Stoper & Cohen, 1989; Von Helmholtz, 1866). Retinal signals are generally unaffected by exposure to altered GIFs, except at the high levels of headwards acceleration ($+Gz$) that reduce blood flow to the eyes and the brain, causing blackout or loss of consciousness (Doolittle, 1924). In contrast, extra-retinal signals may be altered dramatically by even relatively minor perturbations in the magnitude of GIFs, and changes in these signals are probably responsible for the elevator illusion.

Specifically, it has been proposed that normal otolith-oculomotor reflexes are calibrated for the terrestrial environment, and that altered GIFs change this reflex activity (Cohen, 1973a; 1981; Whiteside, Graybiel, & Niven, 1965). These changes in otolith-oculomotor activity, in turn, can lead to illusory displacements of visual targets. According to this explanation, exposure to hypergravity stimulates the otolith organs to drive the natural resting position of the eyes downward. As a result, light from a visual target that was at the normal line of regard under terrestrial conditions now strikes the retina at a point below the fovea. This point was formerly associated with an object in the upper portion of the visual field. Thus, when viewed in hypergravity, the target appears to be elevated.

It is important to note that no net downward displacement of the eyes is necessary to achieve this illusion. With only a single spot of light as a visual target, increased stimulation of the otolith organs could initiate a vertical nystagmus, with slow phase downward (Marcus & Van Holten, 1990). To maintain foveal vision of a target that is objectively at the level of the eyes during exposure to hypergravity, an "upward" command to the oculomotor muscles would have to be initiated. Efferent monitoring of this "upward" command would be sufficient to generate an apparent upward displacement of the target. The same mechanism, but with directionally opposite effects, is suggested for the events that occur during exposure to hypogravity.

The important conclusion to draw from the preceding analysis is that the directional effects of muscle loading/unloading and the elevator illusion are opposite to one another, and thus tend to alter the magnitude, and perhaps even the direction, of the initial target-pointing errors that would result if only one of these effects were present.

Perceptual-Motor Adaptation to Altered GIFs
Major Research Findings

As described in the preceding section, initial exposure to hypogravity or hypergravity is accompanied by both perceptual and perceptual-motor disturbances, with serious implications for performance. However, anecdotal observations indicate that pilots and astronauts adjust fairly rapidly to these unusual conditions (Leonov & Lebedev, 1973; Schmitt & Reid, 1985). The nature and the time-course of this adaptive process have been examined in a variety of experiments, with special emphasis on hand-eye coordination. For research purposes, centrifuges have most commonly been used to produce indefinitely long periods of hypergravity, and aircraft, flying parabolic maneuvers, have been used to produce brief (30-second) alternating periods of 0 G and hypergravity.

One of the earliest published reports of visual-motor adaptation to altered GIFs was by Canfield, Comrey, and Wilson (1953). In this experiment, 48 subjects served as their own controls in exposure conditions of 1, 3, and 5 Gz, produced by centrifugation, and presented in various orders. The task was to reach out ballistically and touch 5-in. square targets located above, below, to the left, and to the right of the subjects' eyes.

The results indicated a tendency in the hypergravity condition to point too low and toward the center of the target array. The under-reaching is presumably caused by the muscle loading effect described previously; because the environment was both well-structured and fully illuminated, the reported under-reaching was probably not attenuated by the elevator illusion. Although adaptation was not systematically examined in this study, it was noted that whenever the 1-Gz condition immediately followed the 3-Gz or the 5-Gz condition, subjects frequently reached too high (at which they often voiced surprise). This is clearly an aftereffect of exposure to increased muscle loading, and suggests that at least some adaptation had occurred.

In addition to causing target-pointing errors, the hypergravic conditions were also associated with an increase in the time required both to initiate and to complete movements of the limb, an observation in concert with previous findings of these same investigators (e.g., Canfield, Comrey, & Wilson, 1949). It was not indicated in the present study if the increased response time ever diminished under hypergravity, although evidence of this sort of adaptation was reported by Canfield, Comrey, Wilson, and Zimmerman (1950) for discriminative reaction times.

Harald Von Beckh (1954) tested the bait-striking accuracy of aquatic turtles (*Hydromedusa tectifera* and *Chrysemis ornata*) during the weightless phase of parabolic flight, and reported that the animals' initial errors (direction not specified) decreased over 20-30 parabolas, although they never reached normal levels.

In a second experiment, pilots were required to reach out and draw a cross in each square of an array during a series of parabolas. With their eyes open, the pilots were only slightly less accurate during the 0-G phase of the parabolas than during 1-Gz horizontal flight; with their eyes closed, however, they made large errors, placing the crosses too high. These errors of over-reaching were probably due to muscle unloading. As in the study by Canfield et al. (1953), the presence of a well-structured visual field (immediately prior to each pointing response) probably attenuated or eliminated effects due to the elevator illusion. The small errors for the eyes-open condition demonstrate that, with visual feedback and non-ballistic reaching movements, individuals can make mid-course corrections while the hand is moving to the target. The initial errors for the eyes-closed condition decreased substantially with repeated flights, indicative of adaptation. Further, Von Beckh noted that the pilot with the longest record of instrument flying underwent very dramatic adaptation after only the second flight.

Lo Monaco, Strollo, and Fabris (1957; cited both by Von Beckh, 1959, and Leonov & Lebedev, 1973) created momentary periods of hypogravity in one of the earliest recorded accounts of "bungee cording." Subjects sat in a chair suspended on elastic links at the base of a 14-m tall "subgravity tower." When released from its constraints, the chair sprang rapidly upward toward the top of the tower, fell toward the ground, and oscillated repeatedly, with decreasing amplitude, until it came to rest. The period of hypogravity on the first "flight" was only 2-3 s, during which, the subjects, with eyes open, repeatedly attempted to hit a 15-cm diameter target with a pencil. Given the number of responses and the brief periods of hypogravity, the subjects probably made ballistic, rather than visually guided, responses. According to Von Beckh's (1959) description, the subjects initially reached too high (the muscle unloading effect), but adapted quickly on subsequent trials.

Gerathewohl, Strughold, and Stallings (1957) required their subjects to point ballistically with a metal stylus at a bull's eye while they continuously viewed their hand and the target. The subjects performed this task repeatedly during 10-s exposures to hypogravity produced by a steep dive in an aircraft, as well as during 4-5-s exposures to 3 Gz hypergravity produced by the pullout from the dive. The subjects initially reached too high during the dive, and too low during the pullout. Although the errors caused by hypogravity in the dive declined very rapidly, the errors caused by hypergravity during the pullout were only partially overcome.

Whiteside (1961) tested the hand-eye coordination of two subjects in a variety of altered gravitational-inertial environments by having them point at the image of a target viewed in a mirror that blocked sight of the hand. With this technique, originally developed by Held and Gottlieb (1958), subjects were able to reach for a continuously presented visual target without obtaining visual feedback regarding the position of their hand, and without having to point ballistically to avoid opportunity to make mid-course corrections. Whiteside examined the effects of both hypogravity (10-20 s of 0 G during parabolic flight) and hypergravity (2-3 min of 2 Gz produced by centrifugation). He also simulated the muscle unloading effects of hypogravity by submerging his subjects to the neck in water.

Surprisingly, in the 0-G phase of parabolic flight, the subjects initially pointed *below* the target, in contrast to the over-reaching responses reported in previous studies. Whiteside proposed that his subjects were seeing the target as lower than it actually was, due to the downward-acting elevator illusion, and were simply pointing to that location. No change in this pointing error occurred during subsequent trials, presumably because no error-corrective visual feedback was provided.

When the subjects were exposed to 2 Gz, they tended to point too high, an error Whiteside again attributed to the elevator illusion, now acting in the upward direction.   And, as with the O-G condition, no adaptation was observed.

Because the otolith organs are stimulated normally in the water-submersion condition, only muscle unloading effects would be expected. Indeed, Whiteside reported that his subjects pointed too high under this condition.   Performance improved markedly over trials, indicative of adaptation based solely on proprioceptive/kinesthetic error-corrective feedback from the limbs.

Cohen (1970a) used subjects as their own controls in three conditions of exposure to 2 Gz produced by centrifugation. In the Visual, Kinesthetic and Proprioceptive (VKP) feedback condition, subjects were allowed to view the preferred hand as they pointed at a visual target.   In the Kinesthetic and Proprioceptive (KP) feedback condition, subjects pointed at the image of the target as seen through a mirror that concealed direct view of the hand. Finally, in the Passive (P) condition, subjects sat motionless and supported by the chair during exposure to hypergravity.

An "adaptation curve" was obtained only for the KP condition, and revealed that most subjects initially pointed below the target. However, within 3-5 trials, these errors disappeared and were replaced by errors of pointing too high; the over-reaching errors persisted for the remainder of the trials in hypergravity. About 20 s after returning to 1 Gz, target-pointing accuracy for both the practiced and the non-practiced hand was measured for all three exposure conditions. For these measurements, the apparatus was set as it was in the KP condition. Substantial over-reaching was initially obtained following exposure under both the VKP and KP conditions, but not following exposure under the P condition. Thus, passive exposure to hypergravity appears to be insufficient to produce changes in hand-eye coordination. This finding is in general agreement with studies of adaptation to other sensory-motor rearrangements, such as those produced by prisms, as reviewed by Welch (1978).

Partial transfer of adaptation from the practiced to the non-practiced hand occurred for both VKP and KP conditions (63% and 22%, respectively), leading Cohen to infer the existence of a central process of correction for the hypergravity-induced errors, rather than some sort of simple muscular post-contraction phenomenon, or one based on unilaterally altered limb proprioception. However, because intermanual transfer was not complete in either condition, it remains possible that a portion of the aftereffects for the practiced hand can be explained by these (or other) peripheral mechanisms.

The relatively greater intermanual transfer for the VKP condition over the KP condition in this experiment may be due to an adaptive reduction in the strength of the elevator illusion which, presumably, would affect pointing for both the practiced and non-practiced hands. Studies of prism adaptation (e.g., Cohen, 1973b; Redding & Wallace, this volume, Uhlarik & Canon, 1970) have indicated that providing subjects with error-corrective visual feedback from the target-pointing hand will yield what appears to be a partial visual adaptation to the prismatic displacement (especially if this feedback occurs at the terminus of each response). It is possible, then, that analogous effects were obtained in the present context.

In a subsequent study, Cohen (1970b) examined the effects of KP feedback on target-pointing accuracy during exposure to terrestrial (1 Gz) conditions and to conditions of 1.5 and 2.0 Gz, as produced by centrifugation. Post-exposure aftereffects were not assessed in this study.

As expected, no significant change in pointing accuracy occurred over trials in the 1-Gz (control) condition. In the 2-Gz condition, subjects pointed too low for the first few trials, and then too high for the remaining trials. Interestingly, in the 1.5-Gz condition, subjects initially pointed too high, and continued to do so throughout the exposure period.

Thus, at levels of 2 Gz or higher, it appears that muscle loading initially has a greater effect on target pointing (prior to adaptation) than does the elevator illusion (because the two opposing mechanisms are initially resolved in the direction predicted by the effect of muscle loading). In part, this may be because the elevator illusion was not at full strength at the outset of the target-pointing period (e.g., Cohen, 1973a). Clearly, it will be important to measure both muscle loading and the elevator illusion separately to clarify their respective roles in hand-eye coordination during exposure to altered GIFs.

In a recent study by Watt, Money, Bondar, Thirsk, Garneau, and Scully-Power (1985), the hand-eye coordination of two payload specialist astronauts was tested during orbital shuttle mission 41-G. The astronauts were seated and held in position by means of an elastic cord stretched across their legs. Their task was to point, with eyes covered, at the remembered positions of five targets seen 5 minutes previously. Each pointing response was initiated at the chest, and visual feedback was provided when each response was completed. Testing occurred 3 weeks preflight (1 Gz), beginning at 11 hours in-flight (OG), and at 2.2 hours postflight (1 Gz).

In comparison to their high degree of accuracy preflight, both subjects were very inaccurate during the initial in-flight test. Their errors showed both a compression and a distortion of their reaching responses; attemps to point to targets directly ahead, above, below, to the left, and to the right of the subjects all tended to cluster together below the central target position. The authors reported that one of the astronauts improved during the course of the flight (although no data were presented), while the other did not. When pointing in-flight, the astronauts expressed great surprise that the targets were not where they remembered them to be. Post-flight responses were as accurate as the preflight responses.

For several reasons, the results of this study are difficult to interpret. First, pointing responses were below the targets, which is not the direction of error predicted from muscle unloading, although it is the expected result if the downward-acting elevator illusion influenced pointing responses to previously seen spatial positions. Second, the pointing task used in this study was handicapped by its high memory demands. Third, the errors were not simply ones of reaching too low, but also of reaching toward the midline. Fourth, the first in-flight tests occurred after 11 hours of exposure to weightlessness, which provided more than enough time for the subjects to adapt to the reduced muscle loading. Finally, the post-tests were conducted after 2.2 hours of experience in 1-Gz, and presumably the subjects engaged in sufficient hand-eye coordinations during this period to have recovered from any adaptation to the reduced muscle loading.

This review of previous studies has revealed several factors that should be, but often are not, taken into consideration when examining perceptual-motor adaptation to altered GIFs. The failure of investigators to eliminate, control, or otherwise assess some of these factors may explain some of the apparent contradictions in the literature. In the next section, we will discuss these factors in detail.

Important Experimental Considerations
Speed of limb movements. Under "everyday", uncontrolled conditions of exposure to hypogravity or hypergravity, pilots and astronauts frequently fail to reveal any obvious problems in their hand-eye coordination, event at the outset (Ballinger, 1952; Schmitt & Reid, 1985). One likely explanation for this observation is that most of these limb movements are made relatively slowly and under visual guidance.

Thus, one variable of crucial importance is whether subjects are allowed to make their reaching responses slowly, with visual guidance and proprioceptive feedback, or ballistically, without adequate time for visual guidance or the use of proprioceptive feedback (see Carlton, this volume). If subjects reach out slowly, the initial errors of under-reaching or over-reaching can usually be corrected enroute to the target, as we saw in Von Beckh's (1954) study of pilots. Alternatively, if reaches are truly ballistic, requiring less than 100 ms for their completion (Carlton, 1981; Zelaznik, Hawkins, & Kisselburgh, 1983), as in some research in this area, an initial misdirection of the hand is likely to remain uncorrected during the reaching response, resulting in substantial initial target-pointing errors.

It can be seen, then, that to reveal the immediate effects of altered gravitational-inertial environments on hand-eye coordination, ballistic pointing responses are to be preferred over slower, visually guided, movements. However, there may be some problems with this approach, because mid-trajectory corrections may still be possible on some trials if the movements are not truly "ballistic". Also, ballistic pointing responses tend to be more variable than responses made at slower speeds (Abrams, this volume; Meyer, Smith, & Wright, 1982; Meyer, Smith, Kornblum, Abrams, & Wright, 1990; Schmidt, Zelaznik, Hawkins, Frank, & Quinn, 1979).

Distance of limb movements. Because the time to complete a reaching movement depends on the distance as well as the speed of the movement, it is very important to specify and/or control the distance that the limb must move to the target. Reach distance has a pronounced influence on dynamic parameters of arm reaching movements; not only can greater distances result in longer reach times and higher peak velocities, but greater reach distances also can require a greater expenditure of energy and more rapid fatigue (Aume, 1973). Failure to specify the distance to the target can yield uncontrolled measures of both the accuracy and the precision of reaching responses, as well as a subsequent inability to explain experimental results.

Restraint and support of limbs. If subjects' limbs are not restrained and supported before and between pointing trials, changes in the apparent weight of the limbs might inform the subjects of the nature of the gravitational-inertial environment, perhaps allowing them to overcome any muscle loading/unloading effects and to point accurately on their very first attempt. Thus, it is important to restrain and support the limbs whenever they are not being used. The anomalous results of some earlier studies may be due to this problem. For example, this artifact may explain Whiteside's (1961) finding of initial under-reaching during exposure to the 0-G condition, and initial over-reaching during exposure to the 2 Gz condition; his findings are consistent with errors of hand-eye coordination that are due solely to the elevator illusion.

Type of feedback. Visual guidance of reaching movements can be eliminated by devising a situation in which subjects can reach out relatively slowly to a visual target without being able to see the hand or the arm. The technique developed by Held and Gottlieb (1958), and used recently by Cohen and Welch (1988), achieves this end by requiring subjects to point at a mirror-reflected target. This technique has the added advantage of allowing subjects to view the target continuously while pointing, rather than only at the start, and thus does not suffer from the potential memory limitations of other techniques.

As we have noted repeatedly, when useful visual feedback is precluded by a mirror-reflected target, or by ballistic pointing to the remembered position of a previously viewed target, subjects can overcome the muscle loading or unloading effects by means of kinesthetic/proprioceptive feedback from the limbs. In contrast, subjects cannot overcome the errors caused by the elevator illusion under these conditions. The pointing errors caused by the elevator illusion result from a discrepancy between the perceived location and the true location of the visually presented target. If no information is provided to indicate that the apparent and the true locations of the target differ, there is no basis for adaptation to these errors.

If subjects see the pointing hand at the termination of each response (rather than continuously viewing the hand as it moves to the target), several things can be expected. First, there are likely to be initial pointing errors because, in the absence of visual feedback, and with only proprioceptive information available, subjects have little opportunity to detect errors while the hand is moving to the target. Second, on subsequent trials, the subjects probably will overcome their errors through practice. Third, with repeated practice at target pointing, the subjects may experience a reduction in the elevator illusion. This third prediction is the most speculative, and is based on the assumption that the processes involved in adaptation to the visual effects of altered GIFs are similar to those involved in adaptation to the visual effects of wearing prisms.

The adaptation of visual-motor control measured in previous studies of altered GIF magnitude has generally involved an elimination of the muscle loading/unloading effects based on proprioceptive/kinesthetic feedback from the limbs. The subjects in these studies were prevented from experiencing visual feedback that could have revealed to them the specific portion of their target-pointing errors due to the elevator illusion alone. As a consequence, they were unable to correct behaviorally for these particular errors. Without specific sensory information, adaptation of the illusion is presumably impossible. Clearly, an attempt to produce adaptation to the visual effects of exposure to altered GIF magnitude represents an important goal for future research.

Structure of the visual field. As indicated previously, the elevator illusion is weak or nonexistent in the presence of a structured visual field. Thus, under such conditions, errors in pointing due specifically to this illusion should be reduced or abolished, leaving primarily the effects of muscle loading/unloading. The adaptive shifts in hand-eye coordination reported by Von Beckh (1954), Canfield et al. (1953), Gerathewohl et al. (1957), and Lo Monoco et al. (1957; as cited by Von Beckh, and by Leonov and Lebedev, 1973) all appear to result from a reduction of the muscle loading/unloading effects. With minimal visual structure, or where the target is the only visible object, as in the studies by Cohen (1970a, 1970b) and Cohen and Welch (1988), the elevator illusion should remain and contribute to errors in pointing.

Head orientation. The magnitude of the elevator illusion depends on the orientation of the subject's head with respect to the direction of the altered GIFs (Cohen, 1973a; Correia et al., 1968; Schöne, 1964). When the head is pitched forward by only 15°, the magnitude of the elevator illusion is reduced to about half of that produced when the head is erect; when the head is pitched forward by 30°, the elevator illusion is virtually eliminated (Cohen, 1973a). Thus, the orientation of the subject's head must be considered in any studies involving hand-eye coordination where the task involves reaching for a visual target whose apparent location can be changed by the elevator illusion.

Presentation of altered GIFs. Studies of adaptation, both to prismatic displacement (e.g., Lackner & Lobovits, 1978) and to rotating environments (e.g., Graybiel & Wood, 1969; Hu, Stern, & Koch, 1991), indicate that adaptation is more complete if the sensory rearrangement is introduced gradually, rather than presented all at once. One might expect the same to be true for the presentation of altered GIFs.

Distribution of practice. Studies of adaptation to prisms indicate that adaptation is greater when exposure is distributed over a series of trials with intervening "rest periods" than when it is "massed" (Cohen, 1973b; Taub & Goldberg, 1973). The same may be true for visuomotor adaptation to altered GIFs, although this variable has yet to be investigated.

Accuracy and precision as dependent variables. Most studies of hand-eye coordination measure the size and direction of the subjects' target-pointing errors (and the change in these as a function of practice). Another potentially important dependent variable is the trial-to-trial variability in pointing, as assessed, for example, by the within-subject standard deviation (which Schmidt et al. (1979) term "effective target width"). This measure is indicative of the *precision* of eye-hand coordination and, in theory, is independent of *accuracy*. Note, however, that if substantial changes in pointing occur during a given set of trials as a result of adaptation, the adaptation itself will produce a large dispersion in pointing response. Indeed, this seems to have occurred in the studies by Gerathewohl et al. (1957) and by Ross (1991); it would be misleading to interpret the increased variability obtained in these studies as a reduction in the precision of hand-eye coordination.

Independent measures of muscle loading/unloading and the elevator illusion. Because the effects of altered GIFs on muscle loading/unloading and on the elevator illusion work in opposite directions, they tend to cancel each other by their respective influences on hand-eye coordination. Therefore, it is very important to isolate and measure these effects individually; some ways to do this are suggested here.

A measure of the elevator illusion that is probably unaffected by muscle loading/unloading can be obtained by having the subject use a hand-held joystick to set a small motor-driven visual target to apparent eye-level in an otherwise dark environment (e.g., Cohen, 1973a). Because no target-reaching limb movements are involved in this task, and because the limbs are fully supported on an arm rest while the joystick is manipulated, muscle loading/unloading effects should not influence the results.

One way proposed to measure the effects of muscle loading/unloading in the absence of the elevator illusion is based on Cohen's (1973a) finding that the elevator illusion is eliminated when the head is pitched forward by approximately 30°. Thus, under these conditions, visual target-pointing responses should be influenced only by the muscle loading/unloading effects. Another suggestion, soon to be implemented in our laboratory, entails having the subject, with head erect, point at body-referenced positions (e.g., the level of the shoulders) in total darkness. Clearly, because the subject is not pointing to a visually presented or visually referenced target, the initial reaching errors should result exclusively from the effects of muscle loading/unloading.

Finally, to know precisely what the individual and combined contributions of muscle loading/unloading and the elevator illusion are during studies of hand-eye coordination under altered GIFs, it would be extremely helpful to measure each of these effects simultaneously. Operationally, this could be closely approximated by having the subject rapidly alternate between providing measures of muscle loading/unloading and measures of the elevator illusion. By counterbalancing the order in which we obtain the two measures, we should be able to control for order effects, and thus determine the time-course of both muscle loading/unloading and the elevator illusion as they independently influence visual-motor control.

Summary

The initial reaching responses in both hypogravity and hypergravity environments are subject to errors, especially when the task involves ballistic movements or is carried out in the absence of visual control. These errors are due to the combined (and opposite) effects of muscle loading/unloading and the elevator illusion. Finally, a number of important variables have been isolated from our review of the research literature, and should be incorporated, or addressed in some manner, in future studies.

Overview of Current Research by the Authors
Because muscle loading/unloading and the elevator illusion are of both theoretical and practical interest, we have undertaken a series of investigations to increase our understanding of their effects and to provide the means by which they can be ameliorated.

The specific goals of this research program are threefold: to achieve a more complete understanding of the effects of altered GIFs on perception and performance, to delineate the roles of intersensory interactions and feedback mechanisms in perceptual and behavioral adaptation to altered GIFs, and to develop analytic, descriptive, and predictive techniques that can be used to further our understanding of the mechanisms underlying adaptation to altered sensory environments in general.

Strategy for Investigating Spatial Perception and Behavior
Our more encompassing goal is to understand the interrelationships among the spatial senses under virtually all environmental conditions, both normal and abnormal. We assume that this understanding is greatly facilitated by examining situations where the normal environment is rearranged or otherwise altered, as is the case with hypogravity or hypergravity. As we have shown, these conditions can cause organisms to experience serious perceptual and perceptual-motor difficulties. Thus, it is our belief that, through careful observation of perceptual and behavioral responses to both acutely and chronically presented anomalous sensory environments, we will achieve a more complete understanding of the physiological mechanisms and principles that subserve normal spatial perception and behavior. The basic idea, eloquently articulated long ago by Bernard (1865), is that both normal and abnormal functioning are governed by the same rules, differing only because of the special conditions under which these rules manifest themselves.

Anomalous organismic conditions can be produced by surgical interventions such as deafferentation, extirpation, or ablation. However, alternative nonsurgical and non-invasive methods are also available, and are much preferred for a variety of reasons. Thus, rather than observing the functioning of an anomalous organism in a normal environment, we can examine the functioning of a normal organism in an anomalous environment. The effects of employing some of these techniques, especially those involving optical rearrangements of visual space, have been discussed extensively elsewhere (Kornheiser, 1976; Pick & Hay, 1966; Rock, 1966; Ross, 1975; Welch, 1978).

The Importance of Motor-Generated Sensory Feedback

Spatially directed responses that depend on precise visual-motor coordination have both motor and sensory components. The spatial direction of the response involves the localization of stimulus sources in the physical environment, as well as the precise specification of motor programs to direct movements of the relevant body parts. Because motor outputs have sensory consequences (e.g., the sight of the moving hand), they can serve as "probes" of the environment, allowing organisms to evaluate the relationships between actions and their spatial effects. Thus, sensory consequences of motor acts (motor-generated sensory feedback) can provide information to recalibrate subsequent motor activity. This concept, articulated in the context of adaptation to altered sensory environments by Held and Hein (1958), is at the heart of the hypotheses tested in many of our current research efforts.

Recent studies by Soechting and Flanders (1989a,b) have elucidated how visually derived representations of target locations are converted into appropriate motor commands. They have argued that errors in pointing result from inaccurate transformations from extrinsic to intrinsic coordinates. In this context, it would appear that altered GIFs provide distortions of both visually derived representations of spatial locations (due to the elevator illusion) and the spatial consequences of formerly appropriate motor commands (due to muscle loading/unloading). Detailed analyses of visual-motor coordination in altered gravity based on this approach remain to be accomplished.

Adaptation of Hand-Eye Coordination in Parabolic Flight

In a recent study (Cohen & Welch, 1988) human subjects were flown through a series of five consecutive parabolas in a Learjet aircraft. The task was to point rapidly and repeatedly at a visual target during each of the hypogravity phases of the parabolas, *or* during each of the hypergravity phases; ten responses were made during the selected phase of each parabola. The target was the image of a spot of light viewed in a mirror that precluded sight of the reaching limb. By engaging in active visual-motor responses during either one or the other of the two altered G conditions, the subjects would experience motor-generated sensory feedback specific to that condition alone. It was hypothesized that this specificity would cause them to adapt only to that form of altered GIF, while remaining unaffected by the other.

The outcome of the experiment was that, on the first few trials of the first parabola, the subjects reached approximately at the target in either of the conditions. On the remaining trials of the first parabola, and throughout the subsequent parabolas, they reached too high in the hypergravity condition and too low in the hypogravity condition.

We interpreted our results as follows: Initially, during a given phase of a parabola, subjects are confronted with the effects of muscle loading (or unloading) and an upward (or downward) acting elevator illusion. Because these effects work in opposite directions, the initial pointing responses were nearly accurate. However, the kinesthetic/proprioceptive feedback from the pointing hand was sufficient to overcome the effects of muscle loading/unloading over successive trials, thus causing the subjects to point at the illusorily (upward or downward) displaced visual target. Because visual feedback from the hand was precluded, this latter error, due to the elevator illusion, remained uncorrected throughout the remainder of the trials.

When the aircraft returned to straight and level flight, the effects of the adaptation persisted briefly, and the subjects initially reached below the target if they had responded only during the hypogravity phase of the parabolas, or above the target if they had responded only during the hypergravity phase of the parabolas.

Of particular interest in this study was the demonstration that, by limiting the subjects' activity to the hypogravic phase of parabolic flight, the *functional* duration of exposure to hypogravity can be extended well beyond the 30-s length of each parabola. Because there is no earthly way to provide extended periods of microgravity (which can only occur in orbital or interplanetary flight), our present findings may have significant applied value for the training of astronauts in parabolic flight. For example, a subject allowed to engage in active body movements only during the 30-s hypogravity phases of a set of 20 consecutive parabolas, while remaining immobile during the alternating hypergravity phases, could be exposed to a full 10 minutes of *functionally* continuous hypogravity (i.e., 20 x 30 s = 600 s). Not only is the cost of parabolic flight appreciably less than that of orbital flight, but the notion that the effects of hypogravity in parabolic flight are always destroyed by the intervening periods of hypergravity appears to be in error.

General Summary and Conclusions
Human perception and performance are affected by changes in the magnitude of the gravitational-inertial environment. The extra weight of the limbs during exposure to hypergravity (muscle loading) causes initial (ballistic) reaching responses to be too low, while the reduced weight of the limbs during exposure to hypogravity (muscle unloading) causes them to be too high. If a visual target is viewed in an otherwise dark setting under altered gravitational-inertial conditions, the elevator illusion will also occur, and will induce target-pointing errors opposite in direction from those caused by muscle loading/unloading.

Anecdotal observations indicate that the errors of hand-eye coordination that occur in either hypogravity or hypergravity are overcome quite rapidly by pilots and astronauts, a conclusion amply confirmed by a series of controlled experiments going back more than 40 years. To date, no systematic attempts have been made to modify the elevator illusion or its concomitant effects on performance; thus far, the only adaptive changes of hand-eye coordination that have been demonstrated unequivocally are those that result from the reduction or elimination of muscle loading/unloading effects.

We have suggested that it might be possible to induce adaptation to the elevator illusion if subjects are provided with visual feedback about their target-pointing errors caused specifically by this illusion. Studies of prism adaptation indicate that the ideal condition for this form of adaptation would be *terminal* visual exposure of the pointing limb.

Important considerations in studies of adaptation to altered GIFs include the following: (a) speed of limb movements, (b) distance of limb movements, (c) restraint and support of limbs, (d) type of feedback, (e) structure of the visual field, (f) head orientation, (g) presentation of altered GIFs, (h) distribution of practice, (i) accuracy and precision as dependent variables, and (j) independent measures of muscle loading/unloading and the elevator illusion.

We examined the effects of alternating hypogravity and hypergravity, as produced by parabolic flight, on open-loop target-pointing. Our results demonstrated that adaptation was specific to those conditions under which motor-sensory feedback was provided (see also Proteau, this volume); the subjects who pointed at the target during the hypogravic phase, but were passive during the hypergravic phase, adapted only to the hypogravic condition; likewise, the subjects who pointed at the target only during the hypergravic phase adapted to that condition. In both cases, the adaptation was based on elimination of the effects of muscle loading/unloading.

To what extent do the results of the research described in this chapter generalize to the "real-life" environment of the astronaut or the pilot in altered G environments? First, it can be concluded that the elevator illusion and its effects on hand-eye coordination will probably not occur when the environment is well-lighted and well-structured. Only when background lighting is poor or nonexistent should this illusion appear and have an influence on target-reaching behavior. Thus, in a visually structured environment, the loading or unloading of the limbs will be the major cause for initial errors in reaching. Second, it should be clear by now that if one reaches for objects *slowly* and with continuous visual guidance of the limb, there should be little or no difficulty in bringing the hand to the target, even at the outset of exposure to the altered gravitational-inertial conditions. Thus, it would appear that the astronaut or pilot is most at risk of mis-reaching when making ballistic movements and/or when looking elsewhere immediately after initiating a response. Even if errors in hand-eye coordination do occur at the beginning of exposure to altered GIFs, rapid adaptation will take place when sensory feedback is provided. Thus, it may be argued that, during the early stages of exposure, astronauts or pilots should avoid making very rapid and/or non-visually guided movements when reaching for objects.

A number of questions about the effects of altered gravity on motor control remain unanswered. First, what are the *individual* quantitative effects of muscle loading/unloading and the elevator illusion on hand-eye coordination? We are currently addressing this question by exposing subjects to 1.75 to 2.0 Gz in a human centrifuge and using *independent* measures of each of these two phenomena, as described in a previous section of this chapter.

Another important issue is the intriguing question of whether it is possible to develop simultaneous adaptations to both altered and normal gravitational-inertial environments. A good deal of anecdotal evidence (e.g., Von Beckh, 1954; Leonov & Lebedev, 1973), and a few controlled experiments scattered through the literature (e.g., Lackner & Graybiel, 1982) suggest that such "dual adaptations" are possible. Our strategy is simply to alternate subjects between adapting to hypergravity in the centrifuge and "readapting" to normal gravity. We anticipate that, with repeated alternations, it may become possible for subjects to make the switch between these conditions with little or no difficulty, in contrast to the typical disturbance of perception and perceptual-motor control that generally characterizes the initial transitions.

A number of subsidiary questions may also be raised here, such as what the best conditions are for creating dual adaptations, and what kinds of cues subjects use to distinguish the two environments and thereby invoke the appropriate adaptation. The practical implications of this research are substantial. For example, if dual adaptations can reliably be produced and the conditions for their creation clearly delineated, it is possible to envision astronauts being successfully trained on "preflight adaptation trainers" (e.g., Parker, Reschke, Arrott, Homick, & Lichtenberg, 1985). If the astronauts could retain this adaptation training over long periods of time, they could be expected to function with full effectiveness, even when they first enter space.

References

Aume, N.M. (1973). An exploratory study of arm-reach dynamics under several levels of gravity. *Ergonomics, 16*, 481-494.

Baker, J.T., Nicogossian, A.E., Hoffler, G.W., & Johnson, R.L. (1976). Measurement of a single tendon reflex in conjunction with a myogram:

The second manned skylab mission. *Aviation Space, & Environmental Medicine, 47,* 400-402.

Ballinger, E.R. (1952). Human experiments in subgravity and prolonged acceleration. *Journal of Aviation Medicine, 23,* 319-321.

Bernard, M.C. (1865). Introduction à l'étude de la médecine expérimentale. [Introduction to the study of experimental medecine]. Paris: J.B. Baillière et fils.

Canfield, A.A., Comrey, A.L., & Wilson, R.C. (1949). A study of reaction time to light and sound as related to increased positive radial acceleration. *Journal of Aviation Medicine, 20,* 350-355.

Canfield, A.A., Comrey, A.L., & Wilson, R.C. (1953). The influence of increased positive g on reaching movements. *Journal of Applied Psychology, 37,* 230-235.

Canfield, A.A., Comrey, A.L., Wilson, R.C., & Zimmerman, W.S. (1950). The effect of increased positive radial acceleration upon discrimination reaction time. *Journal of Experimental Psychology, 49,* 733-737.

Carlton, L.G. (1981). Processing visual feedback information for movement control. *Journal of Experimental Psychology: Human Perception and Performance, 7,* 1019-1030.

Cohen, M.M. (1970a). Sensory-motor adaptation and aftereffects of exposure to increased gravitational forces. *Aerospace Medicine, 41,* 318-322.

Cohen, M.M. (1970b). Hand-eye coordination in altered gravitational fields. *Aerospace Medicine, 41,* 647-649.

Cohen, M.M. (1973a). Elevator illusion: Influences of otolith organ activity and neck proprioception. *Perception & Psychophysics, 14,* 401-406.

Cohen, M.M. (1973b). Visual feedback, distribution of practice, and intermanual transfer of prism aftereffects. *Perceptual and Motor Skills, 37,* 599-609.

Cohen, M.M. (1981). Visual-proprioceptive interactions. In R.D. Walk & H.L. Pick (Eds.), *Intersensory perception and sensory integration* (pp. 175-215). New York: Plenum.

Cohen, M.M., & Welch, R.B. (1988). Hand-eye coordination during and after parabolic flight. *Aviation, Space, & Environmental Medicine, 59,* 68.

Correia, M.J., Hixson, W.C., & Niven, J.I. (1968). On predictive equations for subjective judgments of vertical and horizon in a force field. *Acta Oto-laryngologica, Monograph Supplement 230.*

Doolittle, J.H. (1924). *Accelerations in flight.* In Tenth Annual Report of N.A.C.A. (Report no 203, pp. 373-388). Washington: Government Printing Office.

Gauer, O.H., & Zuidema, G.D. (1961). *Gravitational stress in aerospace medicine.* Boston: Little Brown and Company.

Gazenko, O.G. (1983). Man in space, an overview. *Aviation, Space, & Environmental Medicine, 54,* 53-55.

Gerathewohl, S.J. (1952). Physics and psychophysics of weightlessness visual perception. *Journal of Aviation Medicine, 23,* 373-395.

Gerathewohl, S.J., Strughold, H., & Stallings, H.D. (1957). Sensorimotor performance during weightlessness. *Journal of Aviation Medicine, 28,* 7-12.

Gilson, R.D., Guedry, F.E., Jr., Hixson, W.C., & Niven, J.I. (1973). Observations on perceived changes in aircraft attitude attending head movements made in a 2-g bank and turn. *Aerospace Medicine, 44,* 90-92.

Glenn, J.H., Jr. (1962). The Mercury-atlas-6 space flight. *Science, 136,* 1093-1095.

Graybiel, A., & Brown, R.H. (1951). The delay in visual reorientation following exposure to a change in direction of resultant force on a centrifuge. *Journal of General Psychology, 45,* 143-150.

Graybiel, A., Clark, B., & MacCorquodale, K. (1947). The illusory perception of movement caused by angular acceleration and by centrifugal force during flight. I. Methodology and preliminary results. *Journal of Experimental Psychology, 37,* 170-177.

Graybiel, A., Clark, B., & Zarriello, J.J. (1960). Observations on human subjects living in a "slow rotation room" for periods of two days. *A.M.A. Archives of Neurology, 3,* 55-73.

Graybiel, A., & Wood, C. (1969). Rapid vestibular adaptation in a rotated environment by means of controlled head movements. *Aerospace Medecine, 40,* 638-643.

Held, R., & Gottlieb, N. (1958). Technique for studying adaptation to disarranged hand-eye coordination. *Perceptual and Motor Skills, 8,* 83-86.

Held, R., & Hein, A. (1958). Adaptation to disarranged hand-eye coordination contingent upon reafferent stimulation. *Perceptual and Motor Skills, 8,* 87-90.

Hernandez-Korwo, R., Kozlovskaya, I.B., Kreydich, Y.V., Martinez-Fernandez, S., Rakhamanov, A.S., Fernandez-Pone, E., & Minenko, V.A. (1983). Effect of seven day spaceflight on structure and function of human locomotor system. *USSR Report: Space Biology and Aerospace Medicine, 17,* 50-58.

Homick, J.L., Reschke, M.F., & Miller, E.F., II. (1977). The effects of prolonged exposure of weightlessness on postural equilibrium. In R.S. Johnson & L.F. Dietlein (Eds.), *Biomedical results from skylab* (NASA SP-377, pp. 104-112). Washington: NASA.

Hu, S., Stern, R.M., & Koch, K.L. (1991). Effects of preexposures to a rotating optokinetic drum on adaptation to motion sickness. *Aviation, Space, & Environmental Medicine, 62,* 53-56.

Kornheiser, A.S. (1976). Adaptation to laterally displaced vision: A review. *Psychological Bulletin, 83,* 783-816.

Lackner, J.R., & Graybiel, A. (1982). Rapid perceptual adaptation to high gravitoinertial force levels: Evidence for context-specific adaptation. *Aviation, Space, & Environmental Medicine, 53,* 766-769.

Lackner, J.R., & Lobovits, D. (1978). Incremental exposure facilitates adaptation to sensory rearrangement. *Aviation, Space, & Environmental Medicine, 49,* 362-364.

Leonov, A.A., & Lebedev, V.I. (1973). Psychological characteristics of the activities of cosmonauts. *NASA TT F727.* Washington: NASA.

Marcus, J.T., & Van Holten, C.R. (1990). Vestibular-ocular responses in man to +Gz hypergravity. *Aviation, Space, & Environmental Medicine, 61,* 631-635.

Matin, L. (1976). A possible hybrid mechanism for modification of visual direction associated with eye movements - the paralyzed-eye experiment reconsidered. *Perception, 5,* 233-239.

Matin, L., & Fox, C.R. (1989). Visually perceived eye level and perceived elevation of objects: Linearly additive influences from visual field pitch and gravity. *Vision Research, 29,* 315-324.

Meyer, D.E., Smith, J.E.K., & Wright, C.E. (1982). Models for the speed and accuracy of aimed movements. *Psychological Review, 89,* 449-482.

Meyer, D.E., Smith, J.E.K., Kornblum, S., Abrams, R.A., & Wright, C.E. (1990). Speed-accuracy tradeoffs in aimed movements: Toward a theory of rapid voluntary action. In M. Jeannerod (Ed.), *Attention and performance XIII* (pp. 173-226). Hillsdale, NJ: Erlbaum.

Nicogossian, A.E., & Parker, J.F. (1982). *Space physiology and medicine.* Washington, D.C.: NASA.

Parker, D.E., & Reschke, M.F. (1983, November). Changes in self motion perception following prolonged weightlessness. *NASA JSC Memorandum STS-8 DSO-0433.*

Parker, D.E., Reschke, M.F., Arrott, A.P., Homick, J.L., & Lichtenberg, B.K. (1985). Otolith tilt-translation reinterpretation following prolonged weightlessness: Implications for preflight training. *Aviation, Space, & Environmental Medicine, 56,* 601-606.

Pick, H.L., Jr., & Hay, J.C. (1966). The distortion experiment as a tool for studying the development of perceptual-motor coordination. In N. Jenkin & R.H. Pollack (Eds.), *Perceptual development: Its relation to theories of intelligence and cognition.* Research Programs in Child Development. Chicago: Institute for Juvenile Research.

Reschke, M.F., Anderson, D.J., & Homick, J.L. (1984). Vestibulospinal reflexes as a function of microgravity. *Science, 225,* 212-214.

Rock, I. (1966). *The nature of perceptual adaptation.* New York: Basic Books.

Ross, H.E. (1975). *Behavior and perception in strange environments.* New York: Basic Books.

Ross, H.E. (1991). Motor skills under varied gravitoinertial force in parabolic flight. *Acta Astronautica, 23,* 85-95.

Schmidt, R.A., Zelaznik, H.N., Hawkins, B., Frank, J.S., & Quinn, J.T., Jr. (1979). Motor output variability: A theory for the accuracy of rapid motor acts. *Psychological Review, 86,* 415-451.

Schmitt, H.H., & Reid, D.J. (1985). *Anecdotal information on space adaptation syndrome.* Houston: NASA/Space Biomedical Research Institute, USRA Division of Space Biomedicine.

Schöne, H. (1964). On the role of gravity in human spatial orientation. *Aerospace Medicine, 35,* 764-772.

Soechting, J.F., & Flanders, M. (1989a). Sensorimotor representations for pointing to targets in three-dimensional space. *Journal of Neurophysiology, 62,* 582-594.

Soechting, J.F., & Flanders, M. (1989b). Errors in pointing are due to approximations in sensorimotor transformations. *Journal of Neurophysiology, 62,* 595-608.

Stoper, A.E., & Cohen, M.M. (1989). Effect of structured visual environments on apparent eye level. *Perception & Psychophysics, 46,* 469-475.

Taub, E., & Goldberg, I.A. (1973). Prism adaptation: Control of intermanual transfer by distribution of practice. *Science, 180,* 755-757.

Uhlarik, J.J., & Canon, L.K. (1970). Influence of concurrent and terminal exposure conditions on the nature of perceptual adaptation. *Journal of Experimental Psychology, 91,* 233-239.

Von Beckh, H.J.A. (1954). Experiments with animals and human subjects under sub- and zero-gravity conditions during the dive and parabolic flight. *Journal of Aviation Medicine, 25,* 235-241.

Von Beckh, H.J.A. (1959). Human reactions during flight to acceleration preceded by or followed by weightlessness. *Aerospace Medicine, 30,* 391-409.

Von Helmholtz, H. (1962). *Treatise on physiological optics.* (J.P.C. Southall, Trans.). New-York: Dover. (Original work published in 1866)

Watt, D.G.D., Money, K.E., Bondar, R.L., Thirsk, R.B., Garneau, M., & Scully-Power, P. (1985). Canadian medical experiments on shuttle flight 41-G. *Canadian Aeronautics and Space Journal, 31,* 215-236.

Welch, R.B. (1978). *Perceptual modification: Adapting to altered sensory environments.* New York: Academic Press.

Whiteside, T.C.D. (1961). Hand-eye coordination in weightlessness. *Aerospace Medicine, 32,* 719-725.

Whiteside, T.C.D., Graybiel, A.A., & Niven, J.I. (1965). Visual illusions of movement. *Brain, 88,* 193-210.

Young, L.R., Oman, C.M., Watt, D.G.D., Money, K.E., & Lichtenberg, B.K. (1984). Spatial orientation in weightlessness and readaptation to earth's gravity. *Science, 225,* 205-208.

Zelaznik, H.N., Hawkins, B., & Kisselburgh, L. (1983). Rapid visual processing in single-aimed movements. *Journal of Motor Behavior, 15,* 217-236.

Part 2:
Prehension and gesturing

VISION AND MOTOR CONTROL
L. Proteau and D. Elliott (Editors)

CHAPTER 8

EYE, HEAD AND HAND COORDINATION DURING MANUAL AIMING

HEATHER CARNAHAN

Department of Physical Therapy, Elborn College
University of Western Ontario, London, Ontario, N6G 1H1

When we reach toward visual targets there is a complex interaction between movements of the eyes, head and arm. It may be possible to make inferences about the role of vision in this process, by monitoring how these various body parts move together. This chapter will discuss eye, head and hand coordination, and will address what this organization can reveal about the role of vision in movement control. It will begin with a general review of the research that has evaluated the coordination of the eyes, head and hand, as well as the role of head movements in pointing accuracy. Data will then be presented that deals with how visual strategies can be inferred from evaluating the temporal organization of eye, head and arm movements. Finally, differences between how visual information is used during pointing and grasping will be discussed.

Eye and Head Organization Alone

There has been a plethora of research conducted on the coordination of eye and head movements alone (see Guitton, 1988; Guitton & Volle, 1987; Laurutis & Robinson, 1986). Researchers have shown that saccadic eye movements and head movements may be initiated at quite different times relative to each other, depending on factors such as target amplitude, predictability and visibility (Barnes, 1979; Bizzi, Kalil, & Morasso, 1972; Funk & Anderson, 1977; Guitton & Volle, 1987; Zangmeister & Stark, 1982). For example, when monkeys were required to move only their eyes to an unexpectedly appearing target, the initiation of eye movement, preceded that of the head. There appears to exist an interaction between eye lead time and target eccentricity such that the amount that the eyes lead the head decreases as target eccentricity increases (Guitton & Volle, 1987). Conversely, when monkeys fixated on a predictable target, head movements began before a saccade was generated (Bizzi, 1974; Bizzi, Kalil, & Tagliasco, 1971). This pattern of movement in which the head leads the eyes is also found when human infants fixate on visual targets (Regal, Ashamead, & Salapatek, 1983).

There have only been a handful of studies, that have examined the coordination of free, eye and head movements during *manual* aiming (Biguer, Jeannerod, & Prablanc, 1982; Biguer, Prablanc & Jeannerod, 1984; Carnahan & Marteniuk, in press; Herman, Herman, & Malucci, 1981). It has not yet been clearly established whether the relationship between the eye and head is altered when the task also involves an arm movement, however preliminary data (which will be discussed later in this chapter) suggests that this is the case (Herman et al., 1971).

What Is The Best Way To Compare
Eye, Head and Limb Movements?

Contrasting eye and head movements is relatively straightforward because both body parts rotate about a centrally located axis. Although the true axes of rotation are slightly different, movements of the eyes and head are generally compared directly. To calculate gaze (which is where the visual axis lies with respect to the environment), eye displacement relative to the head, and head displacement relative to the world are simply added. However, measuring and comparing movements of the eyes, head *and hand* is not as straightforward. It is understandable why there exists a paucity of literature on eye, head and hand coordination, since measuring and comparing these three systems is technically and methodologically very difficult.

A unique problem arises when direct amplitude comparisons are required for eye and limb movements. The eyes and head rotate about a centrally fixed axis, and limbs are capable of translation as well as rotation. It is difficult to find a common metric of comparison so that coordination can easily be described and quantified. However, some researchers are now quantifying limb movements in terms of rotations, and are using this approach to make direct comparisons with limb, eye and head movements (Hore, Goodale, & Villas, 1990; Strauman, Hepp, Hepp-Reymond, & Haslwanter, 1990). This procedure should prove fruitful.

Another approach to the problem of comparing eye and head rotations, with limb translations is to compare temporal information about landmark events in the eye, head and arm kinematic profiles. For example, comparing the time to reach peak velocity, start, or stop, movements of the eyes, head and limb, is one way (although limited) to describe visuomotor coordination.

A potential difficulty in directly comparing the displacement of eye, head and limb movements is that the ways in which space is calibrated for these systems may be distinct. It has been suggested that eye movements are spatially mapped with the origin of the map being an imaginary location between the centre of the eyes (Howard, 1982). Limb movements however, may be calibrated in terms of a cartesian system (Jeannerod, 1986, 1988; Soechting & Flanders, 1989). This may be one reason why the eye and head system reorganizes when limb movements are added. The eye and head system, may be mapped relative to body centre coordinates. However, once hand movements are introduced, a new mapping, which is based on a cartesian environmentally-based system, must be integrated with the eye map in order to control the system.

Coordination of Eye, Head and Hand Movements
Biguer et al. (1982) were among the first to examine eye, head and hand control in human subjects. They found that when an individual directs his/her eyes, head and arm to a visual target the eyes start to move first, followed by the head and then the finger. However, the EMG for all three body parts is initiated essentially simultaneously. This delay of the head and finger was probably due to the longer electromechanical delays, and larger inertias, for these body parts. In a followup study (Biguer, Prablanc, & Jeannerod, 1984), subjects were required to manually point to targets with either a fixed or free head. Results showed that in the free head condition, pointing accuracy was enhanced, even though the head movement was hypometric to the target. However, these researchers found no correlation between the amplitude of the head movement, and finger accuracy. It thus appears that having the head move freely enhances finger accuracy, but the extent of the free head movement is not critical.

Role of Head Movement in Pointing
It is obvious that vision is one of the dominant sources of information that individuals use to accurately guide their limbs to visual targets (see this book). However, other sources of information are used in conjunction with this visual information. For instance the Biguer et al. (1984) findings suggest that an important source of information for enhancing finger pointing is *head* position. Researchers have previously addressed this issue and their collective findings suggest that the proprioceptive information resulting from head position (static and dynamic) may provide spatial cues that can be utilized to facilitate accurate visually directed pointing (Gazzaniga, 1969; Jeannerod, 1988; Marteniuk, 1978; Taylor & McCloskey, 1988).

In chapter 6 of this volume, Abrams discusses the influence of extraretinal information on pointing accuracy. This discussion of the role of eye position can be extended to include the role of head position in the spatial mapping and the facilitation of finger pointing accuracy. Marteniuk (1978) showed that static head position plays a role in pointing accuracy in humans. In this study, subjects generated pointing movements in two different conditions: with the head fixed away from the target, and the eyes fixated toward the target, or with both the eyes and the head oriented toward the target. Results showed that pointing accuracy was enhanced when both the eyes and head were directed toward the target, eye position alone was not enough to maximize pointing accuracy. Marteniuk concluded that the orientation of the head toward a visual target facilitates the spatial coding of the object into the spatial reference system that is used when limb movements are generated.

The question still remains regarding the type of information (proprioceptive or vestibular) that individuals gain from head movements in order to facilitate their pointing accuracy. To address this issue Cohen (1961) sectioned the dorsal roots in monkeys, effectively abolishing all proprioceptive information from the neck muscles. After this procedure the animals were no longer able to successfully generate reaching movements toward objects, even when their eyes were fixated on them. Thus, Cohen demonstrated that the neck muscles serve as a major source of information for spatially coding visual information. As well, Taylor and McCloskey (1988) investigated the role of neck proprioception and vestibular input, in spatially coding head and body position. They had human subjects, point at the position of their big toe with their head, either by turning the body to align the toe with the head, or by turning the head to line up with the toe. In the condition in which the head remained stationary and the body turned, head position sense was gained only from neck proprioception, since vestibular input (caused by head rotation) was minimized. The subjects were found to be equally accurate in both the head stationary and head moving conditions, indicating that proprioceptive sensation from the neck plays an important role in positioning the head relative to a target.

Biguer and associates also examined the contributions of head position signals to pointing accuracy. They created illusory movement of the head through neck muscle vibration (Biguer, Donaldson, Hein, & Jeannerod, 1986). The vibration artificially stimulates the muscle spindles, resulting in the false perception that the muscles are stretching, and the false conclusion reached by the subjects is that the body part must be extending (Goodwin, McCloskey, & Matthews, 1972). When subjects' neck muscles were vibrated, and they were required to reach out and point to a visual target, they consistently overshot the target. Thus, the incorrect information regarding head position was translated into inaccurate visuomotor behaviour, again emphasizing the strong influence of head position on finger pointing accuracy.

What Movement Initiation Patterns
Reveal About Visual Strategies

In a recent study, the effect of limb movements on eye and head organization was examined, in an attempt to evaluate the complex relationship between movements of all three body parts (see Carnahan & Marteniuk, in press, for more details). The purpose of this study was to observe the influence that generating limb movements had on eye and head temporal organization, and to evaluate the interactions between these parts of the visuomotor system.

Subjects were required to point to or visually fixate unpredictable targets of varying eccentricities either with fast *or* accurate limb movements. Thus, there were four conditions (fast *vs.* accurate eye fixation, fast *vs.* accurate finger pointing). The influence that the fast and accurate limb movements had on eye and head timing was monitored. As well, the timing of eye and head movements in the absence of limb movements was evaluated.

The results showed different movement initiation patterns of the eye, head and hand movements, for the finger and eye pointing conditions. For *finger pointing*, initiation patterns interacted with experimenter's instructions (i.e., speed *vs.* accuracy), as well as with the target eccentricity. As depicted in the upper panel of Figure 8.1, in the *fast* condition, the eyes and head started to move simultaneously and both began before the finger, when subjects pointed to targets close to the midline (26°). However, when subjects were pointing to relatively eccentric targets (43°), the head started to move first, the finger was second and the eyes started last. As illustrated on the lower panel of Figure 8.1, different initiation patterns were found for the *accurate* finger pointing condition. When subjects pointed to the close target (26°), the eyes started first, followed by the head and the finger. However, when subjects pointed to the far target (43°), the eyes and head started at the same time and the finger started to move last. These variable initiation patterns could be considered an example of coarticulation, similar to that reported when people type, and the fingers of each hand move toward their respective keys, even while other keys are being struck (Rumelhart & Norman, 1982). Such a strategy of coarticulation might be adopted because it is an effective method of improving response speed when the individual effectors move slowly (Rosenbaum, 1991).

For the *gaze fixation only* condition (not shown), the eyes started to move before the head regardless of the experimenter's request for speed or accuracy, or the target eccentricity. Thus, eye and head initiation patterns were altered when limb movements were added to the system; the introduction of the limb into the visuomotor system caused the patterns to become more variable.

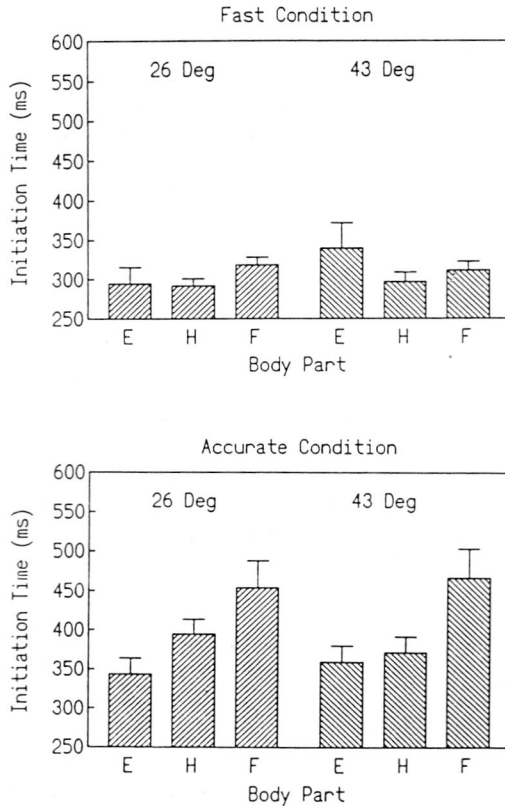

Figure 8.1. Initiation times for the fast and accurate finger pointing conditions. Note that in the fast condition, the head starts to move before the eyes, and for the accurate condition, the eyes start to move first. (Reproduced with permission from Carnahan and Marteniuk, in press).

It was demonstrated in this study that the timing between the eyes and head was altered significantly when hand movements were added to the system. The organizational properties of the visuomotor system were altered when the task was changed from eye pointing to finger pointing. Why did this happen? One possibility is that there exists some sort of head and arm synergy. The nature of this synergy could be similar to the postural synergies proposed by Nashner and colleagues (McCollum, Horak, & Nashner, 1985; Nashner &

McCollum, 1985). The synergy could be triggered by the postural disturbances caused by the extending arm. It is possible that in order to offset the perturbed centre of gravity caused by an arm movement, a corresponding head movement in the opposing direction must be made. Although it was not quantified, the observation was made that along with head rotation, a small head translation backward also occurred. It is possible that this small backwards head movement was made to offset the change in the centre of body mass, caused by the extending arm. Moore and Brunt (1990) have shown that postural adjustments (which act in the opposite direction of the intended movement) are made when seated subjects engage in a reaching task.

Alternatively, the basis of the synergy could be related to neck proprioception and the spatial mapping necessary for accurate limb movements. The previous discussion on the role of head movement in pointing accuracy suggests that the nature or timing of head and eye movements may change when limb movements are introduced. More specifically, it could be proposed that the proprioceptive information derived from the head movements will facilitate finger pointing accuracy, because immediate visual fixation of the target is no longer the sole source of target information.

Movement Termination and Visual Strategies

Even though it has been established that head position plays an important role in modulating manual aiming, the role of vision in the process is still dominant. The importance of vision in directing manual aiming is reflected in the temporal organization of eye, head and limb movements. The Carnahan and Marteniuk data demonstrated that movement initiation is quite variable, and can change as a function of task related variables. Despite this variability in initiation order, the manner in which these body parts stop moving was extremely consistent. Regardless of what the initiation patterns were for the various body parts (in the finger pointing condition), and as illustrated in Figure 8.2, the eyes always reached the target first, followed by the finger and then the head. Moreover, the eyes reached the target at least 200 ms before the finger, regardless of the initiation pattern. A similar pattern of results has also been reported across a series of studies that have manipulated task related variables such as speed and accuracy requirements, target eccentricity, and even target movement (Carnahan, 1989). Thus, the visuomotor system can be variable in initiating movement while still successfully and consistently achieving the specified movement termination goals.

In these studies, termination order did not vary across subjects, or experimental conditions. One explanation for this consistency is that the eyes can simply move so much faster than the arm, and also had a smaller distance to travel. However, it seems reasonable that an objective of the motor system is to get the eyes on the target in enough time in order to make at least one visually based correction, to accurately guide the limb to the target.

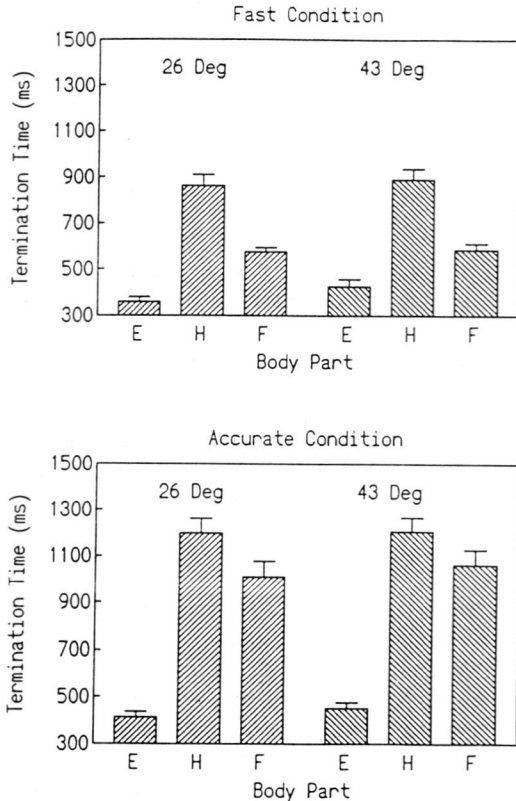

Figure 8.2. Termination times for the finger pointing condition. Note that the eyes always stop moving first, followed by the finger, and then the head. (Reproduced with permission from Carnahan and Marteniuk, in press).

It is compelling that the amount of time that the eyes lead the finger to the target is close to the minimum visual correction times that have previously been reported (Carlton, 1981; Elliott & Allard, 1985; Keele & Posner, 1968; Zelaznik, Hawkins, & Kisselburgh, 1983). Generally, the minimum time for using visual information to make trajectory corrections (approximately 200 ms) is based on a time measurement that began at the initiation of finger movement (see Carlton, this volume), whereas in the Carnahan and Marteniuk (in press) study it is defined as duration from the end of eye movement (start of visual fixation) to the termination of finger movement.

In the typical Keele and Posner (1968) type design, subjects' eyes are fixated on the target before the trial begins. However, in the Carnahan and Marteniuk (in press) study it was necessary for the target to be acquired by the eyes before visual information regarding the target could be used to control finger accuracy. Thus, the visual processing time was not estimated from finger movement initiation but, from eye movement termination to finger movement termination.

The measurement of the use of visual information from the end of eye movement, instead of from the beginning of finger movement, could be a more natural way of thinking about using visual information to accurately reach targets. This is so because we often don't rely on central vision of a target when we initiate our reaches. That is, we may either have some sort of memorial representation of where the target is, or we may initiate our movements based on peripheral information (see Elliott, and Sivak & MacKenzie chapters in this book). We may be well into our movement trajectory before we achieve central fixation of the target. For example, we may initiate a reach toward our calculator on a desktop, while simultaneously trying to locate it visually.

The fact that the 200 ms correction time is estimated from the end of movement implies that visual information is used in the latter stages of movement production to facilitate finger accuracy. This is consistent with other studies which have shown that visual information can be used late in a movement trajectory to correct finger terminal error (Poulton, 1974; Pélisson, Prablanc, Goodale, & Jeannerod, 1986; Prablanc, Pélisson, & Goodale, 1986; Welford, 1968). For example, Prablanc et al. (1986) had subjects point as fast and accurately as possible to a small illuminated target which could be displaced (they used a double-step paradigm) when subjects made an eye saccade toward it. They found that subjects were not consciously aware that the target has moved, yet upon termination of the saccade they were able to accurately point to the new target position, even though there was no evidence of corrections it the arm movement velocity profiles. As well, Georgopoulos, Kalaska, and Massey (1981) showed that monkeys could successfully amend finger movements to perturbed visual targets even when the targets were moved at varying latencies. However, the duration of the movement trajectories directed toward the original target increased as a function of the perturbation latency. This suggests that visual information is used intermittently throughout the entire process of finger movement generation, to facilitate finger accuracy (Prablanc & Pélisson, 1990).

Grasping versus Pointing and Visuomotor Coordination

When we consider how vision is used to control reaching movements we generally do not specify the intended goal of the movement. However, there is preliminary evidence that the goal of the movement does affect the characteristics of the reach (Marteniuk, MacKenzie, Jeannerod, Athenes, & Dugas, 1987). These researchers demonstrated that the intent of the action influences the kinematics of the trajectory. They observed that subjects spend a larger proportion of their reaches decelerating when they were reaching toward a fragile object such as a light bulb as opposed to a sturdy object like a tennis ball. Similarly, we should consider the possibility that the way in which visual information is used may vary as a function of movement goal. We often do not differentiate between grasping and pointing when we generalize about how vision is used when generating limb movements. It is possible, that how individuals use vision may vary as a function of whether they are generating pointing or grasping movements, and that some principles of how vision is used during reaching and pointing is not generalizable to grasping.

A very recent study was conducted to compare how rapidly reaching trajectory corrections can be made when pointing and grasping (Carnahan, Goodale, & Marteniuk, 1991). Subjects were required to reach toward targets, which were suddenly illuminated. Unpredictably, when subjects initiated their movement, the target they were reaching toward was unexpectedly extinguished and a new target either to the left or right of this target was illuminated, giving the impression that the target had suddenly changed position. Subjects were required to reach toward and either touch and point, or grasp the target in its new position. Results showed that reaction times (RTs) to initiate the movements to the first target, were longer for the pointing trials, than for the grasping trials, suggesting that subjects actually spent a longer time preprogramming the pointing movements, than the grasping movements.

The time to peak velocity analysis showed that when subjects were grasping, the peak velocity of their wrist trajectory (indicative of the transport component) was reached sooner when the target was perturbed, than when it remained in the same position. However, when subjects were generating pointing movements, time to peak velocity remained unaffected by the target perturbation. These results, which are presented in Figure 8.3, indicate that subjects were more readily able to amend trajectories during grasping than during pointing. The RT results along with these peak velocity findings suggest that when grasping, less time is spent "preprogramming" the movement and corrections are made "online" while the movement is being generated. However, when pointing, longer time is spent planning the movement during RT, and subjects are less able to modify the trajectory online.

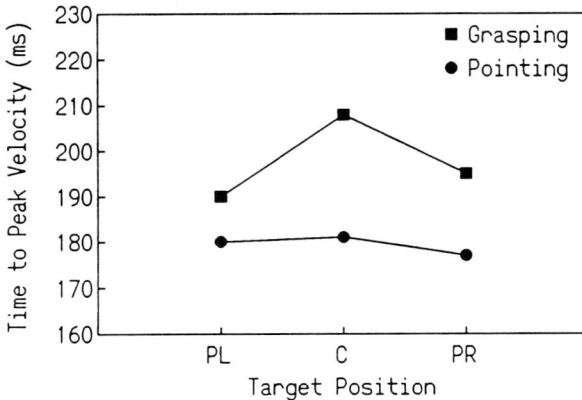

Figure 8.3. Time to peak velocity for the pointing and grasping perturbed trials. For the pointing trials, there is no difference in time to peak velocity between the trials in which the target remains stationary in the centre (C) position, or is perturbed to the left (PL) or right (PR). However, for the grasping condition, peak velocity occurred sooner on the trials in which the target was perturbed.

This question of whether the grasping system is better able to deal with target movement was also tested using a different approach (Carnahan et al., 1991). In a second experiment, subjects were required to either point to, or grasp objects that were either stationary or moving (rolling down a ramp). If the grasping system is better adapted for dealing with moving targets, then reaching MT might be expected to decrease when the targets are moving, and when subjects are required to grasp. In contrast, considering that pointing movements are often directed at stationary targets, MT for a pointing movement directed at a moving target might be expected to increase.

The predicted task by speed interaction was reliable. In the *grasping* condition, MT was shorter for the moving target trials when compared to the stationary trials. However, as illustrated in Figure 8.4, the converse was true for the *pointing* condition. These data support the hypothesis that the visuomotor system is more efficient in dealing with target movement when grasping is intended as opposed to simply pointing.

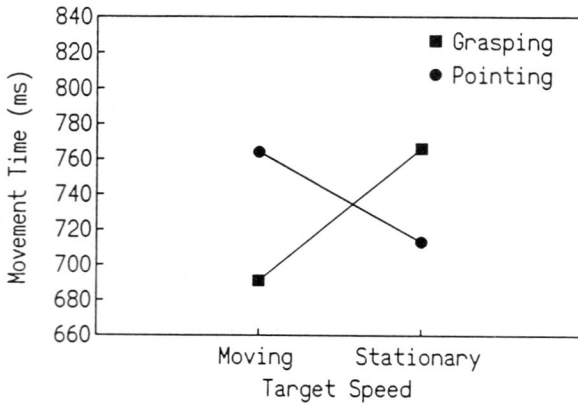

Figure 8.4. Reaching MT increased when subjects were pointing to moving objects, and MT decreased when subjects were grasping moving objects (compared to the stationary conditions).

Why would the motor system be designed in this manner? It is possible that the reaching and grasping system evolved as a system that's function was to acquire targets (for food, prey, defence or locomotion) in a dynamic environment. Thus, it would be necessary for this system to have evolved so that it is possible to rapidly alter visually guided reaches in accordance with target movement. Pointing however may have developed having a functionally different role (i.e., communication), more adapted to the static environment. In this regard, Georgopoulos and Grillner (1989) have proposed that reaching may have evolved from the neural system responsible for the positioning of the lower limb during locomotion. These authors also discuss that some animals use a grasping motion with the paw to improve locomotor performance while climbing on swinging branches, suggesting that grasping may be viewed as a system that has adapted to deal with a dynamic environment.

Eye, Head, and Hand Coordination During Grasping

Now it has been established that there may be differences in how vision is used in grasping and pointing movements I would like to discuss how these visual processing differences may manifest themselves in the ways that movements of the eyes, head and hand are coordinated. The few studies that have examined eye and head coordination during aiming have looked only at pointing movements. In this following study, eye, head and hand movement initiation and termination times were measured during reaching and *grasping* movements (Carnahan, 1989).

Subjects began each trial with their eyes fixed straight ahead and the finger resting directly in front of them. They were instructed to pick up a 3 cm dowel (located 34 cm and either 20 or 50 degrees to the left of the hand start position) whenever they were ready. After a trial began the eyes and head were free to move. No time constraints were placed on the movement.

For the time to initiate movement, there was a main effect for body part. The head (322 ms) started to move before the eyes (408 ms), and before the finger (419 ms). However, there was no significant difference between the start of the eyes and the start of the finger. In Figure 8.5, an EOG trace shows the head moving toward the target before the eye saccade is initiated. In this example, the eyes actually remain fixated on the central fixation point, for as long as 250 ms before the saccade is initiated toward the target.

When termination time was used as the dependent measure there was a main effect for body part. The eyes finished first (504 ms), followed by the finger (1209 ms) and finally by the head (1411 ms). These data support the movement termination findings discussed earlier.

These initiation results are not typical of the pattern of results generally reported when subjects engage in pointing (the eyes usually lead the head and finger). The head led the eye on average by 85 ms. This initiation pattern appears to be unique to the grasping situation. It has already been demonstrated that eye and head organization changes when limb movements are introduced for the purpose of pointing. Following this, it appears that the pattern of eye, head and limb organization is altered once again when grasping movements are intended.

This pattern of coordination may reflect the visual processing requirements of the system when grasping is intended. If it is true that the grasping system can use visual information to generate on-line movement amendments more easily, then perhaps the initiation patterns are reflecting this. That is, the eyes start to move after the head, because vision is not needed immediately. As long as the target is acquired before the hand reaches it, there

H. Carnahan

is enough time to comfortably use visual information to amend the reaching movement. Perhaps with pointing, since visual information is not used as easily online, the eyes must reach the target as soon as possible to optimize the amount of time available to make trajectory corrections.

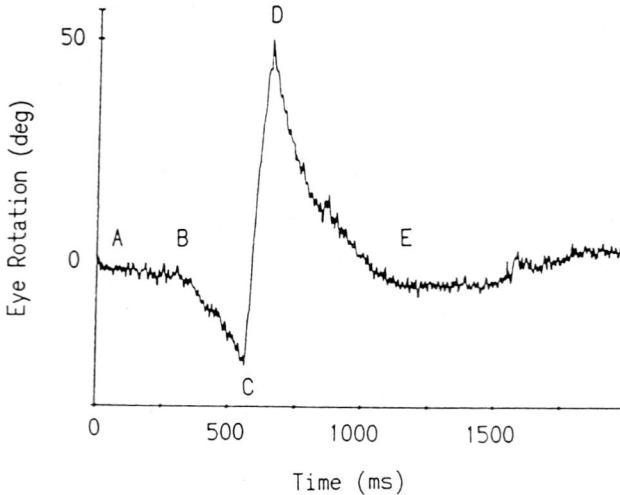

Figure 8.5. This EOG profile (which shows eye displacement relative to the head) for a grasping condition shows that the head starts to move toward the target, before the eyes start to move. From A to B the head and eyes are both stationary. From B to C the eyes remain centrally fixated, and the head rotates toward the target. At C, the saccade is initiated toward the target and reaches the target at D. From D to E the eyes maintain fixation on the target while the head continues to rotate toward the target, thus resulting in the eye rotating in the opposite direction relative to the head.

Summary

It has been demonstrated that vision is not the only source of information that individuals can use to facilitate manual aiming. Proprioceptive information from the neck also enhances pointing accuracy. As well, data were presented that show eye and head temporal organization is altered when limb movements are added to the visuomotor system. The role of head movement in limb pointing accuracy may play a role in this reorganization.

Another finding reviewed in this chapter, is that despite variability in the movement initiation patterns of the eyes, head and hand, the eyes still reach the target in enough time to use visual information regarding target position, to amend the reaching response. Thus, it appears that the coordination of the eyes, head and hand may be based on the visual processing requirements of the manual task. Finally, initiation timing (across experiments), appeared to be affected by the goal of the intended reach (pointing or grasping). This may reflect the ability of the grasping motor system to utilize feedback online more efficiently, and is another example in which the coordination of the visuomotor system is influenced by the visual processing demands of the task.

The limb reaching out to a visual target is just one element of the visuomotor system. We might be more successful in trying to understand the nature of goal directed aiming by examining how *all* the parts of the visuomotor system (i.e., eyes, head, and arm) interact. As well, the goal of the reaching task (pointing or grasping) should be considered, when making conclusions about how visual information is used, because it appears that movement intent has an impact on visuomotor coordination.

References

Barnes, G.R. (1979). Vestibulo-ocular function during coordinated head and eye movements to acquire visual targets. *Journal of Physiology, 287*, 127-147.

Biguer, B., Donaldson, I.M.L., Hein, A., & Jeannerod, M. (1986). La vibration des muscles de la nuque modifie la position apparente d'un cible visuelle. [Vibration of the nape of the neck muscles modifies the perceived location of a visual target]. *Compte-Rendus de l'Académie des Sciences de Paris III, 303*, 43-48.

Biguer, B., Jeannerod, M., & Prablanc, C. (1982). The coordination of eye, head and arm movements during reaching at a single visual target. *Experimental Brain Research, 46*, 301-304.

Biguer, B., Prablanc, C., & Jeannerod, M. (1984). The contribution of coordinated eye and head movements in hand pointing accuracy. *Experimental Brain Research, 55*, 462-469.

Bizzi, E. (1974). The coordination of eye-head movements. *Scientific American, 231*, 100-106.

Bizzi, E., Kalil, R.E., & Tagliasco, V. (1971). Eye-head coordination in monkeys: Evidence for centrally patterned organization. *Science, 173*, 452-454.

Bizzi, E., Kalil, R.E., & Morasso, P. (1972). Two modes of active eye-head coordination in monkeys. *Brain Research, 40*, 45-48.

Carlton, L.G. (1981). Processing visual feedback information for movement control. *Journal of Experimental Psychology: Human Perception and Performance, 7*, 1019-1030.

Carnahan, H. (1989). *Eye, head and hand coordination in prehension and pointing tasks.* Unpublished doctoral dissertation, University of Waterloo, Waterloo.

Carnahan, H., & Marteniuk, R.G. (in press). The temporal organization of hand, eye and head movements during reaching and pointing. *Journal of Motor Behavior.*

Carnahan, H., Goodale, M.A., & Marteniuk, R.G. (1991) *Grasping versus pointing and the differential use of visual feedback.* Manuscript submitted for publication.

Cohen, L. (1961). Role of eye and neck proprioceptive mechanisms in body orientation and motor coordination. *Journal of Neurophysiology, 24*, 1-11.

Elliott, D., & Allard, F. (1985). The utilization of visual feedback information during rapid pointing movements. *Quarterly Journal of Experimental Psychology, 37A*, 407-425.

Funk, C.J., & Anderson, M.E. (1977). Saccadic eye movements and eye-head coordination in children. *Perceptual and Motor Skills, 44*, 599-610.

Gazzaniga, M.S. (1969). Cross-cueing mechanisms and ipsilateral eye-hand control in split-brain monkeys. *Experimental Neurology, 23*, 11-17.

Georgopoulos, A.P., & Grillner, S. (1989). Visuomotor coordination in reaching and locomotion. *Science, 245*, 1209-1210.

Georgopoulos, A.P., Kalaska, J.F., & Massey, J.T. (1981). Spatial trajectories and reaction times of aimed movements: Effects of practice, uncertainty and change in target locations. *Journal of Neurophysiology, 46*, 725-743.

Goodwin, G.M., McCloskey, D.I., & Matthews, P.B.C. (1972). The contribution of muscle afferents to kinaesthesia shown by vibration induced illusions of movement and by the effects of paralysing joint afferents. *Brain, 95*, 705-748.

Guitton, D. (1988). Eye-head coordination in gaze control. In B.W. Peterson & F.J. Richmond (Eds.), *Control of head movements* (pp. 196-207). New York: Oxford University Press.

Guitton, D., & Volle, M. (1987). Gaze control in humans: Eye-head coordination during orienting movements to targets within and beyond the oculomotor range. *Journal of Neurophysiology, 58*, 427-459.

Herman, R., Herman, R., & Malucci, R. (1981). Visually triggered eye-arm movements in man. *Experimental Brain Research, 42*, 392-398.

Hore, J., Goodale, M.A., & Villas, T. (1990). The axis of rotation of the arm during pointing. *Society for Neuroscience Abstracts, 16II*, 1087.

Howard, I.P. (1982). Visual direction with respect to head and body. In I.P. Howard (Ed.), *Human visual orientation* (pp. 275-340). Toronto: Wiley.

Jeannerod, M. (1986). Mechanisms of visuomotor coordination: A study in normal and brain damaged subjects. *Neuropsychologia, 24*, 41-78.

Jeannerod, M. (1988). *The neural and behavioural organization of goal-directed movements.* Oxford: Claredon Press.

Keele, S.W., & Posner, M.I. (1968). Processing of visual feedback in rapid movements. *Journal of Experimental Psychology, 77*, 155-158.

Laurutis, V.P., & Robinson, J.A. (1986). The vestibulo-ocular reflex during human saccadic eye movements. *Journal of Physiology, 373*, 209-233.

Marteniuk, R.G. (1978). The role of eye and head positions in slow movement execution. In G.E. Stelmach (Ed.), *Information processing in motor learning and control* (pp. 267-288). New York: Academic Press.

Marteniuk, R.G., MacKenzie, C.L., Jeannerod, M., Athenes, S., & Dugas, C. (1987). Constraints on human movement trajectories. *Canadian Journal of Psychology, 41*, 365-378.

McCollum, G., Horak, F.B., & Nashner, L.M. (1985). Parsimony in neural calculations for postural adjustments. In J. Bloedal, J. Dichgans, & W. Pratt (Eds.), *Cerebellar functions* (pp. 52-66). Berlin: Springer Verlag.

Moore, S., & Brunt, D. (1990). Anticipatory postural adjustments in seated subjects. *Society for Neuroscience Abstracts, 16II*, 1318.

Nashner, L.M., & McCollum, G. (1985). Organization of human postural movements: A formal basis and experimental synthesis. *The Behavioral and Brain Sciences, 8*, 135-172.

Pélisson, D., Prablanc, C., Goodale, M.A., & Jeannerod, M. (1986). Visual control of reaching movements without vision of the limb. II Evidence of fast unconscious processes correcting the trajectory of the hand to the final position of a double-step stimulus. *Experimental Brain Research, 62*, 303-311.

Poulton, E.C. (1974). *Tracking skill and manual control.* New York: Academic Press.

Prablanc, C., & Pélisson, D. (1990). Gaze saccade orienting and hand pointing are locked to their goals by quick internal loops. In M. Jeannerod (Ed.), *Attention and performance XIII* (pp. 653-676). Hillsdale, NJ: Erlbaum.

Prablanc, C., Pélisson, D., & Goodale, M.A. (1986). Visual control of reaching movements without vision of the limb. I Role of retinal feedback of target position in guiding the hand. *Experimental Brain Research, 62*, 293-302.

Regal, D.M., Ashamead, D.H., & Salapatek, P. (1983). The coordination of eye and head movements during early infancy: A selective review. *Behavioural Brain Research, 10*, 125-132.

Rosenbaum, D.A. (1991). *Human motor control.* San Diego: Academic Press.

Rumelhart, D.E., & Norman, D.A. (1982). Simulating a skilled typist: A study of skilled cognitive-motor performance. *Cognitive Science, 6*, 1-36.

Soechting, J.F., & Flanders, M. (1989). Sensorimotor representations for pointing to targets in three-dimensional space. *Journal of Neurophysiology, 62*, 582-594, 1989.

Strauman, D., Hepp, K., Hepp-Reymond, M.C., Haslwanter, T. (1990). Human eye, head and arm rotations during reaching and grasping. *Society for Neuroscience Abstracts, 16II*, 1087.

Taylor, J.L., & McCloskey, D.I. (1988). Proprioception in the neck. *Experimental Brain Research, 70*, 351-360.

Welford, A.T. (1968). *Fundamentals of skill.* London: Methuen.

Zangmeister, W.H., & Stark, L. (1982). Types of gaze movement: Variable interactions of eye and head movements. *Experimental Neurology, 77*, 563-577.

Zelaznik, H.N., Hawkins, B., & Kisselburgh, L. (1983). Rapid visual feedback processing in single-aiming movements. *Journal of Motor Behavior, 15*, 217-236.

VISION AND MOTOR CONTROL
L. Proteau and D. Elliott (Editors)

CHAPTER 9

FUNCTIONAL CHARACTERISTICS OF PREHENSION: FROM DATA TO ARTIFICIAL NEURAL NETWORKS

MARC JEANNEROD* and RONALD G. MARTENIUK**

*Vision et Motricité, INSERM U94, 69500 Bron, France

**Department of Kinesiology, University of Waterloo
Waterloo, Ontario, N2L 3G1

The present chapter deals with the generation and control of goal directed arm and hand movements during the action of prehension. This is a fundamental aspect of visuomotor behavior and combines two different types of motor behavior. Reaching as a behavior for interacting with the environment, is mostly effected by the proximal joints of the arm, and deals with spatial relations of the objects and the body. Its function is to carry the hand to an appropriate location within extrapersonal space and as such must involve mechanisms for computing distance and direction of the point in space where to move. Grasping is a highly refined skill due to specialization of the hand in primates and humans, it is effected by the distal joints of the fingers, and deals with intrinsic qualities of objects. The coordinate system in which grasping movements are generated therefore does not relate to the body, it relates to the object and the hand.

Reaching and grasping are functionally interrelated, such that reaching is a precondition for grasping. However, the posture that is assumed by the hand before contact with objects represents the end result of a motor sequence which starts well ahead of the actual grasp. The fingers begin to shape during transport of the hand to the object location and contribute to a final spatial configuration where the relative positions of the hand and the object to be grasped are specified, at least in part, by the objects properties. Preshaping is a highly stable motor pattern, involving a progressive opening of the grip with straightening of the fingers, followed by a closure of the grip until object contact. The current model for the coordination between the components of prehension specifies that the mechanisms for achieving a correctly oriented and sized grip operate in parallel with other mechanisms for hand transport to the object location. Parallel distributed processing is a concept that is not only present in several sensory and motor theories of human behavior but also is a chief characteristic of the human's central nervous system. Applied to prehension, a parallel distributed explanation postulates the existence of functional ensembles (the visuomotor channels), characterized by specific input-output relationships, and specialized for generating each component of the action (Jeannerod, 1981; Paillard, 1982). Thus, the input of the visuomotor channel specialized for grasping, for example, is tuned to visual object size, and its output is connected to distal muscles for generating a precision grip. The other channel specialized for the proximal aspect of the movement obviously must have a different structure and mode of activation.

Association of the two components of prehension in most everyday life movements makes their separate and combined study a fascinating one in that one studies an act that makes humans distinct from other animals and at the same time can serve as a window into the structure and organization of the central nervous system. This latter purpose also serves to help formulate theories of motor control that could generalize to other types of well learned skill.

In terms of the development of theories of prehension, recent advances in artificial neural networks may help to conceptualize the underlying mechanisms of prehension. In general (see Jordan, 1990; Jordan & Rosenbaum, 1989; Rumelhart & McClelland, 1986) artificial neural networks exhibit certain characteristics in common with a movement skill like prehension. To begin with, they are based on parallel, distributed control principles which appears to be important neurological and behavioral bases for prehension. As part of this type of control, neural nets can exhibit rapid access and implementation of stored information, not unlike the rapidity observed in the execution of highly skilled movements. Finally, given certain constraints within neural nets, they can exhibit motor equivalence which, as will be pointed out below, is an important characteristic of prehension. Thus, the issues of parallel, distributed control, fast processing of information, and motor equivalence are held in common between the skill of prehension and the operation of artificial neural networks.

The purpose of the chapter will be to review behavioral literature relevant to understanding prehension and, where applicable, understanding motor skills in general. In addition, relevant literature on the recent developments of artificial neural networks will be reviewed and applied to possible mechanisms involved in prehension which may help to change the way we think about these issues and, possibly, lead to new insights.

Reaching

Guiding the hand to the location of an object involves complementary mechanisms which combine for achieving the goal of the movement. One of these mechanisms is directional coding involved in orienting the arm toward a target while another is amplitude coding, the function of which is to generate the proper dynamics along the spatial path selected for that movement. Direction and amplitude of a reaching movement may be specified in parallel, as shown by Favilla, Henning, and Ghez (1990) who measured reaction times of force impulses in response to targets of different amplitudes appearing in two possible directions. Their hypothesis was that, if specification of response amplitude required the prior specification of response direction, amplitude specification would be delayed until direction specification were complete. The results were not in favour of this hypothesis, but were supportive of parallel specification, in that the time course of amplitude specification was not affected by the concurrent need to specify direction. These findings imply separate mechanisms underlying directional coding and kinematic or kinetic coding.

One behavioral way of studying directional control of goal directed movements is to introduce perturbations (for example, an unexpected change in target location) during their execution. The way the kinematics and the trajectory of the movement are reorganized in order to reach the new target, as well as the timing of this reorganization, should reflect the activity of central mechanisms involved in the achievement of the movement goal.

This approach, often referred to as the "double step" stimulation paradigm, has recently become quite popular for the study of sensorimotor coordination. It has been speculated that the perturbation probes the system at the level of preparation or execution of the movement, which is currently activated at the time where it occurs. If this hypothesis is correct, then perturbing the movement at different times should produce different effects (see Jeannerod, 1990). Indeed, the description of behavioral effects produced by double step stimulation experiments will reveal strong differences whether the second step (i.e., the change in target location) is presented prior to the movement, or during the movement itself.

Behavioral Effects of Perturbations Occurring During the Reaction Time of the Reaching Movement

In experiments where the change in target location is made to occur prior to the movement, that is during the reaction time of the response to the first step, the second step is usually presented at fixed intervals with respect to the first interstimulus interval. The conceptual framework for this approach was set 15 years ago by Megaw (1974) and by Gottsdanker (1973). According to the former author, the beginning of the movement should be influenced only by the first target step, and the response to the second step should appear later on, so that the corrected movement would be composed of two successive responses overlapping in time. Alternatively, the latter author postulated that the two responses combine and produce a corrected movement which would reflect from its onset the contribution of the second target step.

The more recent experiments have not completely solved this problem, though they have raised other critical issues on the nature of the corrective mechanism. Georgopoulos, Kalaska, and Massey (1981), in the monkey, used interstimulus intervals of 50-400 ms. They found that, when the movement started, it was first oriented in the direction of the first target and reoriented in the direction of the second target after a period of time which was a function of the duration of the interstimulus interval. In other words, the later the second step occurred, the longer the movement toward the first target lasted. According to the authors, the initial movement to the first target was a fragment of the control response that would have occurred if the target location had not changed. In addition, the time taken from the appearance of the second target to the change in direction of the movement was similar to the control reaction time for that target (i.e., between 200 and 300 ms).

These results were confirmed and expanded to humans by at least two groups of experimenters. Soechting and Lacquaniti (1983) found that the initial part of the movement was oriented in the direction of the first target, and that the change in direction occurred after a time that corresponded to one reaction time. These authors thus rightly adopted the Georgopoulos et al.'s conclusion that, the movement is effected in a continuous, ongoing fashion as a real-time process that can be interrupted at any time by the substitution of a new target. The fact that the effects of this change on the ensuing movement appear without extra delays beyond the usual reaction time is an important result, because it implies that there is no refractory period: a movement trajectory can be corrected even if the second stimulus is given very soon after the first. Van Sonderen, Denier Van der Gon, and Gielen (1988) also confirmed that the reaction times to the change in target location was in the same range as those for single steps. Concerning the initial orientation of the movements, however, they reported that, for short interstimulus intervals (e.g., 25-50 ms), the orientation of the first movement was a combination of the two target positions. For longer interstimulus intervals the arm was initially directed toward the first target, and then shifted abruptly in the direction of the second, as found by the previous authors.

The common finding in all three studies is that the perturbation generated by changing the location of the target during preparation of a movement directed toward the target cannot be corrected before a substantial delay has elapsed. This delay, which approximates the duration of one reaction time, is in any case longer than the minimum time required for vision to influence movements. Possibly it is related to the time needed by the visual map to compute the position of the new target and to establish a new goal for the action. If this is correct, the the amount of time needed to change the trajectory of the movement should relate to the timing of the transformations occurring at the level of this map. This point was illustrated by an experiment of Georgopoulos and Massey (1987). These authors requested subjects to perform reaching movements at various angles from the direction of a visual stimulus and found that reaction times of these movements increased linearly with the size of the angle. They proposed that this increase in reaction time was related to mentally rotating the movement vector until the angle of rotation corresponded to the angle of the required size. The rate of rotation was estimated by the authors to be around 400 degrees per sec.

Changes to movement involving the visual map, where targets for visually-directed movements are represented, thus rely on a relatively slow mechanism involving computation of the location of the new target and the specification of a new motor output. These processes seem to rely, at least partly, on the awareness of the subject. By contrast, as will be shown below, other mechanisms responsible for correcting movements after being perturbed at the time of the movement itself have a much shorter time constant, at the detriment of awareness of the corrections.

Behavioral Effects of Perturbations Time-locked with the Reaching Movements

At variance with the experiments reported above, changes in target location can be made to occur time-locked with the onset of the reaching movements. In this situation, where the second step is conditional to the appearance of the movement triggered by the first step, the interstimulus interval is equal to the reaction time to the first step and therefore varies from trial to trial.

Such an experiment was performed by Pélisson, Prablanc, Goodale, and Jeannerod (1986; see also Goodale, Pélisson, & Prablanc, 1986). Subjects task consisted in pointing at visual targets with their unseen hand. The initial target steps were from midline to targets located between 20 and 50 degrees. The second steps were of a smaller amplitude (e.g., from 30 to 32 degrees, or 40 to 44 degrees) but sufficiently large to make the target change clearly visible. The second steps, when present, were triggered by the onset of the ocular

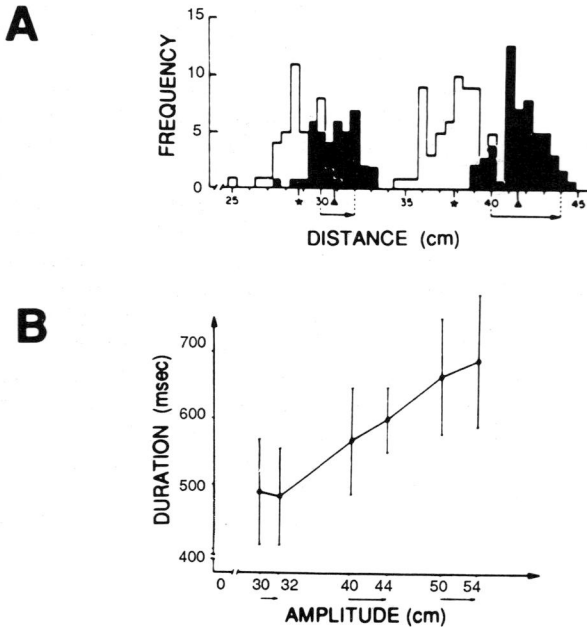

Figure 9.1. On-line corrections of arm direction during a double-step pointing experiment. (a) Distribution of pointings directed at stationary and displaced targets. For a single step to 30 cm, the distribution of pointings represented by the white histogram (mean indicated by the star) undershoots the target. When the target is further displaced from 30 to 32 cm (double step), the distribution of pointings represented by the dark histogram (mean indicated by the filled triangle) undershoots the 32 cm target. Notice, however, that the distance between the star and the triangle is equal to the target shift, i.e., subjects fully correct their ongoing response. The same applied for a target shift from 40 to 44 cm, and for a target shift from 50 to 54 cm (not shown). (b) Duration of hand pointing movement (mean and standard deviations) versus amplitude of the target final position. When a target is displaced for instance from 40 to 44 cm, the duration increases as compared to responses to the 40 cm single step, but the total duration corresponds (by extrapolating from the observed durations to 40 and 50 stationary targets) to the predicted duration of pointings to a stationary target appearing at 44 cm. Thus, no additional reaction time or extra processing time appears within the total duration to the double step stimulation, nor is it associated with a higher variability of movement duration. (Reproduced with permission from Pélisson et al., 1986).

saccades which are normally present shortly before the hand begins to move (Prablanc, Echallier, Komilis, & Jeannerod, 1979; Biguer, Jeannerod, & Prablanc, 1982). As demonstrated by Figure 9.1a, the results showed that, when second steps occurred, subjects invariably corrected their hand trajectories in order to reach the final target positions. Yet, the duration of the corrected movements was not influenced by the perturbation, and was the same as that of control single-step movements directed at targets of the same locations. In addition, the kinematic analysis of the movements showed that double-step trial movements did not differ in duration from single-step trial movements (Figure 9.1b). Specifically, double-step trial movements showed no secondary accelerations (Figure 9.2), which suggests that corrections were generated without delay, as an on-line process. It would be interesting to know, by triggering the second step later and later in the movement, at which point secondary movements would begin to be generated; that is, at which point subjects would shift to a different mode of correction involving programming of a new movement. In addition, the procedure used in this experiment for triggering the second steps (at the onset of eye movements) was such that subjects were not aware of when the changes in target location occurred and that the corrections were generated unconsciously (see also Bridgeman, Kirch, & Sperling, 1981).

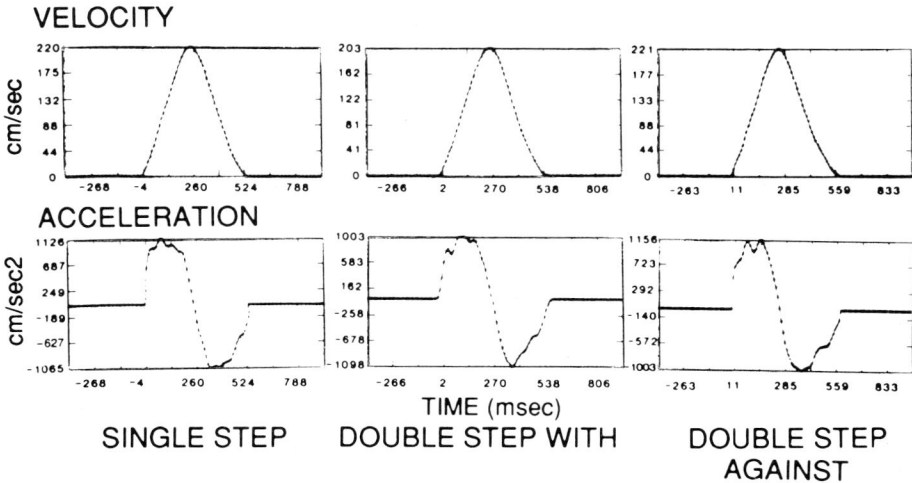

Figure 9.2. Averaged velocity (upper row) and acceleration (lower row) profiles of pointing movements during single-step or double-step trials. Note lack of reacceleration in double-step trials. (Reproduced with permission from Pélisson et al., 1986).

Another example of fast corrections is provided by the experiment of Paulignan, MacKenzie, Marteniuk, and Jeannerod (1990) on prehension movements. In this case, the targets consisted of graspable dowels which were placed on a concentric array at a fixed distance from the starting position of the hand. The dowels were made of transparent material, and were illuminated from below. In some trials the light could be suddenly shifted at the time of onset of the reaching movement, from the initially illuminated dowel to another one located 10 degrees to the right or to the left (perturbed trials). The spatial paths of the wrist, of the tip of the index finger and of the tip of the thumb were monitored. In addition, the kinematics of the wrist movement, as well as the distance between the two fingertips (the size of the finger grip) were computed. When the target was displaced at the onset of the movement (by shifting the light from one dowel to another) a complex rearrangement of the wrist and finger spatial paths was observed, so that the fingers were finally placed in the correct position for an accurate grasp. This rearrangement was effected with only a relatively small increase in total movement time (about 100 ms). Inspection of the wrist kinematics (Figure 9.3a) showed that, following the perturbation, the initial acceleration of the movement aborted and that a secondary acceleration took place in order to direct the hand at the new target position. Interestingly, the first peak in acceleration occurred earlier in the perturbed trials, about 105 ms following movement onset, instead of about 130 ms in the unperturbed ones (Figure 9.3b). The latter result means that within roughly 100 ms the visuomotor system began to react to the change in target location. The duration of 100 ms is less than one reaction time, and approximately corresponds to the minimum delay needed by visual and/or proprioceptive reafferents to influence an ongoing movement (see below). In addition, the fact that the overall movement was completed without an increase in duration beyond this value of 100 ms, means that the pattern of motor commands initially programmed for reaching the target could be rearranged on line. Such a rearrangement implied not only redirecting the wrist toward the new target location, but also introducing new commands for keeping the orientation of the finger grip invariant with respect to the object.

Another result of this experiment, similar to that found by Pélisson et al. (1986) and Goodale et al. (1986), was that, whereas the subjects were aware that the location of the target dowel had changed during the trial, they made an erroneous estimate of the time at which the change occurred: instead of seeing the change occurring immediately after movement onset, they saw it near the end of the movement trajectory, as their hand was coming close to the target. As in the Pélisson et al. experiment, the early motor reaction to the perturbation was dissociated from its perception by the subject. In both experiments, the visuomotor system was able to detect the perturbation and to correct the movement trajectory, even though the information it used for this

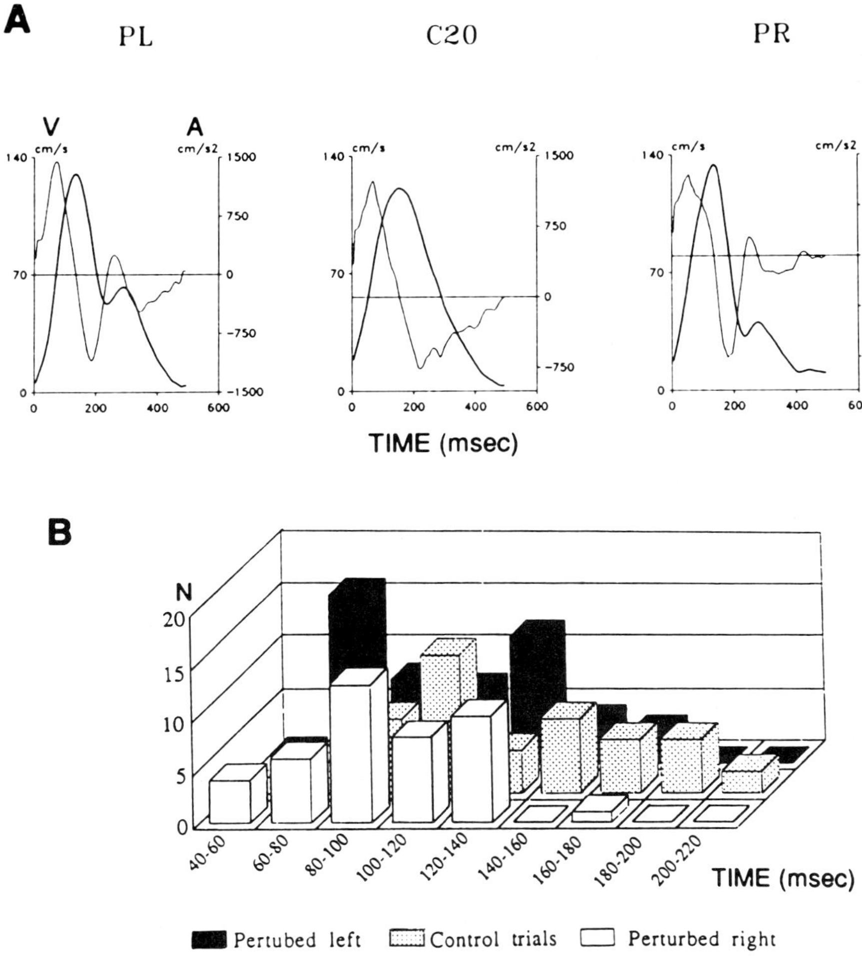

Figure 9.3. (a)   Kinematic pattern of prehension movements in perturbed trials. Three individual movements are shown. The movement shown in the middle of the figure corresponds to a movement directed at the 20 degree dowel in the control (unperturbed) condition (C20). The other two movements correspond to perturbed movements, when the dowel position is shifted to the left (PL) or to the right (PR). The thick line represents the arm velocity, the light line, the acceleration. (b)   Time to peak acceleration in control and perturbed trials. Note shorter time to peak in perturbed trials. (Reproduced with permission from Paulignan et al., 1991).

correction was not consciously available at the time the correction was made (see also Castiello, Paulignan, & Jeannerod, in press).

There are other examples, in a different context, of similar fast corrections in ongoing complex movements. Abbs and Gracco (1984) have described rapid compensation of perturbations applied to articulators during speech, in an experiment where the lower lip was unexpectedly pulled down during production of the phoneme /ba/. They showed that lip closure was nevertheless achieved and that the phoneme was correctly pronounced in spite of the perturbation. Such a compensation, which implied that lip closure was performed by a larger lowering of the upper lip, was thus effected within a delay compatible with the correct production of the phoneme. They suggested that this correction reflected an open-loop adjustment, independent of sensory feedback in the usual sense, but nevertheless relying on proprioceptive signals generated by the ongoing movement. Such a conception clearly fits into the definition of feed-forward mechanisms given by Arbib (1981): "A strategy whereby a controller directly monitors disturbances to a system and immediately applies compensatory signals to the controlled system, rather than waiting for feedback on how the disturbances have affected the system" (p. 1466).

The above experimental results, and particularly the fact that in the Paulignan et al. (1990) study the first detectable difference between perturbed and non perturbed movements occurred some 100 ms after the perturbation (at the time of the first acceleration peak) emphasizes the rapidity of processing of movement-related information for on-line corrections. A duration of 100 ms seems to correspond to the minimum delay needed for visual and/or proprioceptive reafferents to influence an ongoing movement. In the visual modality, evidence for fast processing exists in the context of reaching for targets. Zelaznik, Schmidt, Gielen, and Milich (1983) found that presence of visual feedback during the movement had a positive effect on accuracy, as compared to absence of visual feedback, even for movements of a duration as short as 120 ms. In addition, the same authors reported another experiment where accuracy was measured for movements of very short duration (70 ms; see also Carlton, this volume). They found no effect of presence or absence of visual feedback on errors in distance, although errors in direction appeared to be reduced when visual feedback was present (see also Elliott & Allard, 1985). There are other examples of short visuomotor times; for instance in the case of postural adjustments following perturbation of stance (Nashner & Berthoz, 1978). In the proprioceptive modality, the time for processing kinesthetic reafferent signals was estimated to 70 to 100 ms (e.g., Lee & Tatton, 1975; Evarts & Vaughn, 1978). Finally, similar delays were found for the tactile modality (Johansson & Westling, 1987).

A possible explanation for fast corrections can be proposed. This explanation postulates that the level of representation of the action activated by perturbations arising during the movement is a "dynamic" one, as opposed to the relatively "static" level which accounts for corrections with reprogramming that are generated during the reaction time (see Prablanc & Pélisson, 1990). The difference between the two situations is that perturbations arising during the movement create a discrepancy between the visual signals related to the new target position and the incoming proprioceptive signals generated by the movement itself, which is not the case for perturbations occurring before the movement has started. In this context a dynamic representation can thus be considered as a network that would permanently monitor the movement-related reafferent input and compare it with the efferent motor signals. Any change in the target-limb spatial configuration would be encoded by the network as a deviation from the represented action, and would automatically (and therefore unconsciously) trigger reorganization of the command sent to the execution level.

This mechanism, which represents a revised version of the classic corollary discharge model (e.g., Held, 1961), must operate at a level where the visual and the proprioceptive maps are interconnected, and where a comparison of position of the target relative to the body and position of the moving limb relative to the target can be made.

The Kinematic Coding of Reaching Movements

Another aspect of the control of reaching movements is specification of the correct kinematics along the path selected for getting to the target. Data concerning this aspect come almost exclusively from experiments in humans, which are part of a long and fruitful tradition in psychology (for reviews, see Georgopoulos, 1986; Jeannerod, 1988; Meyer, Smith, Kornblum, Abrams, & Wright, 1990).

Specification of kinematics determines the timing of the movements and the respective durations of their acceleration and deceleration phases. Emphasis on the speed of the movements, as in pointing for example, will result in relatively symmetrical (bell-shaped) velocity profiles. By contrast, emphasis on accuracy, as in prehension, will produce slower movements with a longer deceleration phase and therefore, an asymmetrical velocity profile. In addition, the kinematic pattern also depends on the task in which the subject is involved. This point is illustrated by the results of experiments where reaching movements are compared in different conditions of task constraint. A constraint is defined here as a variable that limits the manner in which movement control occurs. As an example, the target size can be considered as a movement constraint, such that movements aimed at small targets have velocity profiles distinctly different from those produced in attaining larger targets (Soechting, 1984).

The effects of such constraints on movement kinematics were tested by Marteniuk, MacKenzie, Jeannerod, Athenes, and Dugas (1987). In their experiment the goal of reaching movements, as well as the required movement extent and accuracy, were systematically varied. Subjects were asked either to hit a target using the index finger, or to grasp a disk between the thumb and the index finger. Both targets and disks had the same sizes (2 cm or 4 cm) and were placed at distances of 20 cm or 40 cm from the resting position of the right hand. Movement time was longer during grasping than during hitting. Furthermore, according to Fitts' law, it increased in both cases as a function of difficulty of the task (as quantified by combining the target/disk width and the movement amplitude, see Fitts, 1954). Both types of movements had similar values of maximum velocity, and in both cases maximum velocity was higher with increasing target/disk width and with increasing movement amplitude. The most interesting difference between the two types of movements was the partitioning of movement time into acceleration and deceleration. The peak velocity during hitting and during grasping was reached at about the same time, and the longer duration of grasping movements was due to a higher percentage of total movement time spent in the deceleration phase.

Marteniuk et al. (1987) also explored further the determinants of the movement kinematics by comparing grasping movements in two different situations. In both, the subject had to grasp the disk as in the previous experiment; in one situation the disk had to be thrown into a large container, and in the other one placed in a tightly fitting well. Only the reaching parts of the movements were analyzed. This comparison was undertaken with the aim of exploring whether the context within which a movement is executed can affects its characteristics. Indeed movement time was found to be longer for grasping movements executed prior to fitting than for movements executed prior to throwing. This lengthening in movement time was due to lengthening of the deceleration phase (Figure 9.4).

These data clearly indicate that planning and control of trajectories do not occur through the use of a general and abstract representation that would adjust movements for different task conditions, by simply scaling a basic velocity profile. Instead, higher order factors such as representation of the goal, the context, and also probably the knowledge of the result of the action seem to be able to influence not only duration and velocity, but also the intrinsic kinematic structure of the movements.

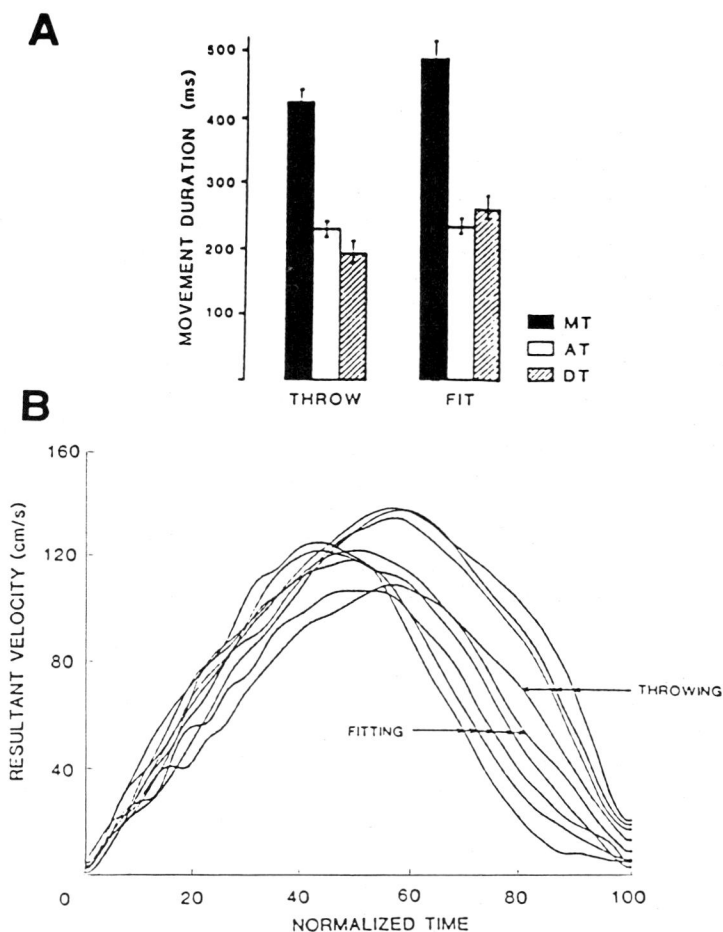

Figure 9.4. Influence of task constraints on the temporal pattern of reaching movement. (a) Movements have been recorded during the action of grasping. On some trials subjects had to grasp the disc and throw it away. On other trials, they had to grasp the disc and fit it in a small box. These task constraints influenced both total movement duration (MT), which was longer in the fit condition, and the velocity profiles. In the throw condition acceleration time (AT) was longer than deceleration time (DT), although the reverse was observed in the throw condition. (b) Representative velocity profiles in the two conditions. Note that time has been normalized for enabling comparison. (Data from Marteniuk et al., 1987).

Grasping

The function of grasping is to acquire objects for the purpose of manipulation, identification and use. Tool behavior is one of the possibilities offered by a fully developed hand, and must also be included into the function of grasping. The anatomy and physiology of the hand and its neural apparatus seem to have evolved across species along two main lines, the development of an opposable thumb and the development of independent finger movements, both properties sometimes collapsed under the terminology of digital dexterity. This section will review studies on the behavioral aspects of grasping with emphasis on sensorimotor, rather than on purely motor, aspects of the function.

Anticipatory Finger Posturing: The Pattern of Grip Formation

The type of grip that is formed by the hand in contact with the object represents the end result of a motor sequence which starts well ahead of the action of grasping itself. The fingers begin to shape during transportation of the hand at the object location. This process of grip formation is therefore important to consider, because it shows dynamically how the static posture of the hand is finally achieved. No systematic investigation of this aspect of grasping (preshaping) seems to have been made until the film study by Jeannerod (1981).

The pattern of grip formation is not an undefinable pattern, it is specified, at least in part, by the type of action in which the subject is engaged. Finger movements during preshaping contribute to the building of a specific spatial configuration, where the relative positions of the hand and the object to be grasped are constrained both by biomechanical factors and by higher order visuomotor mechanisms for object recognition.

Preshaping first involves a progressive opening of the grip with straightening of the fingers, followed by a closure of the grip until it matches object size. The point in time where grip size is the largest (maximum grip size) is a clearly identifiable landmark which occurs within about 60% to 70% of the duration of the reach (see below), that is, well before the fingers come in contact with the object (Jeannerod, 1981, 1984; see also Wing, Turton, & Fraser, 1986; Wallace & Weeks, 1988).

The amplitude of grip aperture during grip formation covaries with object size (Gentilucci, Castiello, Scarpa, Umilta, & Rizzolatti, in press; Jeannerod, 1981, 1984; Marteniuk et al., 1987; Marteniuk, Leavitt, MacKenzie, & Athenes, 1990; Wallace & Weeks, 1988; Wing & Fraser, 1983; Wing et al., 1986). Marteniuk et al. (1990) found that for each increase of 1 cm in object size, the maximum grip size increased by 0.77 cm (Figure 9.5a). Subjects use two different strategies for achieving this pattern. In the first

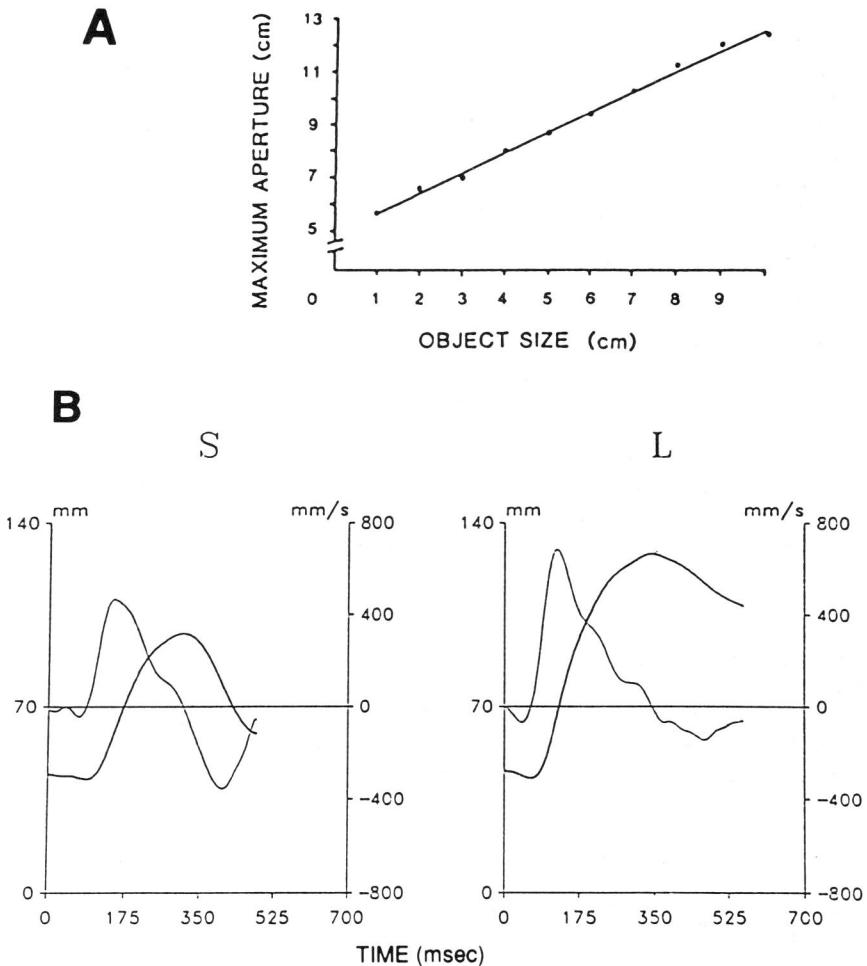

Figure 9.5. Relationship of grip size to object size. (a) Maximum aperture of grip plotted against object size (Reproduced with permission from Marteniuk et al., 1990). (b) Kinematic pattern of grip aperture in two prehension movements. On the left, movement directed to a small dowel (S). On the right, to a larger cylinder (L). Thick lines: change in grip size as a function of movement time. Thin lines, rate of change (velocity) of grip size. Note maximum grip size reached earlier during prehension of small dowel. (Reproduced with permission from Paulignan et al., in press).

strategy, the grip size increases at the same rate whatever the object size (see Figure 9.5b), with the consequence that maximum grip size is reached earlier in movement time for a small object than for a large object (e.g., Gentilucci et al., in press; Marteniuk et al., 1990). By contrast, in other cases, the rate of increase is faster for a large than for a small object, with the consequence that grip size peaks at the same time for both large and small objects. This second pattern corresponds to that described earlier by Jeannerod (1981). It seems that the only determinant for using either one of the two strategies is movement time, so that the subjects tend to equate the time to maximum grip aperture for large and small objects when movements become faster.

One possible explanation for the biphasic pattern of grip formation relates to the thumb-index finger geometry. Because the index finger is longer than the thumb, the finger grip has to open wider than required by object size, in order for the index finger to turn around the object and to achieve the proper orientation of the grip. Indeed, the movement of the index finger contributes the most to grip formation, whereas the position of the thumb with respect to the hand tends to remain invariant (Wing & Fraser, 1983) (Figure 9.6a, b). The outcome of this motor pattern has been formalized by Arbib (1985), Iberall, Bingham, and Arbib (1986) and Iberall and Arbib (1990). Their idea is that the basic elements of the hand are the "virtual fingers", one of them being the thumb (VF1), another one, the finger(s) that oppose to the thumb (VF2), and a third one the unused finger(s) (VF3). In grasping a small object with a precision grip, VF2 will be composed of the index finger only, whereas in power grip, VF2 will include four fingers. During preshaping, vision extracts from the object an "opposition space" to which the opposition space of the hand will be matched by proper selection and orientation of virtual fingers. It is apparent from experimental data that there are preferred orientations for the hand opposition space. For example, Paulignan, MacKenzie, Marteniuk, and Jeannerod (1991) showed that the same orientation of the hand was preserved during prehension of the same object (a vertical cylindrical dowel) placed at different positions in the working space (Figure 9.6c). This implies that, in order to preserve the same hand orientation with respect to the object, different degrees of rotation of the wrist or the elbow are required. In other words, the all limb, and not only its distal segments, would be involved in building the appropriate hand configuration for a given object.

Other, more complex, factors can also account for the pattern of hand configuration prior to, and during grasping. Hand configuration reflects the activity of higher order visuomotor mechanisms involving perceptual identification of the object and those involving generation of appropriate motor commands. Accordingly, a more global approach has been developed by several authors, showing that preshaping, manipulation and tactile exploration

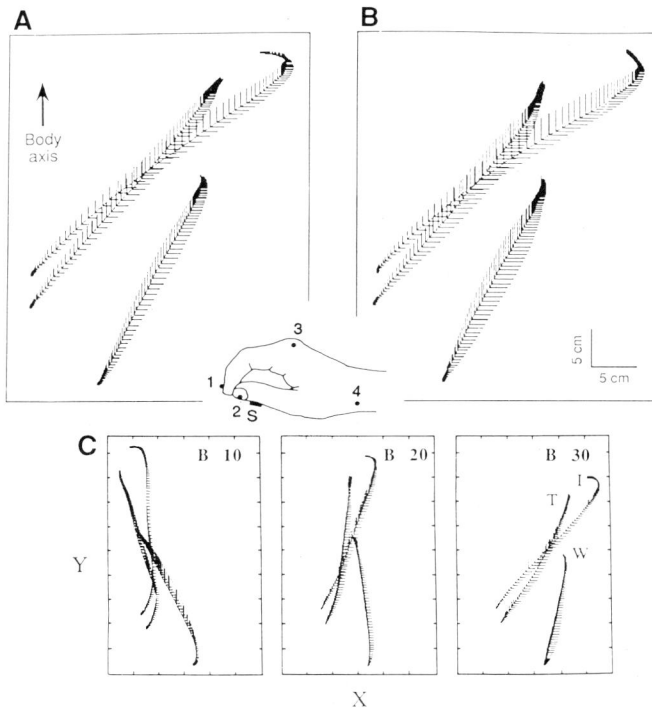

Figure 9.6. Spatial path and variability of wrist, index finger and thumb during prehension movements in one normal subject. (a) Prehension of a small object. (b) Prehension of a larger object. (c) Prehension of the same small object located at 10, 20 or 30 degrees (B10, B20, B30 respectively) to the right of subject midline. W: trajectory of a marker located at the wrist level; T: at the tip of the thumb; I: at the tip of the index finger. Trajectories are sampled at 250 Hz and normalized over 10 trials. The horizontal and vertical bars represent the values of one standard deviation with respect to the mean trajectory, for each of the 100 normalized frames. The drawing of the hand in the middle of the figure shows the starting posture of the subject's hand: 1 and 2 indicate the index and the thumb markers, 4, the wrist marker. S, start switch for triggering perturbations at onset of movement. Note larger contribution of the index finger to the pattern of grip formation during preshaping. Also, note relatively similar orientation of hand opposition in spite of differences in object size or location. (Reproduced with permission from Paulignan et al., 1991 and in press).

of objects are "knowledge-driven". For instance, Klatzky, McCloskey, Doherty, Pellegrino, and Smith (1987) showed that subjects tend to classify usual objects into broad categories, the boundaries of which are determined by the pattern of hand movements these objects elicit when they are to be grasped, used and manipulated. Four main prototypical hand shapes (e.g., poke, pinch, clench, palm) seem to be sufficient for defining the interaction between the hand and most usual objects. In addition, this differentiation of hand shapes according to the forms of objects is retained in preshaping during actual reaching (see also Pellegrino, Klatzky, & McCloskey, 1989). The same idea that characteristics of reach and grasp movements depend on prior knowledge gained from previous real world interactions with objects, has also been emphasized by Athenes and Wing (1990).

These findings seem consistent with the fact that hand movements during grasping are largely predetermined by object-related visual input and by object affordance. Indeed, in normal subjects, both the biphasic pattern of grip formation (opening of the finger grip followed by anticipatory closure) and the coordination of the reaching and grasping components are correctly achieved in situations where the hand remains invisible to the subject. Similarly, the size of the maximum grip aperture correlates with the size of the object in the absence of visual feedback from the hand (Jeannerod, 1981, 1984). The fact that object-related visual input acts proactively in generating appropriate hand configurations is thus a strong indication that cognitive representation mechanisms can directly influence the motor control of the hand. This point is further developed in the next section.

Grip Accuracy

Accuracy, a currently used parameter for assessing motor performance, can be easily quantified in situations like aiming or reaching. This is less true in the case of natural prehensive movements, where the terminal error in finger position during the grasp is likely to integrate the errors intervening at the level of the other segments of the arm. It is thus necessary, in order to determine the accuracy of the grip component *per se*, to use experimental situations where the grip component would be disconnected from, and therefore unperturbed by, the other components of prehension.

Jeannerod and Decety (1990) undertook such an experiment where they examined the accuracy of finger movements in matching the size of visually presented objects. Feedback cues arising from execution of the movement were excluded as much as possible, by preventing both vision of the performing hand and contact of that hand with the target objects. Thus, subjects had to match the size of their finger grip to the size of objects presented through a mirror precluding vision of their hand. The distance between the tip of the index finger and the tip of the thumb was measured.

The results showed that, in spite of a general trend toward overestimation, the mean grip size correlated positively and linearly with target size with high correlation coefficients.

These results raise important issues concerning the accuracy of converting retinal signals related to object size into a finger posture, what is called here visuomotor estimation. First, it is interesting to compare the present results with those obtained in psychophysical experiments testing the subjective scaling of visual length or size. A long time ago, Hering (quoted by Marks, 1978) had proposed that perceived length of a line should be proportional to its physical length, and should not follow a logarithmic function as it is the case in the perception of other physical dimensions. This contention has been substantiated (Stevens & Guirao, 1963). The subjective scaling of area, however, has been found by several authors to be related to physical size by a power function with an exponent of 0.7-0.8 (Ekman & Junge, 1961; Stevens & Guirao, 1963). The discrepancy between subjective scaling of length and area might be related to the type of instructions given to the subjects. Teghtsoonian (1965) showed that, if subjects were required to estimate the "objective area" of circles (How large are the circles?), their judgements followed a linear relation to physical size, as it was found for the lines. In contrast, if the task was to estimate the "apparent size" of the same circles (How large they look to you?), then their judgements followed a power function (exponent 0.76) with respect to physical size. The Jeannerod and Decety results, showing that visuomotor estimation of size is linearly related to target size, therefore suggests that motor programs actually "read" objective size rather than apparent size.

The second point raised by the Jeannerod and Decety results is that the degree of precision attained in visuomotor estimation of size is lower than during the real movement of grasping objects. This difference may be explained by the fact that in the Jeannerod and Decety experiment visual feedback signals were lacking, although these signals are well normally present during prehension. It is a well known general rule in motor control that accuracy of visually goal-directed movements deteriorates in conditions where visual feedback is either suppressed or degraded. Indeed, studies where accuracy of prehension was measured as the global outcome of natural coordinated reach and grasp movements (i.e., where possible errors in grip formation cumulated with errors in execution of the other components of the movement), showed relatively large errors when visual feedback was prevented (Jeannerod, 1984) or in conditions of speeded performance (Wallace & Weeks, 1988). In order to explain the higher accuracy attained in pure visual scaling, one might also suggest that in this task subjects rely on the readout of analytic visual mechanisms (specialized for size, orientation, etc.), which are likely to be quite accurate. In contrast, during visuomotor estimation, subjects have to rely on mechanisms involved in the transformation of visual information into motor commands. It is likely that the complexity of this transformation involves many potential sources of inaccuracy.

The Coordination of Reaching and Grasping

The current model for explaining the coordination of actions involving simultaneously several motor components, like prehension, is that of parallel functional ensembles, characterized by specific input-output relationships, and specialized for generating each component of the action (see Jeannerod, 1981). During prehension, the reaching arm movement carrying the hand to the location of the target-object is executed in parallel with finger movements which shape the hand in anticipation of the grasp.

Theories involving parallel activation, however, are faced with the problem of how the different systems that are separately, and simultaneously, activated can be coordinated with each other in order to achieve the desired action. The fact the two components of prehension are separate movements with different modes of organization, and belonging to different levels of visual processing, implies different timing. They must therefore be bound by a distinct coordination mechanism, so that transport of the hand and shaping of the fingers coincide in time, and the reach terminates exactly at the time where the fingers come in contact with the object. Indeed, most of theories which account for multiple joint actions involve specific mechanisms for ensuring coordination of the components in the temporal and spatial domains (Arbib, 1981; Bernstein, 1967). Arbib's conception (see Iberall & Arbib, 1990) implies that the various motor schemas responsible for prehension (e.g., "Move the arm", "Orient", "Preshape", "Enclose object", etc.) are hierarchically organized within a broader coordinated control program which determines the order and the degree of activation of the schemas.

Such a coordination mechanism cannot be regulated by peripheral input. Activation of cutaneous afferents at the time of contact of fingers with the object would not be sufficient, in spite of the rapidity of tactile-motor loops, to account for a proper timing of prehension. By extrapolating from the data of Johansson and Westling (1987), a delay of 75 ms is needed for tactile signals to influence an ongoing movement. This means that, if the hand was still travelling at 50 cm/s at the time of contact, it would pass the object location by more than 3.5 cm before full stop. This would not be compatible, either with the degree of accuracy required for grasping, or with the observed spatiotemporal organization of the grasp. The alternative hypothesis, that coordination between the two components belongs to a preorganized functional temporal structure, will thus be examined.

Spontaneous Temporal Covariation of the Two Components

One way of studying the degree of coordination between the two components of prehension is to look for possible invariant relationships between kinematic landmarks of the respective trajectories of the involved limb segments, during natural movements. It was shown by Jeannerod (1984) that the main kinematic landmark of the grip component, the time to maximum grip aperture, occurred at a fixed ratio of total movement time. In the Jeannerod paper, this ratio was between 75% to 80% of movement time in most subjects. Very similar values (72%-82%) were found in a sample of adult subjects studied by Von Hofsten and Ronnqvist (1988) and Wallace, Weeks, and Kelso (1990). These latter authors showed that the ratio of time to maximum grip to movement time was remarkably stable in spite of large variations in movement time and speed and in spite of different initial postures of the fingers. Such a temporal invariance is strongly indicative of a functional coupling between the two components.

Another, though more controversial finding is also in favour of functional coupling. Jeannerod (1981, 1984) had observed that the time to maximum grip aperture was systematically correlated with the occurrence of secondary accelerations of the wrist movement. Later work by other authors, however, failed to replicate this result (e.g., Marteniuk et al., 1990). In fact, the explanation to this discrepancy may be that secondary accelerations are not always observed in prehension movements. Whereas they are clearly present during lowering the hand toward the object in movements executed in the vertical plane (as in Jeannerod's experiments), they are absent in other types of prehension movements (e.g., performed in the horizontal plane). Finally, the search for other kinematic covariations between the two components has not been entirely successful until now. A significant correlation was occasionally found between the time occurrence of the velocity peak of the transportation component, and the time occurrence of the maximum grip aperture (Paulignan et al., 1990, Marteniuk et al., 1990). This correlation was not observed in all subjects and, within subjects, was not present in all conditions. Another, more consistent correlation was found between the time occurrence of maximum grip aperture and that of the peak deceleration of the wrist (Gentillucci et al., in press).

Effects of Systematic Manipulation of One Component on the Kinematics of the Other

Another way of understanding the nature of the coordination process between the two components of prehension, is to examine the degree of stability of each given component when the other one is manipulated experimentally. In an experiment where subjects had to grasp a dowel by performing reaching movements either at a normal speed, or as fast as possible, Wing et al. (1986) found that maximum grip aperture was larger when the movement was faster. The authors interpreted this effect of movement speed on the grip component as an anticipatory error correcting behavior. This result was partly confirmed by Wallace and Weeks (1988) in a series of experiments involving manipulation of error tolerance during grasping. Error tolerance boundaries were imposed on the grasp, with the corollary that the subjects had to regulate the duration of their movements in order to minimize their error. As predicted by Fitts' law (1954), the duration of the movement was indeed a function of the accuracy demand. As an example, when the error tolerance was large (e.g., displacements of the object by up to 7.6 cm were allowed during the grasp), the movement lasted 243 ms and the wrist velocity was 165 cm/s. By contrast, when the error tolerance was smaller (maximum object displacement allowed, 1.5 cm), movement duration increased up to 379 ms and the wrist velocity dropped to 117 cm/s. The interesting point was that in the conditions where the movement duration decreased, the grip size tended to become larger by about 10%. Wallace and Weeks were able to demonstrate that the critical factor affecting grip size was movement duration, not movement speed: in manipulating simultaneously the two parameters (by instructing the subjects to perform movements of fixed durations directed at targets placed at different distances), they showed that increasing movement velocity without increasing movement duration had no effect on grip size. These results of Wallace and Weeks, along with those of Wing et al., thus demonstrate that decrease in movement time (whether it is an indirect effect of larger error tolerance, or a consequence of instruction given to the subject) produces larger grip apertures.

Conversely, systematic changes of the grip component may affect the transportation component. Marteniuk et al. (1990) found prehension movements to last longer when the size of the object was smaller, due to lengthening of the deceleration phase of reaching. This finding is consistent both with the fact that movement time is generally a function of task difficulty, and with the fact that changes in total movement time are generally due to changes in duration of the deceleration phase (e.g., Marteniuk et al., 1987; as described previously). Finally, in addition to affecting movement time, object size also affected the timing of the grip. Marteniuk et al. (1990) showed that the time to maximum grip aperture was function of task difficulty, such that maximum grip size occurred earlier for smaller targets.

Effects of Perturbations Applied to One of the Components

Perturbing the input of a sensorimotor system is a commonly used paradigm for probing the various functional levels of control of that system (see Jeannerod, 1990). In this section, the effects of unexpected changes in position or in size of the object prior to or during a prehension movement are examined. Change in position of the object will be considered as perturbing specifically the transportation component of the movement, whereas a change in size of the object will be considered as perturbing the grip component. The problem will be to determine to what extent a perturbation theoretically limited to one of the components will exert its effects on that component only, or whether it will also affect the other.

Perturbation of object location. As mentioned earlier, perturbations affecting the position of the object at the onset of the movement produce rapid corrections of the wrist trajectory, such that the object is correctly grasped with only little increase in movement time (Paulignan et al., 1990). A more complete analysis of these results, however, showed that the effects of object displacement were not limited to corrections affecting the transport component. The grip aperture was also affected. On most perturbed trials, the two components of prehension became kinematically coupled during corrective responses. The alteration of the wrist trajectory for reorienting the movement to the new target position was immediately followed (within about 50 ms) by a brief interruption of grip aperture, not required by the situation since object shape and size remained unchanged (Paulignan et al., 1990, 1991) (Figure 9.7). This finding was recently replicated by Haggard and Wing (in press). In their situation, the subject's arm was suddenly pulled back by a mechanical device during approach to the object. This perturbation triggered a rapid correction of the transport component, such that the arm was reaccelerated in order to reach for object position. In about 70% of perturbed trials, the perturbation applied to the arm also provoked a reversal of grip aperture. As the change in grip formation occurred some 70 ms later than the change in transport, Haggard and Wing interpreted this coupling of the two components as an effect of a coordination mechanism. They proposed that (proprioceptive) information generated by the arm movement was used for stopping grip formation during the correction for the transport. By this way, the temporal coordination between the two components would be restored during the final phase of the movement.

Perturbation of object size. Rapid changes in the size or shape of a graspable object are not easily produced. Paulignan, Jeannerod, MacKenzie, and Marteniuk (in press) used concentric objects made of a central dowel (diameter, 1.5 cm) surrounded by a large cylinder (6 cm). These objects were made of translucent material and were illuminated from below (see Paulignan et al., 1990). The light illuminating the dowel could be shifted below the cylinder, so that the cylinder now appeared to be illuminated. This created the impression of a sudden expansion of the dowel, without change in spatial location. The reverse effect (light shifted from cylinder to dowel), thus giving the impression of a shrinking of the cylinder, could also be produced.

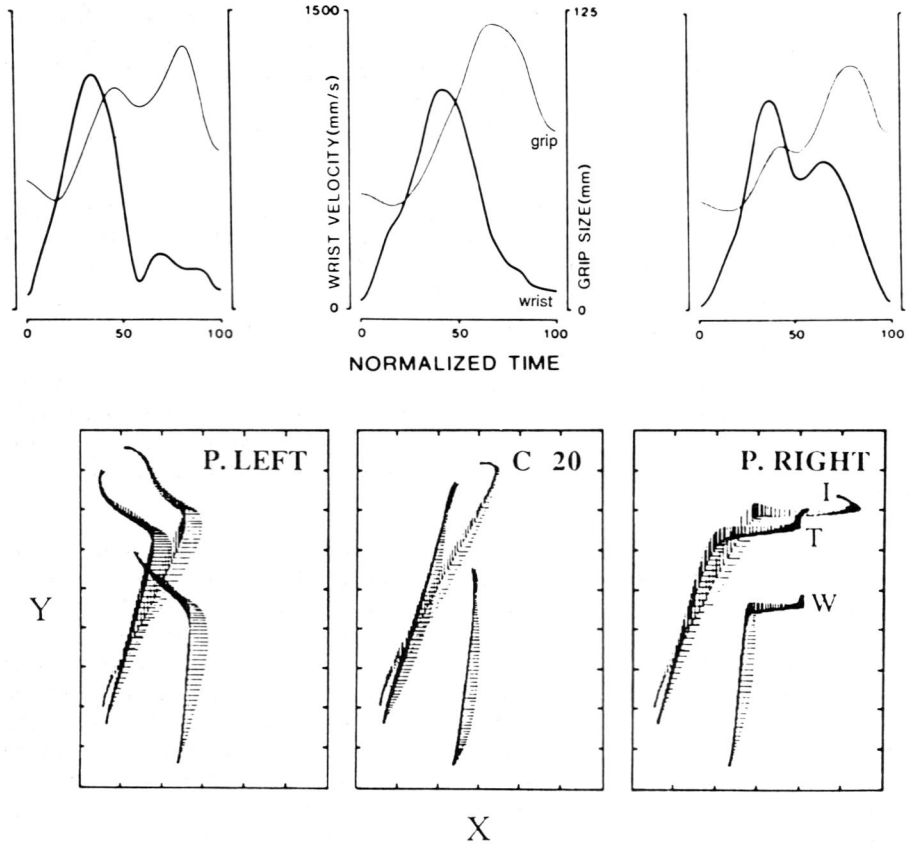

Figure 9.7. Effects of sudden change in object position at the onset of prehension movements. Upper row: Kinematics of prehension during a movement directed at the central (20 degree) dowel (middle diagram) and during a perturbed-left and a perturbed-right movement (diagrams on the left and on the right, respectively). Thick line, velocity of the wrist. Thin line, grip aperture. Data averaged on 6 subjects. Note corrections not only on wrist kinematics, as already shown in Figure 9.3, but also on grip kinematics. (Reproduced with permission from Paulignan et al., 1990). Lower row: Spatial paths and variability of wrist, index finger and thumb in the same experiment. (Reproduced with permission from Paulignan et al., 1991).

Perturbing object size at the onset of the movement clearly affected the grip component. If the perturbation consisted in increasing object size (from dowel to cylinder), the grip formation sized to the small object was interrupted and the grip aperture was reincreased in order to accommodate the large object. This correction in grip size began a relatively long time (330 ms) after the perturbation.

Response to the perturbation in object size, however, was not limited to the grip component. Total movement time was increased by over 170 ms in the small to large perturbation, and by about 85 ms in the large to small case. No change could be seen in the timing of kinematic landmarks of the transport component, at least during the first 300 ms, which means that the additional movement time was spent in the later part of the movement. Indeed, a long-lasting low velocity phase was present in the transport component, so that the wrist virtually stopped at the vicinity of the object before the fingers came in contact with it (Figure 9.8, upper and lower right). This was in contrast to the kinematics of the wrist and grip during unperturbed trials (Figure 9.8, upper and lower left).

Examination of the finger trajectories in the perturbed trials in the Paulignan et al. experiment revealed that the relative contribution of each finger to the grip pattern was changed with respect to unperturbed trials. In the normal condition, the contribution of the index finger to the grip was clearly larger than that of the thumb. In the perturbed trials, it appeared that the contribution of the thumb was increased. In addition, large rotations of the wrist were observed, which were not present in unperturbed conditions. These changes represent a remarkable example of "motor equivalence", a concept which accounts for performing the same action by different means according to the conditions of execution. Other examples of motor equivalence in hand movements have been described by Cole and Abbs (1986) during rapid pinch. These authors showed that, over repetitions of this simple behavior the individual finger and thumb joints varied in their angular positions at the time of contact of the two fingers, while the point of contact remained invariant. This was obtained by corresponding covariations of the finger paths. As already stated above, these quick changes in motor patterns responsible for motor equivalence seem to imply an encoding of the end positions of the limb segments with respect to the goal, rather than an encoding of detailed prescriptions for joint positions.

Taken together, the results of perturbation experiments, showing covariation of movements of the wrist and the fingers during corrections create the impression of a synergy of the two components of prehension. Perturbation of object location, which should, in principle, affect the transport component only, also affects grip formation. Similarly, perturbation of object

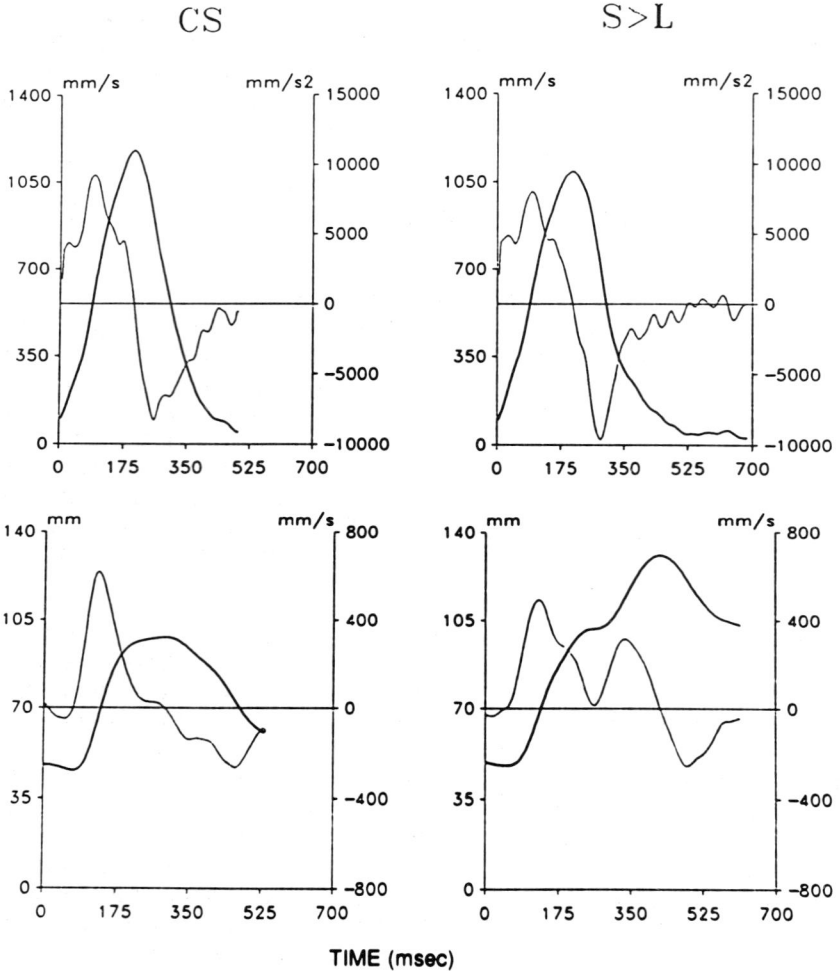

Figure 9.8. Effects of sudden change in object size at the onset of prehension movement. Diagrams on the left, wrist kinematics (upper), and grip kinematics (lower) of unperturbed movements directed at a small object. Diagrams on the right show the effects of a sudden increase in object size (S>L) at onset of movement. Note increase in duration of low velocity phase of wrist movement (upper), and correction in grip size (lower). Also note relatively late occurrence of the grip correction, as compared to early corrections in wrist trajectory during perturbation of object position (Figures 9.3 and 9.7). (Reproduced with permission from Paulignan et al., in press).

size affects not only grip formation, but also transport. A synergy, however, would have implied a strict temporal correlation of the components, which was not confirmed by statistical analysis of the data. Instead, the two submovements that compose prehension appear to be only loosely time-coupled. This pattern of coordination is remindful of the coordinative-structure concept, whereby independent musculoskeletal elements can become functionally linked for the execution of a common task, without implying structural relationships between them (e.g., Kelso, Holt, Kugler, & Turvey, 1980).

Neural Networks and Prehension

Artificial neural net models or neural nets attempt to achieve performance through dense interconnection of simple computational elements. While there is controversy (Roberts, 1989) as to their biological validity, there are advantages for using them to infer possible ways in which the central nervous system can store and process information (Rumelhart & McClelland, 1986). These models are composed of nodes or units that are computationally nonlinear and operate in parallel. There are input and output units, and where there are nonlinear functions to be modelled, hidden units. These nets learn via established learning or training rules and are capable of performing a wide variety of tasks, including sequential movement (Jordan, 1990).

Neural nets and movement "knowledge". One issue that the underlying basis of neural nets helps to conceptualize is how "knowledge" about movement might be stored within the central nervous system and how it might affect movement planning and execution. Earlier in this chapter, we reviewed literature which suggested that object-related visual inputs act proactively in generating appropriate hand configurations. We concluded that this is a strong indication that cognitive representation mechanisms can directly influence the motor control of the hand. Churchland (1989, Chapter 9) describes how neural nets, with a layer of hidden units connecting input and output units, can indeed "represent knowledge" of very complex non-linear processes. Reviewing some of the literature, he describes one such network that was trained to discriminate submarine sonar signals returning from rocks *vs.* mines. Another network takes vector codings for seven-letter segments as inputs and produces vector codings for phonemes as outputs. These outputs, after being fed through a speech synthesizer, allow the network to transform printed words into audible speech.

Churchland makes the point that what is interesting about neural nets that learn tasks like the above, is that the "knowledge" that the networks have acquired and represent consists of a set of connection weights, among the network's units, that were established during the "training up the network" phase. These connection weights, during training, essentially partition the abstract space represented by the possible activation levels of each of the hidden units into activation vectors so that novel inputs can be "recognized" as belonging to a particular subvolume of this abstract space. Thus knowledge representation is seen as the configuration of weighted connections among a set of neuron-like units. Once trained, these networks are very rapid in performing their task. As well, the amount of space that can be used to represent the countless tasks humans encounter in a lifetime is virtually limitless. If, as Churchland speculates, this view of knowledge representation is congruent with how the brain represents knowledge, then we can expect the brain to be able to be partitioned into billions of functionally relevant subdivisions of its internal vector spaces, each responsive to a broad but proprietary range of highly complex stimuli. This is based on the premise that a network the size of the human brain has approximately 10^{11} neurons, and 10^3 connections on each neuron, for a total of 10^{14} total connections, and at least 10 distinct layers of hidden units. The time to "process" or classify an input takes only tens or hundreds of ms because the parallel coded input only has to make its way through up to ten layers of the massively parallel network to the functionally relevant layer that will produce the appropriate information at the output units.

Of course, this kind of structural organization may be ideal for attempting to model some aspects of prehension. As reviewed previously, a parallel distributed explanation of prehension postulates the existence of functional ensembles (the visuomotor channels), characterized by specific input-output relationships, and specialized for generating each component of action (Jeannerod, 1981). How a neural net might represent these components and provide for their interrelationship are empirical questions.

Neural nets and non-linear, self-organizing principles. Another characteristic of neural networks that make them similar to the non-linear, self-organizing principles used to model human movement organization (e.g. Kugler & Turvey, 1987) is that these networks exhibit principles of attractor dynamics with relaxation being their main mode of computation. The brain might be best understood as a kind of relaxation system in which computations are performed which attempt to iteratively seek to satisfy a large number of weak constraints (Rumelhart & McClelland, 1986, chapter 4). The system is thought of as settling into a solution rather than calculating a solution. An example of this is the recent work of Massone and Bizzi (1989) where, using Jordan's (1990) recurrent neural net, they showed that this neural net was capable of learning to control a redundant three-joint limb from an initial posture to several targets. During the testing of this network, it was shown to spontaneously exhibit attractor-dynamics properties with final end-point positions behaving like point attractors. Jordan (1990) has found the same for a network controlling a simulation of a six degree of freedom manipulator. The solution found by the network was stable and was a periodic attractor both in articulatory and endpoint coordinates.

What the above discussion implies is that the brain, through interconnections of these neuron like units, can represent "knowledge" and has considerable "space" to do so. As well, very complex, non-linear "computations" are performed within these neural nets that exhibit the same properties as those computations performed to model movement as a non-linear dynamical system (Kugler & Turvey, 1987). Except in the former case, no formula exists upon which to base the calculations. The calculations are an inherent aspect of the system.

Neural nets and time to reorganize movement. One of the interesting characteristics of prehension reviewed earlier had to do with how rapidly this movement could be reorganized after a perturbation. This issue is related to the conceptual issue of parallel processing which artificial neural nets embody and for which the brain appears to be ideally suited. As Rumelhart and McClelland (1986) assert, given the relatively slow processing rate of neurons (10s of ms), the apparently fast speed of the brain in solving complex problems is due to the fact that single neurons only compute very simple functions and they proceed in massive parallelism. Since there might be as many as 10^{11} neurons with each neuron being an active processor, the parallelism is on a very large scale. Further, as stated above, there may be many distinct layers of hidden units in the human brain which means, given the total number of possible neuronal connections, that its internal vector spaces can be partitioned into many billions of functionally relevant subdivisions. If an input is received that is relevant to one of these divisions, the network is capable of producing the appropriate activation vector in tens or hundreds of ms. The estimate is that for each layer of hidden units only a few tens of ms are necessary because of the parallel coded input. Thus, depending on where the functionally relevant layer of hidden units are, the massively parallel network literally needs only tens or hundreds of ms to traverse through the two to ten layers of hidden units. Finally, since the relevant information is stored as a set of connection weights, the information is automatically accessed by the coded input which means there is no additional time required to search for the information.

From the above, one might envisage information about movement being accessed and implemented in relatively short time intervals. Certainly, the times to amend or modify movements that were reviewed above (under 100 ms) would appear to be entirely within the realm of the capacity of a massively parallel system like the brain.

Neural nets and sensorimotor integration. The work of Held (1961) was cited earlier as utilizing a classical corollary discharge model of control. In fact, we postulated a revised version of this model and suggested that in terms of prehension, it must operate at a level where the visual and proprioceptive maps are interconnected, and where a comparison of position of the target relative to the body and position of the moving limb relative to the target can be made.

MacKenzie and Marteniuk (1985a, b) have argued that at the base of all coordinated movement is an integrated store of sensory and motor information that defines relations among body parts and near body space for the achievement of movement goals. The actual expression of coordinated movement, however, is modified by an interplay of motor and sensory information that tunes movement to the specific context of the situation. Prediction is a trademark of such a system and is derived from past experience where the learned relationship between afferent signals (interoceptive and exteroceptive) and efferent signals allows the prediction of the consequences of a particular action.

While the brain appears to abound in areas concerned with sensorimotor integration, artificial neural networks may be one way of studying the characteristics of this process. Massone and Bizzi (1989) give an example of how visually based targets triggered a network to perform sensorimotor transformations to produce time trajectories of a limb to acquire the targets. In another case, Dean (1990) constructed a neural net that simulated walking in the stick insect. The simulation showed that walking occurred by directly associating the joint configurations in one leg with those of another, in this way demonstrating proprioceptive-motor transformations.

Neural nets and motor equivalence. Jordan's network (1990) achieves motor equivalence through introduction of a smoothness constraint on the network. This allows the same target position to be achieved in different ways by a multiple degree of freedom effector depending on the temporal context. As one of Jordan's simulations shows, performance of a six degrees of freedom manipulator was variable in articulatory coordinates while invariantly achieving to goal of the task. Massone and Bizzi (1989) demonstrated motor equivalence in a different way through the use of Jordan's model and a redundant three joint limb. After the network had been trained to move from an initial position to one of several targets in task space, the target was suddenly moved to a new location once the limb had begun movement towards it (not too dissimilar to the experiment described above by Paulignan et al., 1990). When this happened, the network was still able to successfully accomplish the task even though the initial posture of the limb for the new target position was different from the initial position of the limb during network training.

The demonstration of motor equivalence by neural networks is not surprising given their characteristics as reviewed above. They are capable of solving nonlinear functions and exhibit self-organizing principles, properties that are intrinsic to the phenomenon of motor equivalence. Also inherent in the demonstration of motor equivalence in a human nervous system must be the issue of sensorimotor integration, which would allow rapid transformations between afferent and efferent information. In this way, a movement could not only be constantly monitored for its effectiveness in achieving the movement goal but its components could be covaried in a meaningful manner, so that the goal could be successfully attained with a number of different configurations of the components (e.g., Paulignan et al., 1991).

References

Abbs, J.H., & Gracco, V.L. (1984). Control of complex motor gestures: Orificial muscle responses to load perturbations of lip during speech. *Journal of Neurophysiology, 51*, 705-723.

Arbib, M.A. (1981). Perceptual structures and distributed motor control. In V.B. Brooks (Ed.), *Handbook of physiology, - The nervous system II* (pp. 1449-1480). Bethesda, MD: American Physiological Society.

Arbib, M.A. (1985). Schemas for the temporal organization of behavior. *Human Neurobiology, 4*, 63-72.

Athenes, S., & Wing, A.M. (1990). Knowledge directed coordination in reaching for objects in the environment. In S. Wallace (Ed.), *Perspectives on the coordination of movement* (pp. 285-301). Amsterdam: North-Holland.

Bernstein, N. (1967). *The coordination and regulation of movements.* Oxford: Pergamon Press.

Biguer, B., Jeannerod, M., & Prablanc, C. (1982). The coordination of eye, head and arm movements during reaching at a single visual target. *Experimental Brain Research, 46*, 301-304.

Bridgeman, B., Kirch, M., & Sperling, A. (1981). Segregation of cognitive and motor aspects of visual function using induced motion. *Perception & Psychophysics, 29*, 336-342.

Castiello, U., Paulignan, Y., & Jeannerod, M. (in press). Temporal dissociation of motor responses and subjective awareness: A study in normal subjects. *Brain.*

Churchland, P.M. (1989). *A neurocomputational perspective.* Cambridge: MIT Press.

Cole, K.J., & Abbs, J.H. (1987). Kinematic and electromyographic responses to perturbation if a rapid grasp. *Journal of Neurophysiology, 57,* 1498-1510.

Dean, J. (1990). Coding proprioceptive information to control movement to a target: Simulation with a simple neural network. *Biological Cybernetics, 63,* 115-120.

Ekman, G., & Junge, K. (1961). Psychophysical relations in visual perception of length, area and volume. *Scandinavian Journal of Psychology, 2,* 1-10.

Elliott, D., & Allard, F. (1985). The utilisation of visual feedback information during rapid pointing movements. *Quarterly Journal of Experimental Psychology, 37A,* 407-425.

Evarts, E.V., & Vaughn, W.J. (1978). Intended arm movements in response to externally produced arm displacements in man. In J.E. Desmedt (Ed.), Cerebral motor control in man: Long-loop mechanisms. *Progress in Clinical Neurophysiology* (pp. 178-192). Basel: Karger.

Favilla, M., Henning, W., & Ghez, C. (1989). Trajectory control in targeted force impulses. VI. Independent specification of response amplitude and direction. *Experimental Brain Research, 75,* 280-294.

Fitts, P.M. (1954). The information capacity of the human motor system in controlling the amplitude of movement. *Journal of Experimental Psychology, 47,* 381-391.

Gentilucci, M., Castiello, U., Scarpa, M., Umilta, C., & Rizzolatti, G. (in press). Influence of different types of grasping on the transport component of prehension movements. *Neuropsychologia.*

Georgopoulos, A.P. (1986). On reaching. *Annual Review of Neuroscience, 9,* 147-170.

Georgopoulos, A.P., & Massey, J.T. (1987). Cognitive spatial-motor processes. *Experimental Brain Research, 65,* 361-370.

Georgopoulos, A.P., Kalaska, J.F., & Massey, J.T. (1981). Spatial trajectories and reaction times of aimed movements: Effects of practice, uncertainty and change in target location. *Journal of Neurophysiology, 46,* 725-743.

Goodale, M.A., Pélisson, D., & Prablanc, C. (1986). Large adjustments in visually guided reaching do not depend on vision of the hand or perception of target displacement. *Nature, 320,* 748-750.

Gottsdanker, R. (1973). Psychological refractoriness and the organization of step-tracking responses. *Perception & Psychophysics, 14,* 60-70.

Haggard, P., & Wing, A.M. (in press). Remote responses to perturbation in human prehension. *Neuroscience Letters.*

Held, R. (1961). Exposure-history as a factor in maintaining stability of perception and coordination. *Journal of Nervous and Mental Disorders, 132,* 26-32.

Iberall, T., & Arbib, M.A. (1990). Shemas for the control of hand movements: An essay on cortical localization. In M.A. Goodale (Ed.),

Vision and action: The control of grasping (pp. 204-242). Norwood, NJ: Ablex.

Iberall, T., Bingham, G., & Arbib, M.A. (1986). Opposition space as a structuring concept for the analysis of skilled hand movements. In H. Heuer & C. Fromm (Eds.), Generation and modulation of action pattern. *Experimental Brain Research Series, 15,* 158-173.

Jeannerod, M. (1981). Intersegmental coordination during reaching at natural visual objects. In J. Long & A. Baddeley (Eds.), *Attention and performance IX* (pp. 153-168). Hillsdale, NJ: Erlbaum.

Jeannerod, M. (1984). The timing of natural prehension movements. *Journal of Motor Behavior, 16,* 235-254.

Jeannerod, M. (1988). *The neural and behavioral organization of goal-directed movements.* Oxford: Claredon Press.

Jeannerod, M. (1990). The representation of the goal of an action and its role in the control of goal-directed movements. In E.L. Schwartz (Ed.), *Computational neuroscience* (pp. 352-368). Cambridge: MIT Press.

Jeannerod, M., & Decety, J. (1990). The accuracy of visuomotor transformation. An investigation into the mechanisms of visual recognition of objects. In M.A. Goodale (Ed.), *Vision and action: The control of grasping* (pp. 33-48). Norwood, NJ: Ablex.

Johansson, R.S., & Westling, G. (1987). Signals in tactile afferents from the fingers eliciting adaptive motor responses during precision grip. *Experimental Brain Research, 66,* 141-154.

Jordan, M.I. (1990). Motor learning and the degrees of freedom problem. In M. Jeannerod (Ed.), *Attention and performance XIII* (pp. 796-836). Hillsdale, NJ: Erlbaum.

Jordan, M.I., & Rosenbaum, D.A. (1989). Action. In M.I. Posner (Ed.), *Foundations in cognitive science* (pp. 727-767). Cambridge: Bradford/MIT Press.

Kelso, J.A.S., Holt, K.G., Kugler, P.N., & Turvey, M.T. (1980). On the concept of coordinative structures as dissipative structures. I. Empirical lines of convergence. In G.E. Stelmach & J. Requin (Eds.), *Tutorials in motor behavior* (pp. 49-70). Amsterdam: North-Holland.

Kugler, P.N., & Turvey, M.T. (1987). *Information, natural law, and the self-assembly of rhythmic movement.* Hillsdale, NJ: Erlbaum.

Klatzky, R.L., McCloskey, B., Doherty, S., Pellegrino, J., & Smith, T. (1987). Knowledge about hand shaping and knowledge about objects. *Journal of Motor Behavior, 19,* 187-213.

Lee, R.G., & Tatton, W.G. (1975). Motor responses to sudden limb displacements in primates with specific CNS lesions and in human patients with motor system disorders. *Canadian Journal of Neurological Sciences, 2,* 285-293.

MacKenzie, C.L., & Marteniuk, R.G. (1985a). Motor skill: Feedback, knowledge, and structural issues. *Canadian Journal of Psychology, 39,* 313-337.

MacKenzie, C.L., & Marteniuk, R.G. (1985b). Bimanual coordination. In E.A. Roy (Ed.), *Neuropsychological studies of apraxia and related disorders* (pp. 345-358). Amsterdam: North-Holland.

Marks, L.E. (1978). Multimodal perception. In C. Carterette & M.P. Friedman (Eds.), *Handbook of perception, volume VIII. Perceptual coding* (pp. 321-339). New York: Academic Press.

Marteniuk, R.G., Leavitt, J.L., MacKenzie, C.L., & Athenes, S. (1990). Functional relationships between grasp and transport components in a prehension task. *Human Movement Science, 9,* 149-176.

Marteniuk, R.G., MacKenzie, C.L., Jeannerod, M., Athenes, S., & Dugas, C. (1987). Constraints on human arm movement trajectories. *Canadian Journal of Psychology, 41,* 365-378.

Massone, L., & Bizzi, E. (1989). A neural network model for limb trajectory formation. *Biological Cybernetics, 61,* 417-425.

Megaw, E.D. (1974). Possible modification to a rapid on-going programmed manual response. *Brain Research, 71,* 425-441.

Meyer, D.E., Smith, J.E.K., Kornblum, S., Abrams, R.A., & Wright, C.E. (1990). Speed-accuracy tradeoffs in aimed movements. Toward a theory of rapid voluntary action. In M. Jeannerod (Ed.), *Attention and performance XIII* (pp. 173-226). Hillsdale, NJ: Erlbaum.

Nashner, L.M., & Berthoz, A. (1978). Visual contribution to rapid motor responses during postural control. *Brain Research, 150,* 403-407.

Paillard, J. (1982). The contribution of peripheral and central vision to visually guided reaching. In D.J. Ingle, M.A. Goodale, & R.J.W. Mansfield (Eds.), *Analysis of visual behavior* (pp. 367-385). Cambridge: MIT Press.

Paulignan, Y., Jeannerod, M., MacKenzie, C.L., & Marteniuk, R.G. (in press). Selective perturbation of visual input during prehension movements. 2. The effects of changing object size. *Experimental Brain Research.*

Paulignan, Y., MacKenzie, C.L., Marteniuk, R.G., & Jeannerod, M. (1990). The coupling of arm and finger movements during prehension. *Experimental Brain Research, 79,* 431-436.

Paulignan, Y., MacKenzie, C.L., Marteniuk, R.G., & Jeannerod, M. (1991). Selective perturbation of visual input during prehension movements. I. The effects of changing object position. *Experimental Brain Research, 83,* 502-512.

Pélisson, D., Prablanc, C., Goodale, M.A., & Jeannerod, M. (1986). Visual control of reaching movements without vision of the limb. II. Evidence of fast unconscious processes correcting the trajectory of the hand to the final position of a double-step stimulus. *Experimental Brain Research, 62,* 303-311.

Pellegrino, J.W., Klatzky, R.L., & McCloskey, B.P. (1989). Time course of preshaping for functional responses to objects. *Journal of Motor Behavior, 21,* 307-316.

Prablanc, C., & Pélisson, D. (1990). Gaze saccade orienting and hand pointing are locked to their goal by quick internal loops. In M. Jeannerod (Ed.), *Attention and performance XIII* (pp. 653-675). Hillsdale, NJ: Erlbaum.

Prablanc, C., Echallier, J.F., Komilis, E., & Jeannerod, M. (1979). Optimal response of eye and hand motor systems in pointing at a visual target. I. Spatio-temporal characteristics of eye and hand movements and their relationships when varying the amount of visual information. *Biological Cybernetics, 35,* 113-124.

Roberts, L. (1989). Are neural nets like the human brain? *Science, 243,* 481-482.

Rumelhart, D.E., McClelland, J.L., & the PDP Research Group. (1986). *Parallel distributed processing. Volume 1: Foundations.* Cambridge: Bradford/MIT Press.

Soechting, J.F. (1984). Effect of target size on spatial and temporal characteristics of a pointing movement in man. *Experimental Brain Research, 54,* 121-132.

Soechting, J.F., & Lacquaniti, F. (1983). Modification of trajectory of a pointing movement in response to a change in target location. *Journal of Neurophysiology, 49,* 548-564.

Stevens, S.S., & Guirao, M. (1963). Subjective scaling length and area and the matching of length to loudness and brightness. *Journal of Experimental Psychology, 66,* 177-186.

Teghtsoonian, M. (1965). The judgement of size. *American Journal of Psychology, 78,* 392-402.

Van Sonderen, J.F., Denier Van der Gon, J.J., & Gielen, C.C.A.M. (1988). Conditions determining early modification of motor programmes in response to changes in target location. *Experimental Brain Research, 71,* 320-328.

Von Hofsten, C., & Ronnqvist, L. (1988). Preparation for grasping an object: A developmental study. *Journal of Experimental Psychology: Human Perception and Performance, 14,* 610-621.

Wallace, S.A., & Weeks, D.L. (1988). Temporal constraints on the control of prehensive movements. *Journal of Motor Behavior, 20,* 81-105.

Wallace, S.A., Weeks, D.L., & Kelso, J.A.S. (1990). Temporal constraints in reaching and grasping behavior. *Human Movement Science, 9,* 69-93.

Wing, A.M., & Fraser, C. (1983). The contribution of the thumb to reaching movements. *Quarterly Journal of Experimental Psychology, 35A,* 297-309.
Wing, A.M., Turton, A., & Fraser, C. (1986). Grasp size and accuracy of approach in reaching. *Journal of Motor Behavior, 18,* 245-260.
Zelaznik, H.N., Schmidt, R.A., Gielen, C.C.A.M., & Milich, M. (1983). Kinematic properties of rapid aimed hand movements. *Journal of Motor Behavior, 15,* 217-236.

VISION AND MOTOR CONTROL
L. Proteau and D. Elliott (Editors)

CHAPTER 10

THE CONTRIBUTIONS OF PERIPHERAL VISION AND CENTRAL VISION TO PREHENSION

BARBARA SIVAK and CHRISTINE L. MACKENZIE

*Department of Kinesiology, University of Waterloo,
Waterloo, Ontario, N2L 3G1*

This chapter considers the processing of visual information in reaching and grasping, specifically the roles of central and peripheral vision. First, we review and note the functional implications of the structural differences between central and peripheral regions of the retina and higher visuomotor processing areas. Second, we examine experimental evidence for the functional roles of central and peripheral vision for planning and motor control in aiming, pointing, reaching and grasping tasks. Finally, recommendations for future research are presented.

The motor act of prehension depends on the net output of complex neural processing, including the processing of visual information. Although it is possible to reach and grasp an object without the use of vision, vision adds important dimensions. When vision is used, grasping involves a complex interplay between the visual and motor systems. For example, visual information about the spatial location of objects and object characteristics such as size, shape and texture must be integrated with planning and control information in the motor systems. Grasping an object requires that the hand be transported to an object (transport component) at the same time as the fingers are postured in anticipation of making contact with the object (grasp component). Jeannerod (1981, 1984) suggested that the transport component is organized through a visuomotor channel processing information about extrinsic properties of an object (e.g., the location), and the grasp component is organized through a visuomotor channel processing information about intrinsic properties of an object (e.g., size and shape). However, recent evidence has shown that visual information about object characteristics can also influence the planning and control of the transport component (Marteniuk,

Leavitt, MacKenzie, & Athenes, 1990; Paulignan, MacKenzie, Marteniuk, & Jeannerod, 1990, 1991, in press; Sivak & MacKenzie, 1990).

When considering the visual system, one must acknowledge that visual information is provided through both peripheral and central retinal areas, each specialized for processing specific types of information. Observations of an individual reaching to grasp an object suggest the following possible contributions of peripheral and central vision to prehension. Individuals usually fixate the eyes and turn the head toward an object to be grasped. Before the start of the grasping movement, both the hand and arm are in the peripheral visual field (hereafter referred to as peripheral vision) while the object is in the central visual field (hereafter referred to as central vision). During the reach, the hand and arm remain in peripheral vision while the object to be grasped is in central vision. Near the end of the movement, the hand comes into central vision, along with the object, while the arm remains in peripheral vision. When contact is made with the object, both the hand and object are seen in central vision while the arm remains in peripheral vision. Because visual information is provided in parallel by both central and peripheral regions of the retina during prehension, it is important to understand how these specialized retinal areas contribute to prehension.

Anatomical and Physiological Characteristics of Central and Peripheral Vision: Functional Implications

The functional uniqueness of central and peripheral vision is due to the anatomy and physiology of the retina and higher visuomotor processing areas. The retina, the first site of visual processing, consists of three layers of cells, each of which plays an important role in the processing of visual stimuli. Similar to other sensory systems, the layer that first receives the visual stimuli consists of sensory receptor cells. The distribution of photoreceptors in the retina varies (Davson, 1980). Cones are found mostly in the central foveal regions and their number decreases sharply toward the periphery. Cones function best in bright light conditions and have the pigments and neural connections that allow for the perception of colour and for high visual acuity. Rods are abundant in the periphery, less so in the central regions, and become non-existent in the foveola or centre of the fovea. Rods function optimally at night, or whenever illumination is low (Davson, 1980).

In the foveal region of the retina where cones are abundant, there is often a one-to-one relationship between cones and bipolar cells, found in the next nuclear layer in the retina involved in the processing of visual information. This one-to-one relationship continues to the ganglion cell layer (Davson, 1980). In the fovea, a ganglion cell might receive information from one, two or three cones. In contrast, in the periphery, a ganglion cell might receive information from as many as 200 rods. Functionally, this many-to-one mapping between rods and ganglion cells in the peripheral region of the retina provides low spatial resolution. Conversely, the one-to-one relationship between the cones, bipolar and ganglion cells in the foveal region functionally allows for the mediation of fine visual acuity.

Further differences between peripheral and central retinal regions have been noted at the ganglion cell layer, primarily for cats and non-human primates (Enroth-Cugell & Robson, 1966). Although, X, Y and W cells were first identified in the cat retina (see Lennie, 1980; Rodieck, 1979 for reviews), there is now evidence of the existence of comparable classes in the retina of non-human primates. The terms X-like, Y-like and W-like are used to describe the different classes of ganglion cells in primates. The vast majority of ganglion cells are X-like cells. They are highly concentrated in and near the fovea and project to the lateral geniculate body, not to the superior colliculus. The characterisitc features of X-like ganglion cells such as small receptive fields indicate that X-like cells are likely to be the source of information essential for form or object vision (deMonasterio, 1978a, b; Schiller & Malpeli, 1977).

The Y-like cells project to both the lateral geniculate body and the superior colliculus. They are much fewer in number and are mostly found in the region of the peripheral retina. The characteristics features of Y-like ganglions cells such as large receptive fields and sensitivity to rapidly moving stimuli are consistent with a functional role in the detection and localization of stimulus motion and change (deMonasterio, 1978a, b; Schiller & Malpeli, 1977).

As few as 10% of the ganglion cells are W-like cells. Because these cells have been shown to conduct information slowly and respond poorly to visual stimuli, it has been suggested that the W-like ganglion cells have no important role in either object vision or in the detection or localization of stimulus change (deMonasterio, 1978a, b; Schiller & Malpeli, 1977).

In addition to the above structural differences, neurodevelopmental differences are noted between central and peripheral regions of the retina. The retina is fully developed at birth except for the region of the fovea. The foveal area is not completely functional until approximately 6 months after birth (Hollenberg & Spira, 1972; Mann, 1964; Yuodelis & Hendrickson, 1986). It is interesting to note that control of independent finger movements also develops in the last half of the first year. Although independent finger control develops regardless of the state of the visual system (e. g., blind people can type), it would be of interest to examine more carefully whether and how acquisition of independent finger control in grasping is altered when foveal development is abnormal.

Because the high degree of specialization established at the fovea is maintained over only a small area of retina (less than 1% of the retina by area), we have developed a complex oculomotor control system to keep the image of objects of interest on the fovea of both eyes. For foveation, coordination between the somatic and autonomic motor systems is required for motor control of extraocular and intraocular muscles. Moreover, large proportions of the visual centres of the brain are dedicated to the processing of information coming from the small foveal region of the retina. The above illustrates the complexity of visuomotor integration for processing information projected to foveal regions of the retina.

Continuing along the visual pathway, it is noted that the foveal region of the retina dominates the retinotopic representation in the lateral geniculate body and visual cortex. In contrast, only peripheral retina is represented in the superior colliculus. Beyond the area of the occipital lobe, central and peripheral vision show further processing differences. Central visual pathways predominantly project to areas in the temporal lobe (Ungerleider & Mishkin, 1982). This area has been implicated in the processing of visual information related to form perception or what an object is. Both peripheral and central visual pathways project to the parietal lobe (Ungerleider & Mishkin, 1982). This area has been implicated in the processing of visual information related to the location of an object. In their study of optic ataxia, Perenin and Vighetto (1988) showed deficits both in limb transport and hand posturing with damage to posterior parietal areas.

Thus, based on the anatomy and physiology of the visual system, central vision is functionally specialized for responding to spatial patterning of a stimulus. In contrast, peripheral vision is functionally specialized for responding to location and movement stimuli. These differences provide a rationale for examining the contributions made by peripheral and central vision to prehension since this task involves the processing of visual information related to the movement of the arm, location of the object and the object's characteristics (size, texture etc.).

Contributions of Central and Peripheral Vision to Arm Movements: Evidence from Aiming and Pointing Tasks

Reaching to grasp an object requires that the arm be directed to a specific location in space. This goal is similar to that required in aiming or pointing. There is little doubt as to the importance of visual information in maintaining movement accuracy. Studies that have manipulated "what" is seen during the movement have shown that vision of both the hand and target resulted in the greatest accuracy compared to vision of just the target or hand (Carlton, 1981; Elliott & Allard, 1985; Jeannerod & Prablanc, 1983; Prablanc, Echallier, Jeannerod, & Komilis, 1979; Proteau & Cournoyer, 1990). In these cases, vision of both the hand and target allowed for the best visual error information. It is important to point out that, by allowing subjects to see both the hand and target, information was provided through both peripheral and central vision. As well, Elliott and Allard (1985) showed that without vision of the arm during the movement to the target, subjects undershot the target. Here, visual feedback information about the arm, normally provided through peripheral vision was not available. Prablanc et al. (1979) reported that vision of the hand in its starting position before movement begins also affects movement accuracy. Considering that the hand is normally seen with peripheral vision before movement begins, it is logical to suggest that peripheral vision may provide important information for the planning and control of arm movements to specific locations in space.

Furthermore, the importance of "when" visual information is available during the movement has been shown to be relevant in determining movement accuracy. Paillard (1982) and Beaubaton and Hay (1982) showed that subjects benefitted most from visual feedback information provided toward the final phase of the movement. At this time, both the hand and the target would be seen in central vision, and a comparison could be made between the location of the target and the location of the hand. However, they showed that visual feedback information provided at the beginning of the movement also improved accuracy over conditions when no visual feedback information was provided. At this time, the target would be in central vision and the hand would be in peripheral vision. As suggested above, it appears that when the hand is visible in peripheral vision, information is provided which contributes to the final accuracy of the movement.

The experimental results examined above suggest that peripheral and central vision may provide important information in the planning and control of arm movements directed to a specific location in space. Although it was shown that accuracy was best when the hand and target were both visible throughout the movement (both central and peripheral vision would thus be used), the results indicate that visual information provided at the beginning of the movement (peripheral vision would thus have been used) also contributes to controlling movement endpoint accuracy.

While it has been demonstrated that peripheral and central vision contribute to accuracy in the control of arm movement, the question remains as to what their specific roles are. The question of whether peripheral vision is involved in the directional control of hand trajectory during an aiming task performed at high speed was experimentally addressed by Bard, Hay, and Fleury (1985). This was accomplished by manipulating whether and when the hand was seen during its trajectory to the target, while subjects maintained fixation of the target. Their results indicated that when subjects did not see their hand in the terminal phase of movement, a directional correction of arm trajectory could be made by on-line feedback from peripheral vision, even for movements of 110 ms duration. However, their results also indicate that subjects could adjust the direction of the arm when the hand was seen in central vision near the end of the movement, if the movement was slow enough to allow time for movement adjustments. These results suggest to us that, in addition to peripheral vision, central vision also can be used to monitor and change arm direction.

The question of how accuracy is affected when the target is viewed through either peripheral or normal vision (i.e., includes central and peripheral vision, but with central vision used to view the target) was addressed by Prablanc et al. (1979) and MacKenzie, Sivak, and Elliott (1988). Their results indicated that, with vision of the target but not the hand, subjects were just as inaccurate in pointing to a target when they viewed the target with normal vision as when they viewed the target with only peripheral vision. These results suggest that, for locating a target without vision of the limb, it does not matter whether or not the target is seen clearly (i.e., with optimal visual acuity). The crucial variable that appears to predict the accuracy of a pointing movement is whether or not the arm and target are in view throughout the movement. With vision of the target but not the moving limb, proprioceptive information from the limb is inadequate to accurately move the limb to the visible target. In accordance with Paillard and Amblard (1984) and Bard et al. (1985), peripheral vision may be important for providing necessary information about limb velocity and direction needed for arm movement control. In contrast, it has not been determined if information from central fields is necessary for "homing-in" operations of the hand to the target, i.e., for feedback comparisons of hand and target locations.

Contributions of Central and Peripheral Vision to Arm and Hand Movements: Evidence from Reaching and Grasping Tasks

The above results hinted that peripheral and central vision might play important and perhaps different roles in motor planning and control of arm movements directed at specific locations in space. Aiming and pointing movements require directing the arm to a selected location in space. Reaching to grasp an object also has a similar demand. However, in prehension, the hand also plays an important effector role. While the arm is being transported toward the object, the hand preshapes to some maximum aperture, then encloses so that the object can be grasped. Therefore, consideration to movements of the hand must be included in any comprehensive analysis of reaching and grasping movements.

The experiments described in this section were undertaken in our laboratory for the purpose of quantifying experimentally the specific contributions of peripheral and central vision in reaching and grasping movements (MacKenzie & Van den Biggelaar, 1987; Sivak & MacKenzie, 1990). Considering that central vision is believed to be important in grip formation through processing information about intrinsic object characteristics, we wanted to manipulate object size in these experiments. At the outset, we wanted to compare moving and stationary targets, as peripheral vision appears to be optimally specialized for processing information about object motion and location. Typically, when we grasp an object, we foveate the object (i.e., turn the head and eyes to look at the object so that it is projected onto the fovea of the retina). In our first study, we investigated whether it was important if the object to be grasped was seen in central or peripheral vision. We instructed subjects to look at the object or to look at a stationary marked point, so that the object to be grasped was in central or peripheral vision, respectively. Given the results from this initial study, we were motivated to examine more explicitly the specific contributions of peripheral and central vision on the transport and grasp components, when grasping stationary objects of varying sizes.

In all experiments reported in this chapter, three dimensional kinematic data for reaching and grasping were collected at 200 Hz using the WATSMART (Waterloo Spatial Motion Analysis and Recording Technique, Northern Digital Inc.) system. Infra-red emitting diodes (IREDs) served as markers and were positioned on the wrist, index fingernail and thumbnail as well as on the top centre of the object. The wrist IRED was used to indicate the transport component. The index finger and thumb IREDs provided a measure of the aperture between the index finger and thumb, to indicate the grasp component.

Analyses involved examination of peak kinematic values for both transport and grasp components. We looked at peak velocity and peak deceleration for the wrist IRED and maximum aperture between the thumb and index finger IREDS. The study of trajectories of movement has been used by various researchers to make inferences about central nervous system operations (Atkeson & Hollerbach, 1985; MacKenzie, Marteniuk, Dugas, Liske, & Eickmeier; 1987; Sivak & MacKenzie, 1990; Soechting & Lacquaniti, 1983). Systematic variances and invariances in these kinematic profiles over various experimental manipulations may be a reflection of neural mechanisms subserving motor planning and control processes. As well, we provided a qualitative descriptive analysis based on visual examination of the trajectory profiles.

Reaching and Grasping Stationary *vs*. Moving Targets
At the outset, we designed an experiment to evaluate the effects of visual information, object motion and object size on the transport kinematics of the

wrist IRED, aperture between the index finger and thumb IRED, and kinematic features of the thumb and index finger trajectories. Our reasoning at the time was that if the transport and grasp components are organized through independent, parallel visuomotor channels, then experimental manipulations should selectively affect the transport or grasp components.

Healthy, adult subjects were instructed to grasp and lift a cylindrical disk. They were seated unrestrained at a table. In all conditions, subjects were instructed to grasp and lift the object when it was in front of the start position along the sagittal plane, 27 cm in front of the body midline. They were instructed to grasp between the pads of the thumb and the index finger: a small disk (3 cm diameter, 2 cm height) or large disk (6 cm diameter, 2 cm height).

The three "grasping conditions" involved both target motion and visual information about the target object as follows: (a) stationary target - central vision, in which subjects had normal vision and foveated on the stationary disk, placed on the treadmill surface, 27 cm in front of the body midline; (b) moving target - central vision, in which subjects had normal vision and foveated the disk as it approached on a treadmill moving from left to right (with head and eyes free to move); and (c) moving target - peripheral vision, in which subjects had normal vision but fixated on a point marked on the treadmill frame, 60 cm opposite the body midline); thus they saw the disk moving from left to right in peripheral vision. The last two grasping conditions required coincident timing. Trials on which subjects made saccades to the target were discarded in the moving target - peripheral vision condition. In the two moving target conditions, the treadmill speed was 53 cm/s; subjects saw the disk moving for about 1.5 s before grasping. The two disk sizes were crossed with three grasping conditions, for a total of six experimental conditions. The order of the grasping conditions and target sizes was randomized for each subject.

With respect to the grasp component, consistent with other results (Jeannerod, 1981; Marteniuk et al., 1990; Von Hofsten & Ronnqvist, 1989), maximum aperture was bigger for the large (112 mm) than the small (88 mm) disk. Figure 10.1a shows that the effect of disk size was additive to grasping conditions. The maximum apertures were larger with peripheral vision (108 mm) than with central vision, regardless of whether the object was moving (96 mm) or stationary (96 mm). All subjects showed these effects. Further, compared to the stationary target and moving central vision condition, we observed that almost all subjects tended to use fingers additional to the index finger in opposition to the thumb in the moving target - peripheral vision condition. This strategy may have served to increase the probability of grasping the object and enhancing grasping stability.

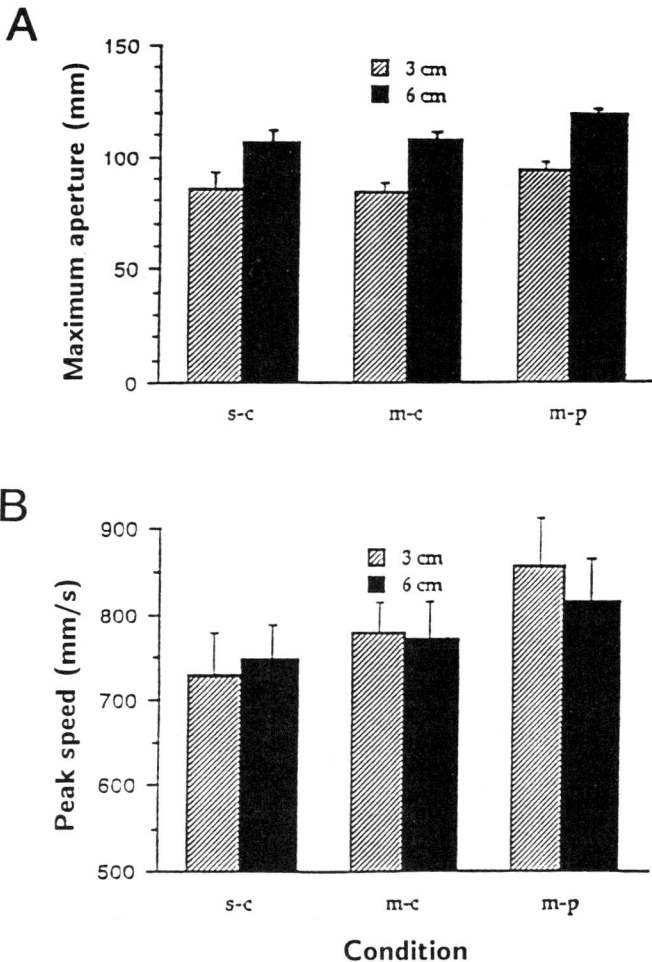

Figure 10.1. (A) Grasp component: Maximum aperture (mm) between the thumb and index finger markers. Maximum apertures were larger in the moving target-peripheral vision condition (m-p) than the stationary target-central vision (s-c) and the moving target-central vision (m-c) conditions. The two central vision conditions did not differ. (B) Transport component: Peak speed or peak tangential velocity (mm/s) of the wrist marker. Peak speed was higher for the moving target-peripheral vision condition (m-p) than the moving target-central vision condition (m-c). Peak speed was lowest in the stationary target-central vision condition (s-c).

Peak speed of the wrist along the path of the trajectory was fastest in the moving target - peripheral vision (838 mm/s) condition, then the moving target - central vision (778 mm/s) condition, then the stationary target - central vision (739 mm/s) condition (see Figure 10.1b). The shape of the time normalized resultant velocity profile was invariant across conditions (i.e., subjects spent about 58% of the movement duration in the deceleration phase). Wrist kinematics were unaffected by target size, i.e., target size affected the grasp but not the transport component. In contrast, target motion affected the transport but did not affect the grasp component.

Separate analyses of the thumb and index finger trajectories revealed that the thumb showed the same effects as the wrist (similar to Wing, Turton, & Fraser, 1986). In contrast, the resultant velocity profile of the index finger was affected by all experimental variables. This appears to contradict the statement above, that target motion affected the transport, but did not affect the grasp component. We think that the result indicates that the index finger is preserving aperture for the task, while taking into account transport changes. Thus, aperture may be a "higher level" control variable than individual finger control.

These results suggested to us that when subjects were using peripheral visual information about the object to be grasped, there were greater demands on aperture calibration, because the maximum aperture was always bigger for peripheral vision than central vision conditions. Whether subjects saw the disk in central or peripheral vision also seemed to affect the transport component, i.e., subjects moved faster when the moving target was in peripheral vision than when the moving target was in central vision.

We conclude that indeed central and peripheral vision are making differential information processing contributions to the control of reaching and grasping. In our vision conditions, subjects had field of view information from both central and peripheral visual fields. Thus, we were unable to make specific conclusions about the relative contributions of central and peripheral vision. At this point, we decided to use a simpler grasping task without target motion, i.e., to examine only stationary targets. Our primary objective for the following experiments was to isolate central and peripheral vision so that we could evaluate the specific contributions of each, in the absence of the other. In order to do this, we developed optical methods to selectively restrict visual information to either central or peripheral regions of the retina.

Restricting Visual Information to Peripheral and Central Regions of the Retina

Examination of the specific contributions of peripheral and central vision to motor planning and control of reaching and grasping was made possible

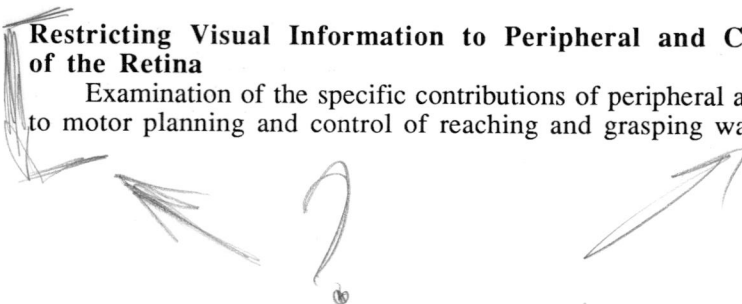

through physical, optical separation of peripheral and central visual fields. A special contact lens system was developed to isolate peripheral vision. This system used a piggyback contact lens design in which a partially painted hard lens was positioned on top of a special soft lens carrier (for details see Sivak, Sivak, & MacKenzie, 1985). The hard lenses were painted black in the center so that when lenses were inserted in both eyes, visual field testing with static perimetry revealed that no visual information was available through the center 10 degrees of the visual field. With the lenses positioned in the eyes, approximately 10 degrees of the central field remained as a scotoma and therefore could not be used to provide visual information even when the eyes or head moved.

To isolate central vision a goggle system was developed that allowed for 10 degrees of binocular central vision (for details see Sivak et al., 1985). This system used swimming goggles that were painted black on the sides to eliminate light. The front of the goggles contained slightly depressed circles into which were fitted circular black opaque disks. Each disk had a 0.5 mm aperture that was displaced 4 mm medially from the centre of the disk. With the goggles on, static perimetry tests confirmed that subjects could use only approximately the central 10 degrees of visual field. The rest of the visual field remained as a scotoma.

The following sections detail prehension experiments conducted using the lenses to restrict information to the peripheral regions of the retina or using the goggles to isolate central vision.

Role of Peripheral Vision in Reaching and Grasping
Using either only peripheral vision or normal vision, subjects used the pads of the index finger and the thumb to reach for, grasp and lift a wooden dowel (11.5 cm high; 2.5 cm diameter) positioned on the table in front of them. The subject's head was supported in a chin rest to stabilize the head in a straight ahead position. However, when wearing the contact lenses, if subjects looked straight ahead they could not see the dowel. Subjects thus needed to turn their eyes to see the dowel with the remaining peripheral visual field of both eyes. All visual conditions were conducted using binocular vision. As a consequence of the experimental procedure to isolate peripheral vision, high resolution vision (mediated by the central regions of the retina) was excluded. This had an important effect on the organization of the transport and grasp components and their coordination.

With respect to the transport component, Figure 10.2a shows a lower peak velocity with peripheral than normal vision. In addition to moving slower, a qualitative visual examination of the time normalized wrist acceleration curves for only peripheral vision and normal vision revealed a

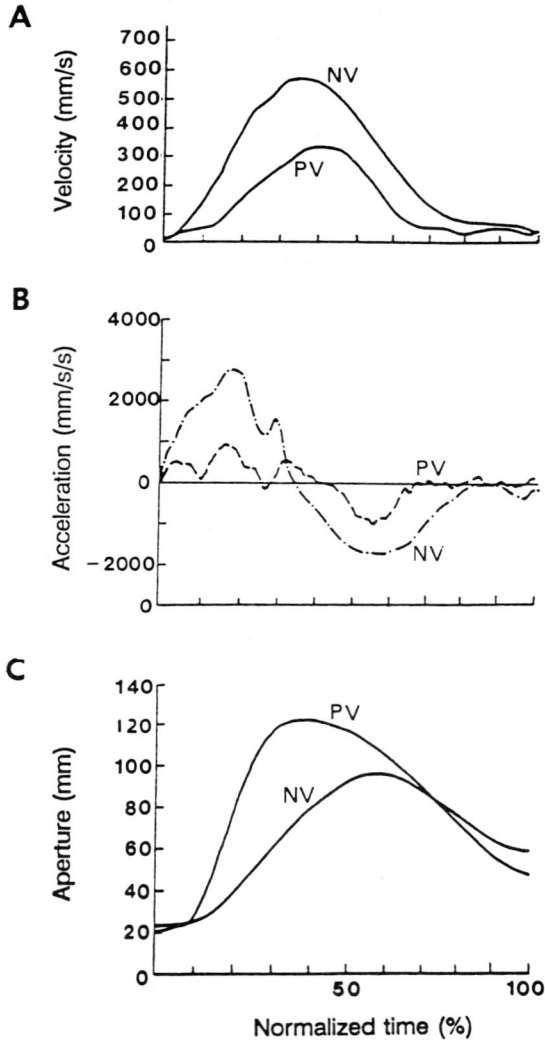

Figure 10.2. Mean curves for one subject for (A) velocity, (B) acceleration and (C) aperture for peripheral vision (PV) and normal vision (NV). In B, note the difference in the shape of the curves after peak deceleration.

difference in the shape of the curves (Figure 10.2b). This difference was particularly evident in the portion of the curve after peak deceleration. In contrast to normal vision, with only peripheral vision subjects appeared initially to decelerate rapidly and then maintained a near 0 acceleration. Further analyses related to this shape difference revealed that the forward movement of the arm toward the dowel stopped earlier with only peripheral vision compared to normal vision. For only peripheral vision, the wrist IRED stopped moving near the dowel when approximately 72% of the movement duration was completed. In contrast, with normal vision, the wrist IRED stopped moving when approximately 91% of the movement duration had been completed. At this point in the analysis, the question was asked "why did the arm stop moving forward earlier with peripheral vision than with normal vision"? The answer to this question lies in understanding what was happening to the hand while the arm was being moved forward toward the object.

To try and answer the above question, the size of the aperture was noted when the wrist IRED stopped moving forward for both peripheral and normal vision. These apertures showed that, at the time that the arm stopped moving, the size of the aperture was wider with only peripheral vision than with normal vision (Figure 10.3). All subjects showed this effect. Thus, subjects appeared to be using a different strategy to complete the actual grasp with only peripheral vision compared to normal vision. With only peripheral vision, subjects were opting to close the hand after the arm had stopped moving. Further analysis revealed that this movement strategy was adopted to allow for tactile information from either the thumb or index finger making contact with the dowel to help in completion of the actual grasp. This initial contact triggered the actual grasp. Johansson and Westling (1987) found that the minimum response latencies of the distal hand muscles to tactile stimuli in the adult was approximately 60-80 ms. As the time from contact with the dowel by either the thumb or the index finger to the time of dowel lift exceeded 80 ms in the peripheral vision condition, this explanation is plausible. In normal vision, this did not seem to be the case; the small aperture at the time that the wrist IRED stopped moving forward suggested that the hand was enclosing while the arm was moving to the dowel. With normal vision, both digits contacted the dowel at approximately the same time (Figure 10.4).

Not only did subjects not adjust the size of the aperture between the index finger and thumb until initial contact was made with the dowel but, in addition, analyses also revealed that subjects opened the hand wider with only peripheral vision (116 mm) than with normal vision (86 mm). Jeannerod (1981) has suggested that the grasp component is organized from visual information about intrinsic properties of an object. For the index finger and thumb precision grasp used by subjects in this study, good visual acuity is probably required.

Therefore, we suggest that the grasp component was affected because high resolution vision was inadequate with only peripheral vision compared to normal vision.

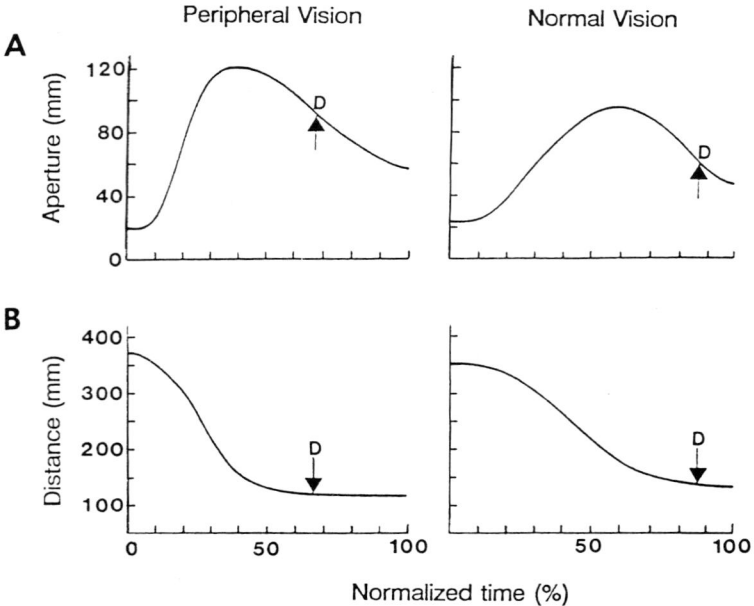

Figure 10.3. Mean curves for one subject for (A) aperture size and (B) distance between wrist IRED and dowel IRED. The point when this distance did not change was used to indicate when the arm had stopped moving forward (D). Note that when the arm stopped moving forward, the size of the aperture is larger for only peripheral vision than for normal vision.

It is interesting to note that the effects on the transport and grasp components did not accommodate to the visual manipulation over the sixteen experimental trials. Subjects repeatedly used the same movement strategy to approach and grasp the dowel when they had only peripheral vision. This is in contrast to experiments with laterally displacing prisms in which subjects accommodate after two or three trials and redirect the arm to maintain

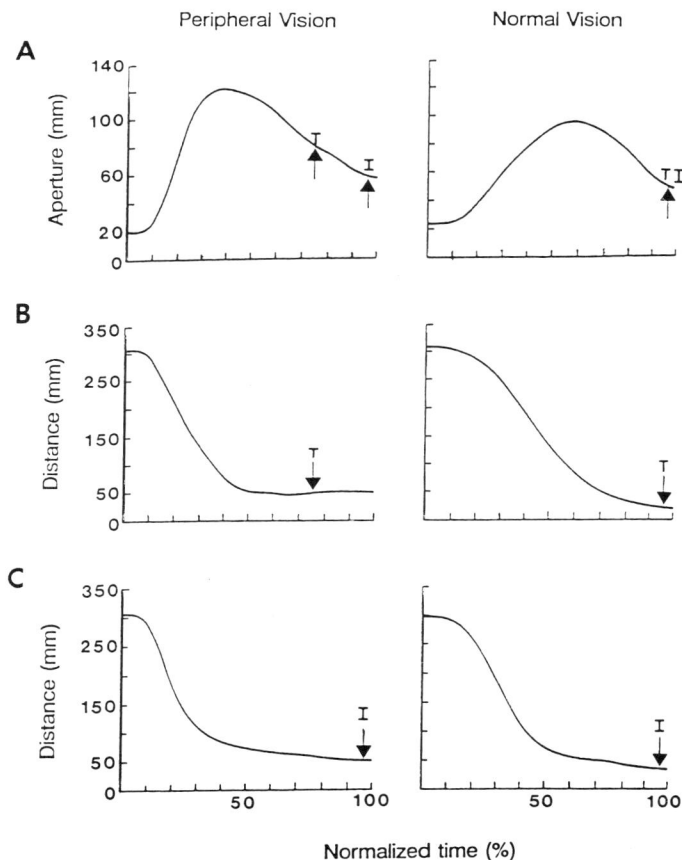

Figure 10.4. Mean curves for one subject for (A) aperture size, (B) distance between thumb IRED and dowel IRED and (C) distance between index finger IRED and dowel IRED. T and I indicate respectively when the distance between the thumb IRED and dowel IRED and between the index finger IRED and dowel IRED did not change. These times were used to indicate when the thumb and index finger made contact with the dowel. Note that with normal vision, both the thumb and index finger contacted the dowel at the same time. In contrast, with peripheral vision, subjects used a tactile control strategy whereby either the thumb or index finger contacted the dowel first. Note the larger aperture at the time of first contact with peripheral vision, compared to normal vision.

movement accuracy (Elliott & Roy; 1981; Held, 1968; Redding & Wallace, this volume). The current results suggests that subjects always used the available information from peripheral vision. They did not accommodate to the size of the object, even after repeated grasps.

Furthermore, we suggest that the organization of the transport component was affected with only peripheral vision because the grasp component was affected. According to Jeannerod (1981), the transport component is organized from information about the location of an object (extrinsic properties) through a visuomotor channel independent from the visuomotor channel used to organize the grasp component. The fact that the organization of the transport component could be affected by what the hand was doing suggests that there may be a communication link between the visuomotor channels for motor planning and control of the transport and the grasp components. It suggests also that the visual information about intrinsic properties of an object may provide important information for motor planning and control of the transport component.

With normal vision, the opening of the hand and the start of enclosing the hand took place while the hand was in peripheral vision. If only peripheral vision is important for the opening and closing of the hand, then the grasp component should not have been affected by the lack of adequate visual information about the characteristics of the dowel. Depressed visual acuity related to the intrinsic features of the dowel appeared to have affected the organization of the grasp component. The information of finger position via peripheral visual fields before movement begins and during the movement appears to be irrelevant to the organization of the grasp component. The data indicate that central visual information about object characteristics obtained both before and during the movement is crucial for planning and control of the grasp component.

The most obvious difference between the visual conditions was the lack of high resolution vision with only peripheral vision compared to normal vision. At this time, we suggest that the effect on the transport and grasp components with only peripheral vision were due to the lack of high resolution vision normally mediated by central regions of the retina. Although similar results might be obtained with central vision using defocusing lenses, it is important to remember that the peripheral retina is incapable of mediating high resolution vision (M.A. Goodale, personal communication, January, 1989). In addition, for grasping tasks that rely less on acuity information (e.g. large available area for finger placement, collective finger grasp types, etc.), the present results may not be achieved.

In summary, both the grasp and transport components were affected with only peripheral vision compared to normal vision. For the index finger and thumb precision-type grasp used in this study, it appears that the organization of the grasp component relied heavily on visual acuity information about intrinsic object characteristics normally mediated via central regions of the retina. The transport component was modified to accommodate the changes in the organization of the grasp component.

Sensitivity of Peripheral Vision to Object Size

The finding that maximum aperture was wider with only peripheral vision compared to normal vision suggested to us that peripheral vision might not be as sensitive to the size of an object as normal vision. We then evaluated how well subjects judged dowel size. Subjects were able to judge dowel size by placing different sized diameter dowels (diameters of 1.0, 2.5 and 5.5 cm) in ascending and descending sizes when presented in a random order, regardless of whether only peripheral or normal vision was used to view the dowels. This indicated that peripheral vision does allow for the ranked judgement of dowel size when the dowels are viewed comparatively. However, it did not explain why subjects opened the hand wider with peripheral vision compared to normal vision. Further investigation of the sensitivity of peripheral vision to object size for grasping was warranted.

Research has shown that changing the size of an object to be grasped, affects visuomotor integration processes subserving reaching and grasping (Von Hofsten & Ronnqvist, 1988; Marteniuk et al., 1990). These studies have shown that, with normal vision, the opening of the hand increases linearly with the size of the object to be grasped. With peripheral vision, how would the size of the hand opening be modulated?

Using either normal vision or only peripheral vision, subjects used the pads of the index finger and thumb to reach for, grasp and lift the 1.0, 2.5, and 5.5 cm diameter dowels presented in a random order. The effect of using only peripheral vision versus normal vision replicated the previous experiment in showing that both the transport and grasp components were affected when subjects used only peripheral vision. Again, for the grasp component, subjects opened the hand wider with peripheral vision than normal vision for all dowel sizes. This strategy may have been used to compensate for the reduced visual acuity of peripheral vision compared to normal vision.

Under normal vision, the size of the maximum aperture increased monotonically with increasing dowel size (see Figure 10.5). This agrees with Marteniuk et al. (1990), Jeannerod and Decety (1990) and Von Hofsten and Ronnquist (1988). In contrast, the results suggest that peripheral vision was not as sensitive to object size as normal vision. There are two reasons for this statement. First, subjects opened the hand wider for all sizes of dowels with peripheral vision compared to normal vision, indicating a problem with gauging the size of the dowel for grasping (Figure 10.5). Second, although subjects showed an increase in the size of the maximum aperture from smaller to larger sized dowels, the increase was small with only peripheral vision compared to normal vision. Aperture size was larger for the 5.5 cm dowel than for the 1 cm dowel; however, the increases in aperture size from the 1 cm dowel to the 2.5 cm and from the 2.5 cm dowel to the 5.5 cm dowel were not significant. This seems to indicate that, with only peripheral vision, small differences in size are not readily detected compared to large differences, for the purpose of calibrating aperture.

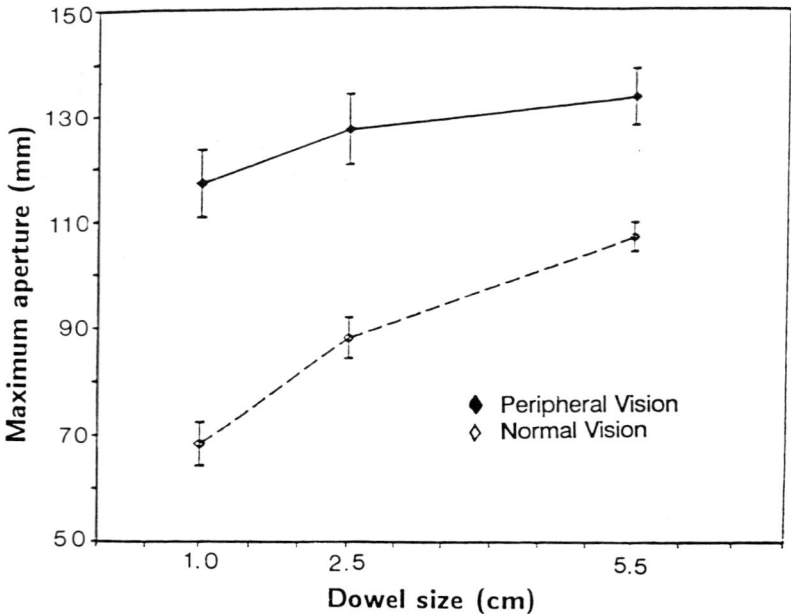

Figure 10.5. The effect of dowel size on maximum aperture for only peripheral vision and normal vision. Note that aperture size appears to increase monotonically for normal vision but to a lesser degree for peripheral vision.

A possible alternative explanation of these results reflects biomechanical constraints of the hand that limit the maximum span between the index finger and thumb. Given that subjects opened the hand wider with peripheral vision than normal vision, the biomechanical limit would have been reached earlier with peripheral vision than normal vision. If peripheral vision was as sensitive to object size as normal vision, then we might have expected to see a significant increase in the size of the maximum aperture between the 1 cm and 2.5 cm dowel but not between the 2.5 cm and 5.5 cm. This is because the biomechanical limit may have been reached for the 2.5 cm dowel. This did not occur. The results did not reflect this biomechanical constraint. Thus, peripheral vision is less sensitive than normal vision in providing size information for calibrating grasping aperture. Perhaps if a greater range of dowel diameters was included in the grasping task (for example .5, 1.0, 1.5, 2.0,..., 10 cm), the decreased sensitivity of peripheral vision to dowel size would be more evident.

Role of Central Vision in Reaching and Grasping

Using the same task, subjects were asked to reach, grasp and lift a wooden dowel (11.5 cm height, 2.5 cm diameter) using either normal vision or only central vision. With the goggles on, the arm and hand were visible neither at the start of the movement nor throughout the reaching movement. The hand came into view when it approached the dowel at the end of the movement. However, the dowel was always visible with the goggles on.

The time normalized wrist kinematic data suggested that the organization of the transport component was affected by the visual manipulation (see Figure 10.6). Subjects spent a greater percentage of movement time after peak deceleration with only central vision (52%) compared to normal vision (39%). Furthermore, pilot observations of subjects reaching for the dowel as fast as possible showed that they consistently undershot the dowel. That is, they always judged the dowel to be closer than it was. The experimental analysis together with these observations suggest that without a large field of view provided by peripheral vision, subjects did not estimate the distance to the dowel accurately. Proprioceptive information from the moving limb provided inadequate information as to where the limb was in relationship to the object. Perhaps proprioceptive information is decalibrated when visual information is disturbed (Soechting & Flanders, 1989a, b). The parameters of the transport component were organized based on information that suggested a nearer target than in reality.

It appeared that subjects gauged the dowel to be closer but when they reached the judged position of the dowel, they had to move further. Possibly these adjustments were based on a comparison type of analysis where the position of the hand relative to the target was deemed incorrect, thus requiring a further forward movement of the arm. Possibly, they simply continued until anticipated tactile contact was achieved. Evidence for these possibilities is

B. Sivak and C.L. MacKenzie

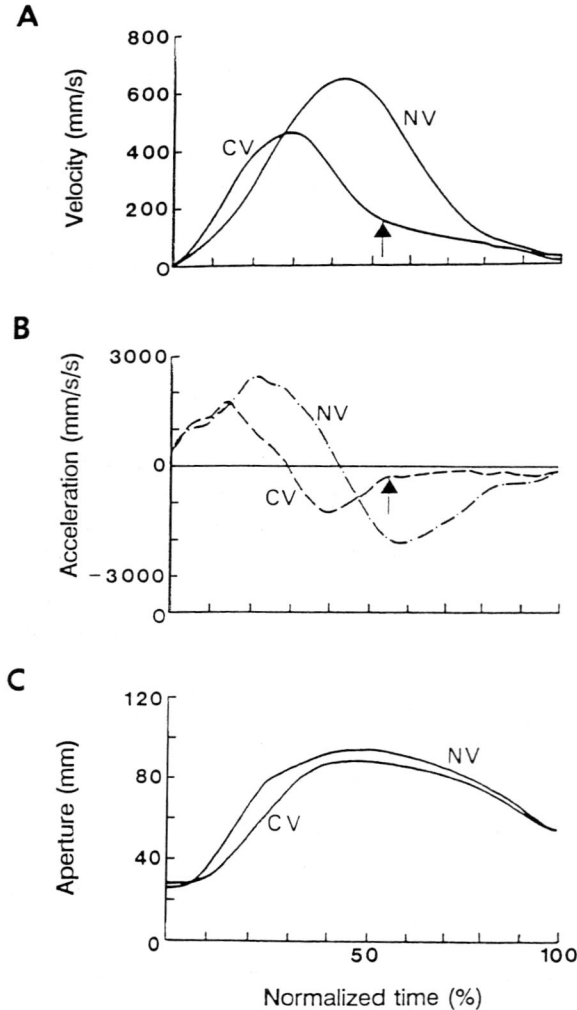

Figure 10.6. Mean curves for one subject for (A) velocity, (B) acceleration and (C) aperture for central vision (CV) and normal vision (NV). Note, that with only central vision, the area of velocity and acceleration curves representing the time spent in deceleration appears to have two phases. The beginning of the second deceleration phase is indicated by an arrow. The two phases are less distinct with normal vision.

provided by the fact that the increase in movement time occurred after peak deceleration, and by the two phases apparent in the deceleration profile (Figure 10.6b). The two phases suggest that subjects planned to end the movement earlier than they did. That is, when subjects did not reach the dowel when they expected to, they continued to decelerate at a constant rate until tactile contact was made with the dowel. It is important to note that these were not submovements indicated by changes in acceleration and deceleration, but rather a constant deceleration. This suggested to us that subjects were moving forward anticipating tactile contact, rather than making a series of corrections.

Consistent with the above is the subjective responses of the subjects. All subjects reported that the dowel appeared closer to them than it was. Surprising to us, however, was that subjects did not adapt to this illusion over the sixteen repeated trials. This is in direct opposition to the findings of prism experiments which showed that subjects adapted to the directional displacing prisms (Elliott & Roy, 1981; Held, 1968; Redding & Wallace, this volume). However, the prism experiments have looked at accuracy as a dependent measure. In this study, subjects were just as accurate with only central vision as with normal vision. That is, they always picked up the dowel. Following up the prism studies with 3-D kinematic analyses of the movements may reveal a different set of results from the previous end-point accuracy results.

In agreement with the suggestion made by Jeannerod (1981, 1984) that the grasp component was organized from visual information about object characteristics, the results showed that the grasp component was *not* affected by having only central vision compared to normal vision (Figure 10.6c). There were no differences in the size of the maximum aperture and the relative time of when maximum aperture occurred between the vision conditions. This is important for two reasons. First, with only central vision, there was no visual information of the position of the fingers both before the movement began and during the movement as is present in normal vision situations. This suggests that visual information for hand and finger posturing is acquired from the object. On-line visual feedback information of finger position while the hand is approaching the dowel is not necessary for planning and control of the grasp component. Peripheral vision appears to be unnecessary for the planning and control of the grasp component.

Second, the fact that the organization of the transport component was affected by only central vision and the organization of the grasp component was not, lends support to the notion that the grasp and transport components are organized through independent visuomotor channels as suggested by Jeannerod (1981). The fact that subjects gauged the dowel to be closer did not affect when, or how much they opened the hand. Furthermore, the transport component showed a two-phase deceleration profile with only central vision.

The grasp component did not demonstrate these two phases, nor was it different with only central vision from normal vision. This means that subjects based the organization of the grasp component on visual information received from the size of the dowel (object characteristics) and did not need the visual information that the transport component needed for its organization.

The major difference between the vision conditions was the small visual field size with only central vision compared to normal vision. The small field size reduced the amount of location information about the dowel and on-line visual feedback information from the moving limb. Although it has been suggested that peripheral vision is used to provide information necessary for directing the limb to the proper location of an object (Bard et al., 1985; Paillard, 1982; Paillard & Amblard, 1985; Paillard & Beaubaton, 1979), subjects in the present experiment had no difficulty with the direction, only the distance of the movement in the absence of peripheral vision. The lack of a significant difference between the vision conditions in deviation from the midline plane attests to this statement. It has been shown that subjects can use information from the peripheral visual fields to adjust the direction of an arm movement (Bard et al., 1985). Because the present study had only one known direction to move, it is not clear that peripheral vision was necessary to assess the direction of movement. In the absence of peripheral vision, eye and head position may provide sufficient information. The results from this study suggest that direction and distance are separable entities of location for grasping an object.

In summary, with only central vision the organization of the transport component was affected compared to normal vision. It appears that the organization of the transport component relies on important cues provided through a wide field of view (both environmental cues and information from the moving limb) which are lacking with only central vision. This information appears to be unimportant for the organization of the grasp component. Central vision appears to provide adequate visual information for organization of the grasp component.

Concluding Remarks and Recommendations for Further Research

It has been established that, compared to normal vision, having only peripheral or only central vision differentially affects reaching and grasping movements. As a result of the low resolution acuity with peripheral vision, it seems that with only peripheral vision, information related to the intrinsic features of a dowel is poorly relayed, thus affecting the planning and control of both the grasp and transport components and their coordination. Because central vision lacks a wide field of view, information related to extrinsic properties of the dowel is inadequate, thus affecting the organization of the transport component (especially the distance), and not the grasp component.

We can speculate on the role of peripheral and central vision when individuals reach to grasp an object with normal vision. Peripheral vision provides important environmental information about object motion and distance, as well as information from the moving limb, needed for the planning and control of the transport component. Central vision provides important information required for the organization and control of both the grasp and transport components. The high resolution image provided by central vision allows for the accurate calibration of grasp aperture for intrinsic object characteristics.

Furthermore, the current view of visuomotor channels, that only visual information about the extrinsic properties of an object are used to organize the transport component, needs to be updated. We suggest that both extrinsic and intrinsic properties about an object can affect the organization of the transport component. That is, processing related to intrinsic properties of an object, such as the clarity of the visual image, may affect the control parameters of the transport component as well as the grasp component. In contrast, visual information about the extrinsic properties of an object does not appear to influence the organization of the grasp component. This does not dispute the existence of independent visuomotor channels but rather suggests that there may be a communication link between them. This communication link may only work one way. That is, when the grasp component is affected, it also affects the organization of the transport component. It is as if the hand "drives" the arm. In contrast, when the organization of the transport component is affected, the organization of the grasp component is not.

More research is needed in the area of visuomotor integration processes. Although it has been shown that peripheral and central vision make specific contributions when reaching to grasp and object, there are many questions that need further empirical exploration. For example, it was shown that tactile information is important when objects are grasped in peripheral vision. Under peripheral vision conditions, how is tactile information used? What information is acquired? How long does it take to achieve stable grasp? What type of grasp is most commonly used? Furthermore, it was shown that under conditions of only central vision, subjects have difficulty estimating the distance to the object but not the direction. Further investigation elucidating the roles of peripheral and central vision in the control of distance and direction of arm movements, especially under conditions of distance and direction uncertainty is needed. Also, as individuals usually foveate the object to be grasped it is important to examine the role of central vision in "homing in" operations before contact is made with the object. An understanding of the nature of these "homing in" operations is of interest.

From a neurodevelopmental and neuropathological perspective, several research questions deserve attention. Congenically blind infants learn to reach and grasp objects. How does reaching and grasping develop in these infants? In the same vein, what is the effect of abnormal development of the eye on the development of reaching and grasping behavior? It was stated earlier that the fovea develops after birth along with the development of independent finger control. How does abnormal development of the fovea affect the emergence of independent finger control for grasping?

Evidence from special populations has shown that lesions in the central nervous system can affect reaching and grasping behavior. It is important to investigate how patients with parietal, temporal and frontal lesions differ in their reaching and grasping behavior.

Finally, it has been documented that aging results in a decrease in the ability to accommodate, a decrease in the sensitivity of mechanoreceptors and frequently the loss of finger flexibility. The question can be asked how are visuomotor integration processes in reaching and grasping affected by these changes.

References

Atkeson, C.G., & Hollerbach. J.M. (1985). Kinematic features of unrestrained vertical arm movements. *Journal of Neuroscience, 5,* 2318-2330.

Bard, C., Hay, L., & Fleury, M. (1985). Role of peripheral vision in the directional control of rapid aiming movements. *Canadian Journal of Psychology, 39,* 151-161.

Beaubaton, D., & Hay, L. (1982). Integration of visual cues in rapid goal-directed movements. *Behavioral Brain Research, 5,* 92-105.

Carlton, L.G. (1981). Visual information: The control of aiming movements. *Quarterly Journal of Experimental Psychology, 33A,* 87-93.

Davson, H. (1980). *Physiology of the eye.* New York: Academic Press.

deMonasterio, F. M. (1978a). Properties of concentrically organized X and Y ganglion cells of macaque retina. *Journal of Neurophysiology, 41,* 1394-1417.

deMonasterio, F.M. (1978b). Properties of ganglion cells with atypical receptive-field organization in retina of macaque. *Journal of Neurophysiology, 41,* 1435-1449.

Elliott, D., & Allard, F. (1985). The utilization of visual feedback information during rapid pointing movements. *Quarterly Journal of Experimental Psychology, 37A,* 407-425.

Elliott, D., & Roy, E.A. (1981). Interlimb transfer after adaptation to visual displacement: Patterns predicted from the functional closeness of limb neural control centers. *Perception, 10,* 383-389.

Enroth-Cugell, C., & Robson, J.G. (1966). The contrast sensitivity of retinal ganglion cells of the cat. *Journal of Physiology, 187*, 517-552.

Held, R. (1968). Plasticity in sensorimotor coordination. In S. J. Freedman (Ed.), *The neuropsychology of spatially oriented behavior* (pp. 457-475). Chicago: Dorsey Press.

Hollenberg, M.J., & Spira, A.W. (1972). Early development of the human retina. *Canadian Journal of Ophthalmology, 7*, 472-484.

Jeannerod, M. (1981). Intersegmental coordination during reaching at natural visual objects. In J. Long & A. Baddeley (Eds.), *Attention and performance IX* (pp. 153-169). Hillsdale, NJ: Erlbaum.

Jeannerod, M. (1984). The timing of natural prehension movements. *Journal of Motor Behavior, 16*, 235-254.

Jeannerod, M., & Decety, J. (1990). The accuracy of visuomotor transformation: An investigation into the mechanisms of visual recognition of objects. In M.A. Goodale (Ed.), *Vision and action: The control of grasping* (pp. 33-48). Norwood, NJ: Ablex.

Jeannerod, M., & Prablanc, C. (1983). The visual control of reaching movements. In J. Desmedt (Ed.), *Motor control mechanisms in man* (pp. 12-29). New York: Raven Press.

Johansson, R.S., & Westling, G. (1987). Signal in tactile afferents from the fingers eliciting adaptive motor responses during precision grip. *Experimental Brain Research, 66*, 141-154.

Lennie, P. (1980). Parallel visual pathways: A review. *Vision Research, 20*, 561-594.

MacKenzie, C.L., Marteniuk, R.G., Dugas, C., Liske, D., & Eickmeier, B. (1987). Three dimensional movement trajectory in a Fitts' task: Implications for control. *Quarterly Journal of Experimental Psychology, 39A*, 629-647.

MacKenzie, C.L., Sivak, B., & Elliott, D. (1988). Manual localization of lateralized visual targets. *Journal of Motor Behavior, 4*, 443-457.

MacKenzie, C.L., & Van den Biggelaar, J. (1987). The effects of visual information, object motion and size on reaching and grasping kinematics. *Society for Neuroscience Abstracts, 13*, 351.

Mann, I.C. (1964). *The development of the human eye.* New York: Grune & Stratton.

Marteniuk, R.G., Leavitt, J.L., MacKenzie, C.L., & Athenes, S. (1990). Functional relationships between grasp and transport components in a prehension task. *Human Movement Science, 9*, 149-176.

Paillard, J. (1982). The contribution of peripheral and central vision to visually guided reaching. In D.J. Ingle, M.A. Goodale, & R.J.W. Mansfield (Eds.), *Analysis of visual behavior* (pp. 367-385). Cambridge: MIT Press.

Paillard, J., & Amblard, B. (1984). Static versus kinetic visual cues for the processing of spatial relationships. In D.J. Ingle, M. Jeannerod & D.N. Lee (Eds.), *Brain mechanisms of spatial vision* (pp. 299-330). La Haye: Martinus Nijhoff.

Paillard J., & Beaubaton, D. (1979). Triggered and guided components of visual reaching: Their dissociation in split-brain studies. In M. Shanani (Ed.), *The motor system: Neurophysiology and muscle mechanisms* (pp. 333-347). Amsterdam: Elsevier.

Paulignan, Y., MacKenzie, C.L., Marteniuk, R.G., & Jeannerod, M. (1990). The coupling of arm and finger movements during prehension. *Experimental Brain Research, 79,* 431-435.

Paulignan, Y., MacKenzie, C.L., Marteniuk, R.G., & Jeannerod, M. (1991). Selective perturbation of visual input during prehension movements: I. The effects of changing object position. *Experimental Brain Research, 83,* 502-512.

Paulignan, Y, MacKenzie, C.L., Marteniuk, R.G., & Jeannerod, M. (in press). Selective perturbation of visual input during prehension movements: II. The effects of changing object size. *Experimental Brain Research.*

Perenin, M.T., & Vighetto, A. (1988). Optic ataxia: A specific disruption in aspects of the deficit in reaching for objects. *Brain, 111,* 643-674.

Prablanc, C., Echallier, J.F., Jeannerod, M., & Komilis, E. (1979). Optimal responses of the eye and hand motor systems in pointing at a visual target. II. Static and dynamic visual cues in the control of hand movements. *Biological Cybernetics, 35,* 113-124.

Proteau, L., & Cournoyer, J. (1990). Vision of the stylus in a manual aiming task: The effects of practice. *Quarterly Journal of Experimental Psychology, 42A,* 811-828.

Rodieck, R.W. (1979). Visual pathways. *Annual Review of Neuroscience, 2,* 193-225.

Schiller, P.H., & Malpeli, J.G. (1977). Properties and tectal projections of monkey retinal ganglion cells. *Journal of Neurophysiology, 40,* 428-445.

Sivak, B., & MacKenzie, C.L. (1990). The integration of visual information and motor output: The contribution of peripheral and central vision. *Neuropsychologia, 28,* 1095-1116.

Sivak, B., Sivak, J., & MacKenzie, C.L. (1985). Contact lens design for lateralizing visual input. *Neuropsychologia, 23,* 801-803.

Soechting, J.F., & Flanders, M. (1989a). Sensorimotor representation for pointing to targets in three-dimensional space. *Journal of Neurophysiology, 62,* 582-594.

Soechting, J.F., & Flanders, M. (1989b). Errors in pointing are due to approximations in sensorimotor transformations. *Journal of Neurophysiology, 62,* 594-608.

Soechting, J.F., & Lacquaniti, F. (1983). Modification of trajectory of a pointing movement in response to a change in target location. *Journal of Neurophysiology, 49,* 548-564.

Ungerleider, L.G., & Mishkin, M. (1982). Two cortical systems. In D.J. Ingle, M.A. Goodale, & R.J.W. Mansfield (Eds.), *Analysis of visual behavior* (pp. 549-586). Cambridge: MIT Press.

Von Hofsten, C., & Ronnqvist, L. (1989). Preparation for grasping: A developmental study. *Journal of Experimental Psychology: Human Perception and Performance, 14,* 610-621.

Wing, A.M., Turton, A., & Fraser, C. (1986). Grasp size and accuracy of approach in reaching. *Journal of Motor Behavior, 18,* 245-261.

Yuodelis, C., & Hendrickson, A. (1986). A qualitative and quantitative analysis of the human fovea during development. *Vision Research, 26,* 847-863.

VISION AND MOTOR CONTROL
L. Proteau and D. Elliott (Editors)
© 1992 Elsevier Science Publishers B.V. All rights reserved.

CHAPTER 11

LIMB APRAXIA: A PROCESS APPROACH

ERIC A. ROY* and CRAIG HALL**

*Department of Kinesiology, University of Waterloo
Waterloo, Ontario, N2L 3G1

**Faculty of Kinesiology, University of Western Ontario
London, Ontario, N6G 1H1

Coordinated movements such as picking up a cup, hammering a nail or greeting a friend with a wave are an important part of our daily interactions with the environment. One facet of our knowledge of these movements can be examined by asking a person to pantomime the movements. Although no one has looked very carefully at how closely the pantomime resembles the actual movement, most normal adults are quite capable of demonstrating these gestures outside the context in which they naturally occur. Persons with neurological disorders, particularly those with a stroke affecting the left hemisphere, may exhibit striking impairments in these pantomime movements (e.g., Roy, 1985; Roy, Square-Storer, Hogg, & Adams, 1991). In looking more closely at how these people are asked to perform these gestures, it seems possible to dissociate between two patterns of performance, depending on whether they are asked to perform to verbal command or to imitate the gesture demonstrated by the examiner. In one pattern, performance is impaired to verbal command but improves dramatically on imitation. In the other, performance is impaired both to verbal command and on imitation (De Renzi, 1985). The observation that performance improves on imitation indicates that in some cases visual information facilitates performance of the gesture. One question to be addressed in this chapter, then, concerns the role that this visual information might play in the elicitation and control of these gestural movements. As such, this question is consistent with the general theme of this book. The other aspect of interest relates to the observation that performance of these gestures is generally impaired to verbal command. One direction we would like to explore is that this means of eliciting a gesture or response requires the person to generate the movement internally based on some representation of the required movement. The difficulty in performing the gesture in this condition may reflect a problem in generating an image of the movement which might be used to control performance.

In order to examine these questions, we will first review the work on apraxia and related movement sequencing disorders to provide a background upon which to base our discussion of the questions of interest. We will then review work on image generation and discuss how this work might help in understanding the impairment in gestural performance when the patient is asked to generate the movement from memory. Our attention will then turn to how visual information provided in the examiner's demonstration might facilitate performance of the gesture and why this performance enhancement to imitation is not observed in all cases. Finally, we will consider directions for future research.

Limb Apraxia: Modality Effects in Gestural Performance

Limb apraxia presents as an inability to perform gestures, for example, show me how to stir coffee, or show me how to salute. Although unable to perform these gestures in a clinical examination, patients are frequently able to do so when in the appropriate environmental situation, for example, when making a cup of coffee at dinner (Heilman & Rothi, 1985; Roy, 1978, 1982, 1983; Roy & Square-Storer, 1990). Limb apraxia is defined both by exclusion and inclusion. By exclusion the disorder is defined as apraxia only if verbal comprehension disorders, visual recognition disorders, basic motor control impairments and generalized impairments to cognitive-behavioral function (dementia) can be ruled out. The definition by inclusion places particular emphasis on the types of errors which are characteristic of apraxia. These errors have been described in detail elsewhere and will not be considered at length here (e.g., Friesen, Roy, Square-Storer, & Adams, 1987; Haaland & Flaherty, 1984; Rothi, Mack, Verfaellie, Brown, & Heilman, 1988; Roy, Square, Adams, & Friesen, 1985).

Limb apraxia, in a general sense, involves two major types: those which affect both limbs (bilateral apraxias) and those which affect only one limb (unilateral apraxias).   Both of these major types have been reviewed extensively elsewhere (Roy, 1982, 1985). Our focus here is largely upon two of the most well-known bilateral apraxias termed ideational and ideomotor apraxia after Liepmann (1900). Both of these apraxias arise from damage to the left hemisphere, usually due to stroke.   Ideational apraxia has been characterized in several ways. One view (e.g., De Renzi & Luchelli, 1988; Poeck, 1983, 1985) has proposed that ideational apraxia is observed primarily in tasks which involve a sequence of actions, particularly those which require the use of multiple objects, such as in making a cup of tea. The individual actions, for example, stirring the tea, are frequently performed fluently, although the object may be used inappropriately. Also, the order of the constituent movements is frequently not correct. Ideational apraxia has also been characterized as an impairment in the use of single objects (e.g., Liepmann, 1900). Both performance to verbal command without actually holding the object and performance with the object are impaired. In contrast, imitating the examiner's demonstration is often much less impaired. This improvement in performance to imitation relative to verbal command led Heilman (1973) to propose a third view of ideational apraxia, a disconnection between the language comprehension area and the more anterior motor control area.

Ideomotor apraxia is often much less severe than ideational apraxia and is often thought to be examination-bound in the sense that it is observed primarily in the clinical setting (De Renzi, 1985; De Renzi, Motti, & Nichelli, 1980). Ideational apraxia may be apparent even in the patient's home environment. Patients with ideomotor apraxia often exhibit less fluency of movement than those with ideational apraxia. They are often able to perform sequences of movements involving real objects in contrast to the ideational apractic, and are most impaired when performing single gestures without the object in hand. The final contrast with ideational apraxia is apparent in the pattern of performance across modalities. In ideomotor apraxia patients are frequently impaired both to verbal command and imitation but often demonstrate marked improvement when using the actual object.

This dissociation between the modalities seen in apraxia has also been observed in movement sequencing (Jason, 1983a, b, 1985, 1986, 1990; Roy, 1981). In a study by Roy, Square-Storer, Adams, and Friesen (1989) three modality conditions were examined. In one condition the sequence was performed without any memory demands in which pictures depicted the sequence of movements to be performed. In the other conditions there were memory demands in which the sequence was performed to imitation or verbal command. In the imitation condition the examiner demonstrated the sequence to be performed, while in the command condition the sequence was presented verbally (e.g., slide, turn, point) to the patient. In both conditions the patient was required to generate the sequence from memory. In this study groups of LHD (N = 28) and RHD (N = 15) patients and normal controls (N = 10) were examined.

Several phases of testing were involved in this study. First the individual movements in each sequence were demonstrated according to the modality condition involved. In the command condition, for example, the movement associated with each verbal label was demonstrated. Following this demonstration, the patient was required to perform five consecutive repetitions of each movement. In the second phase, the individual movements were presented in a random order and the patients had to perform each movement. In order to move on to the third phase, performing the actual movement sequence, the patient had to achieve a criterion of 80% correct in this second phase. In the third phase, seven trials of the sequence were performed and the patient needed to reach a criterion of at least two correct sequences before he could move on to the next higher sequence length. If the patient did achieve this criterion, another movement was added and the three phases were repeated for this next sequence length.

Preliminary analyses (see Roy & Square-Storer, 1990) revealed that in the picture modality there were no differences among the groups as measured by the percentage of correct sequences performed. For the two memory modalities, on the other hand, the LHD patients performed more poorly than the other two groups attaining a significantly lower percentage of correct sequences.

These preliminary findings afford some insight into the nature of this movement sequencing deficit. First, impairments in sequencing in the LHD patients were most apparent when the task placed demands on memory, in accord with the previous work (e.g., Jason, 1983a, b; Roy, 1981). Comparing performance in the second phase of testing in which the patient was required to perform the individual movements in a random order to that in the third phase where movement sequences were involved provides insight into the sequencing deficit. In both the second and third phase demands were placed on memory, although the LHD patients were impaired only in the third phase which involved a sequence of movements. These patients then seemed capable of selecting the appropriate response from memory providing that they do not need to generate a series of movements (phase two performance), and they can perform a sequence of movements providing that there is no need to select the movements from memory (picture modality condition in the third phase). It seems that only when these two aspects are both required in the task does the impairment appear.

Apraxia: Image Generation Disorder

Given the observations discussed above it would appear that damage to the left hemisphere is associated with a striking impairment in performing gestures and movement sequences when these must be generated from memory (i.e., pantomimed). One explanation we would like to explore for this observation, focusing principally upon the performance of single gestures, is that these patients have a problem generating a mental image which could be used in performing the gesture. Before contemplating this possibility, however, it is important to consider alternative explanations.

In many cases this impairment in performing gestures from memory is evident when the patient responds to a verbal command to demonstrate a gesture, suggesting that the problem might arise from a comprehension disorder or from some disruption in processes linking language and movement. Work by Kertesz (1985) and others (e.g., Lehmkuhl, Poeck, & Willmes, 1983; Square-Storer, Roy, & Hogg, 1990) examining the relationship between apraxia and aphasic speech-language deficits has shown some association, particularly between apraxia and comprehension deficits. This relationship as a general explanation for this impairment observed to verbal command seems unlikely since many affected patients are not aphasic and do not have comprehension deficits. As well, many demonstrate a comparable impairment when attempting to imitate the examiner's performance, a condition where verbal comprehension is obviated.

The link between language and movement has been addressed by Heilman (1973) in his suggestion that ideational apraxia, in which the patient shows some improvement in performance to imitation, may involve a disconnection between the language comprehension area in the left hemisphere and the more anterior area responsible for the control of movement. Elliott and Weeks (1990) also considered this language-movement linkage in their work on sequencing in Down's Syndrome. A disruption to some link between language and movement also seems doubtful as a general explication for the impairment observed to verbal command in apraxia as many of those LHD patients demonstrating an impairment to verbal command exhibit a comparable impairment when imitating the examiner's performance.

Work by Roy et al. (1989) alluded to in the previous section also affords evidence against the notion that the impairment in performing to verbal command can be explained based on a comprehension disorder or a disruption to a language-movement link. Recall in this study that the LHD patients performed more poorly in the verbal and imitation conditions, both of which required the patient to generate the sequence from memory. Had the problem been largely language based one would have expected much poorer performance in the verbal condition.

Having shown that the impairment the LHD patients exhibit in performing gestures from memory does not likely arise from some speech-language-based deficits we can return to our consideration that this impairment may involve a problem generating an image of the gesture to be performed. To begin our discussion consider for a moment what is required to pantomime a gesture. Because the gesture is performed outside of the appropriate environmental context and without the object in hand, in the case of transitive gestures (i.e., gestures involving object use), the performer in essence must in some sense create the context based on his past experience. Information about the location of the hand in space, the arm-hand action and the hand posture must be accessed in some way if the gesture is to be performed correctly. All of this information must be compiled into some meaningful unit. One possible candidate would be a motor (action) image.

The notion that there is a relationship between action and imagery has quite a long history and is certainly captured in the writings of William James (1890a, b). That an image often premediates a response was the basis of his ideo-motor mechanisms theory. According to this theory, repeated experiences of stimulus-response chains produce conditioned anticipatory images of response feedback and these images then become anticipatory to actual performance, resulting in discriminative control over overt responses. Today psychologists continue to be interested in the relationship between covert rehearsal (imagery) and skilled performance, especially as it pertains to certain areas such as sport. They feel that similar learning principles, and likely underlying processes, function at both the behavioral and imaginal levels. Therefore, the use of imaginal strategies should aid in the learning and performance of skilled movement activities (Hall, Buckolz, & Fishburne, in press; Silva, 1983). This argument certainly can be extended to the gestures that must be learned and performed by patients in studies of apraxia.

While the bulk of research has focused on visual imagery, in fact imagery is possible in each of the different modalities (see Paivio, 1986, pp. 101-102) and, in considering movement, kinesthetic imagery (the "feel" of the movement) may be just as pertinent as visual imagery. Furthermore, researchers no longer view imagery as a unitary phenomenon, but as consisting of a number of separate processes or components (Kosslyn, Holtzman, Farah, & Gazzaniga, 1985), suggesting that the nature of the imagery task becomes even more critical because different processes will be employed in different tasks.

The notion that imagery and impairments to this process may have some impact on understanding apraxia might arguably require that there be some identifiable correlates of imagery in brain activity and/or that brain damage may impair imagery processes in some way. Evidence for brain correlates of imagery comes from work using a variety of techniques which record brain activity during mental practice. Using regional cerebral blood flow meaures Roland, Larsen, Lassen, and Skinhoj (1980) compared the pattern of metabolic activity when a subject mentally practised a finger sequencing task to the pattern when the task was actually performed. When actually performed two areas in the frontal lobe, the supplementary motor area and the hand area in the precentral gyrus, exhibited increased activity. Only the supplementary motor area was active when mentally practising the task. This finding is supported by a number of reports using other recording techniques which indicate activity in widely distributed brain areas, but particularly in frontal and parietal regions (see Decety & Ingvar, 1990 for a review).

Paivio and te Linde (1982) reviewed the research pertaining to brain mechanisms underlying episodic and semantic memory functions of nonverbal imagery. They concluded that episodic memory for nonverbal stimuli and semantic memory processes involved in image manipulation are more dependent on the right hemisphere. Paivio (1986) extended this review and suggested that nonverbal-visual imagery processes are more strongly based in the right-parietal and occipital lobes than in other regions. These summaries seem to suggest that imagery processing is predominantly a right-hemisphere function; however, they specifically pertain to the role of imagery in episodic memory processes and in such tasks as mental rotation, tactual spatial learning, and perceptual closure. The left hemisphere may play an equally important role in other aspects of imagery processing. Studies in which imagery instructions have been investigated indicate that imagery encoding can occur in either hemisphere (see Paivio, 1986). Furthermore, the research by Martha Farah and her associates (Farah, 1984; Farah, Gazzaniga, Holtzman, & Kosslyn, 1985) suggests that the posterior left hemisphere is critical for visual image generation. This conclusion was originally reached following Farah's review of published neurological case reports of imagery deficits. In this paper, the successful and unsuccessful performances of patients on cognitive and perceptual tasks that had been administered to them were compared. Using Kosslyn's (1980) componential model of visual imagery, Farah was able to infer through a process of elimination which particular component of mental imagery ability was impaired. The patients with a deficit in the image generation process demonstrated a consistent pattern in lesion site, the posterior left quadrant of the brain.

Given that imagery processes can play an important role in movement and that visual image generation seems to be a left hemisphere process, it is reasonable to propose that the impairment in pantomiming gestures associated with damage to the left hemisphere might be related to a loss of the image generation process. We will propose a model shortly to help explain the possible role of image generation in apraxia.

Heilman and Rothi (1985) have to some extent alluded to the role of an image in apraxia. According to them visuokinesthetic engrams are embodied in the region of the supramarginal gyrus in the parietal lobe of the left hemisphere. Evidence for the integrity of these engrams, they argue, comes from studies of gestural recognition (Rothi, Heilman, & Watson, 1985; Rothi, Mack, & Heilman, 1986). They identified one type of ideomotor apraxia which they attribute to a disruption to these visuokinesthetic engrams. The affected apractic patients exhibit an impairment in gestural recognition (Heilman, Rothi, & Valenstein, 1982).

Although this work of Heilman and co-workers points to the potential importance of representations in memory in pantomiming gestures, it is not clear what role these visuokinesthetic engrams play nor how their disruption affects gesturing. From Heilman's account it would seem that these representations are, to some extent, involved in the selection and/or the control of the movements used in performing the gesture. The exact mechanism is unclear, however.

Visual Information in Apraxia

As we have just seen one of the characteristic features of apraxia is that the affected patients exhibit an inability to demonstrate gestures when these must be generated from memory. In some cases, however, the patient displays an improvement to imitation (e.g., Roy et al., 1991). Some have argued that this dissociation between pantomime and imitation serves to distinguish between ideational and ideomotor forms of apraxia with the former showing this facilitation of performance to imitation (e.g., De Renzi, 1985). Why might performance improve when imitating the gesture demonstrated by the examiner? Possibly visually observing the gesture serves to reduce the information load of the task. When the patient must perform the gesture from memory, considerable demands are placed on response selection. When imitating the gesture these demands may be reduced in that the particular gesture from the patient's repertoire of gestures is essentially selected for him, thus diminishing the information load associated with the decision about what gesture to perform.

An alternative explanation is that the model provided by the examiner removes the need for the patient to generate an image, a process necessitated when pantomiming the gesture. The visual information afforded in the examiner's model may serve to aid the patient in generating and controlling the movements involved in the gesture. Spatial (e.g., location and orientation of the hand, plane of movement of the hand) and postural (e.g., hand posture) information as well as information about the characteristics of the action (e.g., repetitive movement at the wrist) afforded in the examiner's demonstration may be used by the patient to control his movement. How might this visual information be used? One possibility is that the patient monitors his own movements visually in an attempt to match the characteristics of his movement to those apparent in the examiner's gesture. Given the plethora of information available to the patient in the examiner's demonstration it is unclear just what aspect may be important. Studies of prehension (e.g., Jeannerod, 1988) have examined the role of vision in the control of different components of the reaching movement. Work by Sivak and MacKenzie (1990), for example, has shown that peripheral vision is important in the control of the transport component (moving the arm) to the object's location in space, while central or foveal vision is critical for detecting information about object properties which is used to control hand shaping in the grasp component. Such studies might be important in apraxia to discern the role of vision in controlling the various dimensions of movement in a gesture.

While this type of visual monitoring and matching process may be involved in performing many gestures, a number of gestures (e.g., brushing teeth, combing hair, saluting) involve movements directed toward the body and performed in body-centred space. Since these gestures are not easily monitored visually, it would seem the patient must perform an intermodal transform in which the visual characteristics of the examiner's movements are translated into a proprioceptive map. The patient would then ostensibly attempt to match the "felt" characteristics of his movement to those observed visually in the examiner's gesture.

Whichever of these or other possible mechanisms might explain how the patient could use visual information from the examiner's movement, it is clear that the patient must visually analyze the examiner's demonstration and use this information in some way to generate the movements in the gesture. Considering these processes one may envisage several reasons for why some patients (e.g., those with ideational apraxia) do not demonstrate a facilitation in performance to imitation. Possibly the patient makes errors in analyzing the visual characteristics of the examiner's movement and, thus, uses incorrect information to match his own movement against. Alternately he may have difficulty monitoring his own movements either visually or proprioceptively. Finally, he may be unable to transform the visual information from the examiner's gesture into a proprioceptive map to control the movements in body-centred gestures. These latter problems with monitoring movements proprioceptively and with intermodal transformations may explain, to some extent, why body-centred gestures have often been found to be more poorly performed than those directed away from the body and performed in allocentric space (e.g., Cermak, 1985).

A Model and Directions for Future Work

Considering our discussion to this point we have suggested that the inability to pantomime a gesture in apraxia may arise from an impairment in the central generation of movement, possibly due to some disruption to an image generation process. Evidence for this proposal comes from the observation that many of the patients who are unable to pantomime a gesture are able to imitate the gesture when it is demonstrated by the examiner. Presumably, the patient must use some visual information in the examiner's demonstration to generate the movement. A number of patients, however, continue to exhibit performance impairments to imitation. What do these patterns of performance suggest about the nature of apraxia and, more generally, about the potential mechanisms involved in gestural performance? What implications do these notions of the role of image generation and vision in gestural performance have for research into apraxia? To provide a framework for considering these questions we have developed a model (Figure 11.1) depicting the processes and mechanisms which seem implicated in gestural performance given our observations above and considering Kosslyn's computer simulation model of mental imagery (Kosslyn, 1980) and the dual coding approach to mental representations (Paivio, 1986).

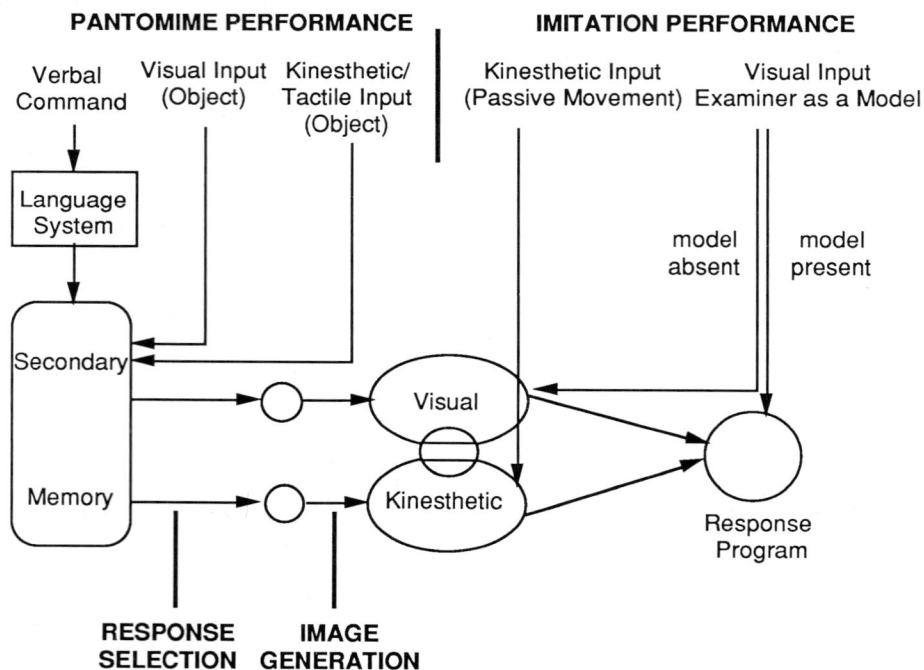

Figure 11.1. A model depicting the processes involved in producing gestures in response to visual and other sources of input.

This model is designed to depict the production of movements which are mediated by imagery processes such as would be required when the patient must pantomime a gesture. In this task the context of the gesture is not present but must in some sense be created mentally. In this situation it is likely that the various movement components must be selected and organized internally as there is no information in the environment on which to rely. It is in these types of processes (selection, organization, context creation) that the imaginal system is assumed to play a major role (see Paivio, 1986, chapter 4 for a discussion of the organizational properties of the imaginal system).

In this model all three modalities are seen to be sources of information for eliciting the gesture. The patient may be asked to demonstrate a gesture (auditory-verbal modality). Alternately, he may be shown an object visually (visual modality) or asked to feel it tactually without vision (kinesthetic modality) and then asked to demonstrate how to use it. In each of these conditions the patient must generate the gesture from memory. One of the first stages in this process must involve response selection. Once selected we envisage that an image of the appropriate response must be generated and retained in working memory. Both visual and kinesthetic information are thought to be represented in this image which is then thought to play an important role in the organization and control of the gestural response. Although the actual mechanism here is not clear, this image may to some extent serve to establish the internal (i.e., body centred) context for the gesture. The kinesthetic component of the image may play a role in the initial stages of response preparation by establishing the postural context for the gesture through a type of pretuning mechanism (Fel'dman, 1966; Frank & Earl, 1990). Following movement initiation both the visual and kinesthetic components may afford information on the expected sensory consequences.

In contrast to the conditions alluded to above, when the patient is asked to imitate a gesture, access to memory and the associated response selection and image generation processes are not necessitated. In this case the visual and kinesthetic modalities are seen to afford the primary sources of information in which the patient attempts to copy the movements apparent in the examiner's demonstration. The patient must be able to analyze the visual and kinesthetic information in order to pick out the features which are critical to accurately imitating what he has seen and/or felt. When visual information from the examiner's demonstration is continuously available, the patient may visually monitor his own performance to ascertain how well it conforms to that evident in the model (for gestures performed in allocentric space) or he may carry out some type of visual-kinesthetic transformation and monitor the felt position of his hand movements (for gestures performed in body-centred space). In either case there is little or no reliance on memory. When the visual information from the examiner's demonstration is not continuously available or when the patient is attempting to imitate a gesture which he has been moved through passively (i.e., kinesthetic information), however, he must retain an image of the gesture and subsequently use this to control his performance. In this case image formation and retention are involved. To some extent performance in this situation shares this image retention process with that involved when the patient must perform the gesture from memory. When performing from memory the image retained is generated from memory, while the image retained in this imitation condition is formed based on information available in the environment.

The final stage in response production in this model is concerned with the mechanisms involved in controlling movement during performance of the gesture. This part of the model is not well articulated in that it is not of particular import in this discussion. Certainly processes involved in multisegmental coordination and in monitoring feedback are important in this stage (see Jeannerod, 1988 for a review).

Given this model how might one envisage that gestural performance could be impaired and what directions for future work on apraxia are suggested? Looking first at the input side it is possible that gestural performance could be impaired with deficits in comprehension. While gestural impairments associated with profound comprehension deficits affecting the patient's understanding of the verbal command (e.g., show how to salute) cannot be considered apraxia given the definition by exclusion alluded to at the outset, a number of studies (e.g., Kertesz & Hooper, 1982) have found some relationship between more subtle comprehension deficits and impairments to gestural performance. Certainly more work needs to be done to discern what the precise nature of this relationship is (see Square-Storer et al., 1990).

Because both vision and touch/kinesthesis can be a source of information relevant to gestural performance, it is possible that visual and tactual/kinesthetic impairments may give rise to deficient gestural performance. In the case of transitive gestures visual or tactual object recognition disorders due to sensory/perceptual deficits or memory impairments may lead to gestural impairments. Again, as with comprehension above, it is unlikely one could refer to such gestural performance deficits as apraxia, given the definition by exclusion. Problems in the analysis of visual or kinesthetic information, on the other hand, as when observing the examiner's demonstration or when feeling the movements of the limb as it is passively moved through a gesture, could lead to impairments in gesturing which could be termed apraxia. Such a definition of apraxia is certainly favoured in the work on developmental dyspraxia where the impairment to praxis in children is seen to be largely due to some basic problem in the analysis and monitoring of visual and particularly kinesthetic information (e.g., Cermak, 1985).

Such a problem in perceptual analysis could lead to a particular deficit in performance of the gesture to imitation. Theoretically, although never reported, it would seem possible that a patient might be more impaired to imitation than to pantomime in the face of such a perceptual analysis problem. Further, it may be that one might observe rather modality specific forms of apraxia, depending on the functional integrity of the visual and kinesthetic perceptual systems. Indeed, De Renzi, Faglioni, and Sorgato (1982) have reported such modality specific apraxias, although they attributed them more to a disconnection between the affected modality (e.g., vision) and the motor system than to a problem in modality specific perceptual analysis deficits. Given our model and the considerations alluded to above it would seem important to carry out assessments to determine the integrity of perceptual analysis in each modality to discern whether the gestural disorder involves problems in such a process. Such an analysis might conceivably reveal that one system or the other (i.e., vision or tactual/kinesthetic) is in tact in which case it might be possible to use the intact system to augment or substitute for the affected one. To our knowledge such approaches have not been reported in the assessment or treatment of apraxia.

Assuming that the speech-language, visual and tactual-kinesthetic systems are intact one might envisage that the problem in gestural performance relates to a deficit in the internal generation of the response from memory if the patient demonstrates an equivalent gestural deficit regardless of the mode of input. For example, in the case of transitive gestures the patient would show an equivalent impairment whether he is asked verbally to show how to use a hammer, is shown the hammer or feels the hammer tactually without vision. In order for the patient to pantomime the gesture this information from each modality must contact a representation in memory from which an image can be generated. As with other forms of memory (e.g., verbal memory) this representation must involve a number of aspects (Roy, 1983). Roy (1983; Roy & Square, 1985) has described two types of knowledge important in object use and gesture which are of interest here: knowledge of objects and tools in terms of the actions and functions they serve and knowledge of actions which may be independent of tools or objects but into which tools may be incorporated.

Knowledge of objects as they relate to their functions may have internalized linguistic referents similar to functional associates (e.g., Goodglass & Baker, 1976) which form part of the semantic field for objects descriptive of the actions performed with or on the objects (e.g., pushing, sawing). In addition to these "internalized" functional referents Roy (1983) suggests that there are also perceptual referents which provide an externalization of knowledge about function performed with objects. The perceptual attributes of objects afford one source of information here. Through experience the performer learns that certain perceptual features enable certain actions. That is, the perceptual attributes afford (e.g., Saltzman & Kelso, 1987) the actions which are possible (e.g., an object which shares perceptual features with a hammer could be used for hammering). The other source of information here concerns the environmental/situational context in which tools and objects are used.

Roy (1983) suggests that these memory structures may be organized into networks which are formed on the basis of shared features. Two objects which share some features may be confused and, so, one may be chosen and used incorrectly, an error frequently seen in normal adults (e.g., Norman, 1981; Roy, 1982). These memory structures may be mapped on to sets of procedures defining various functions performed on objects. These perceptual/contextual attributes may provide descriptions which are used to activate these procedures, thus providing a rather direct link between perception and action.

The second knowledge component which Roy alludes to focuses upon the actions performed in using objects. As with the former knowledge component Roy suggests that there may be an important linguistic aspect to this component. The semantic structure of action verbs (e.g., to hammer) forming the functional referents of objects in the former knowledge component are conceptualized in a logical or propositional framework (e.g., Miller & Johnson-Laird, 1976). Action then may be conceptualized as verbal descriptions of the relationships of the agent (e.g., the hand) and the objects moving in a particular spatiotemporal pattern. Another source of information here, Roy argues, concerns the body-relevant (e.g., the visual and tactual/kinesthetic information experienced by the performer in making the gesture) and environment-relevant (e.g., the swirling liquid seen when stirring a cup of tea) consequences of action. These aspects of knowledge provide the capability for the performer to verbally describe the action he must perform and to recognize another's performance as representing a particular action or gesture.

One might argue that a great deal of information relevant to action is in the environment and does not need to be "in one's head", so to speak. The perceptual attributes of tools and surfaces (affordances) inform the performer as to what actions are possible. The interaction of object surfaces (e.g., the spoon in the cup when stirring tea) constrains the action such that only a relatively small number of movements are possible. Indeed, Arbib's (1990) notion of coordinative control programs argues that these various perceptual aspects in the environment may be processed in specific sensory channels and linked to particular motor systems which provide for distributed control of movement. The very fact that people are able to pantomime actions outside of the appropriate environmental context, however, suggests that there must be some representation of action and gesture which in a sense enables them to recreate the context so as to perform the gesture appropriately.

The integrity of these representations may be examined through various recall and recognition tests. Work by Roy (1983a, b) and Heilman (Rothi, Mack, & Heilman, 1986) examining pantomime recognition in apractics suggests that some of these patients indeed do have a gestural recognition problem. This notion of a recognition disorder raises a question about what the appropriate type of recognition test might be. Considering visual recognition, for example, it may be that the nature of the visual information depicted is important in detecting a recognition disorder. A static depiction of the gesture as in a photograph may not be the best way of presenting the gesture for recognition as it does not capture the dynamic quality of the actual performance. Further, it may be important to focus on particular features of the gesture in the recognition task (e.g., hand posture, location of the hand, the action or movements of the hand) since it seems clear that gestural performance may exhibit selective impairments which affect one of these dimensions and not another (e.g., Haaland & Flaherty, 1984; Roy et al., 1985). Might it be possible that the patient has a selective memory deficit which might be identified in these types of selective recognition tasks? Roy (1983a, b) has shown that some apractics have a particular problem in identifying hand posture errors in the videotaped performance of an actor demonstrating a gesture. Possibly these recognition problems relate to the incidence of postural errors apparent in the performance of these patients. Such selective correlative analyses between gestural recognition and gestural performance may provide new insights into apraxia.

While the patient's memory representation may be unaffected, impairments at the next stage in our model, response selection, may lead to deficits in gestural performance. A number of researchers have alluded to such a response selection problem as a basis for an apractic deficit (e.g., De Renzi, 1985). Ostensibly, some problem in response selection might be implicated if one observed an increasing impairment in the apractic deficit with an increase in the number of potential gestural responses from which the patient must chose. In reality it may be difficult to validly assess the integrity of this stage in the context of a clinical assessment using this logic since the population of gestural responses is not well defined. Possibly, though, one might attempt to organize gestures into categories, the sizes of which might vary. One might then predict that if the problem were to do with response selection patients would exhibit the greatest performance deficit in gestures from categories with a larger population. A number of attempts have been made to categorize gestures (e.g., Hécaen & Rondot, 1985). More work in this area may prove fruitful for understanding the potential importance of response selection deficits in apraxia.

The next stage we have alluded to in the model concerns image generation. One means of assessing the integrity of this stage would be to incorporate an image generation test into the clinical assessment for apraxia. The task most commonly employed is to ask patients to describe or answer questions about the visual appearance of objects (e.g., whether the entrance to your house is in the centre or off to one side of the front). There is evidence from studies with normals that for retrieving information about aspects of appearance not often explicitly discussed, especially spatial, relational information, imagery is used (Eddy & Glass, 1981; Kosslyn & Jolicoeur, 1980). These tasks have focused on visual imagery but analogous tasks could be developed to assess kinesthetic imagery. In essence the hypothesis here might be that if an impairment in image generation does underlay apraxia, one might predict a coincidence between these disorders or a relationship between the severity of the image generation deficit and the severity of apraxia. To this point no one has carried out this type of investigation.

The next stage in the model pertains to working memory. Here the image generated from secondary memory (when pantomiming the gesture) and the image encoded (when imitating the gesture) is stored to be used in organizing and controlling movement involved in performing the gesture. A deficit at this stage of working memory could adversely affect gestural performance in that the patient, being unable to retain the generated or encoded image, would have difficulty in accurately organizing and controlling the gestural response. According to our model evidence that such a deficit might be the basis for the observed apractic impairment would come from comparing pantomime performance to imitation performance when the visual information is not continuously present or when tactual-kinesthetic information forms the information for imitation. Imitation in these latter two conditions necessitates the encoding and retention of an image in working memory. Similarly, the image generated in the pantomime condition must be stored in working memory. Because these performance conditions share this common stage, gesturing should be uniformly impaired in these three conditions if there is a deficit in working memory.

The final stage in the model dealing with the processes involved in organizing and controlling movement may also be a site for disruption which could lead to apractic performance of gestures. It is this stage which has been the principle focus of work on apraxia. Many researcher's argue that disruptions here at a very low level involving the actual control of movement are not to be considered apraxias (e.g., Heilman & Rothi, 1985). Disruptions to the motor pathways resulting in hemiparesis, ataxia or tremor, for example, would adversely affect gestural performance but, according to the exclusionary definition alluded to above with regard to basic sensory and comprehension deficits, should not be considered apraxia. Disruptions at this stage which are thought to be characteristic of apraxia involve processes which affect more the sequencing or transitions through postures in a movement (e.g., Kimura, 1982). Such a problem might conceivably arise because of a problem in timing the activation of various segments or components of the gesture. Charlton, Roy, MacKenzie, Marteniuk, and Square-Storer (1988) reported such a problem in an apractic man who had difficulty in temporally coordinating the transport component of a reaching movement with the grasp component. Work by Poizner, Mack, Verfaellie, Rothi, and Heilman (1990) also showed a somewhat similar effect in gestural responses like those used in the typical apraxia examination. Investigations of this type of disruption to the control of movement might focus on varying the complexity of coordination task by successively adding components to the movement (see Charlton et al., 1988 and Jason, 1986 for examples).

In contrast to this type of sequencing or postural transition disorder the patient may have a deficit in monitoring feedback in the control of the movement. One might attempt to manipulate the amount and quality of feedback the patient receives from his movement in order to investigate the potential impact of this problem to the observed apractic deficit. For example, if the patient is having considerable difficulty processing visual feedback, one might predict that removing visual information about the arm movement should not appreciably degrade performance. While impairments in processing feedback in a single modality may be an important factor in apraxia, it is possible that the major disorder with regard to feedback processing relates to a deficit in intersensory integration. This problem in particular is seen as most important in developmental forms of apraxia (e.g., Cermak, 1985).

Taken together these predictions derived from our model suggest that apraxia may arise from disruptions at a number of stages in the organization and execution of the gestural response. Given that the gestural impairments in apraxia are so multi-determinant it is not surprising that the disorder has proven difficult to understand. Work in this area might now be fruitfully directed toward attempting to delineate various types of apraxia as has been done in work on aphasia (e.g., Goodglass & Kaplan, 1972) and dyslexia (Coltheart, 1984), following the lead set by Heilman et al. (1982) and Roy (1978, 1983a, b). The observation that the nature of the disorder may be dependent on the conditions under which the gestural response is elicited suggests that this work must be sensitive to the notion of context or task dependent mechanisms in apraxia (Roy & Square-Storer, 1990; Roy et al.,

1991). Movement control in this approach (Arbib, 1985, 1990) arises out of constraints which serve to define task demands. Learning what the constraints are in gestural performance and how the motor system adapts to them may afford new insights into the nature of apraxia. For example, as we have seen, relative to pantomime gestural performance frequently improves to imitation, suggesting that some characteristics of the visual information available in the examiner's demonstration may provide constraints which serve to facilitate performance. Work in apraxia might profitably be spent attempting to identify just what these visual characteristics are. Certainly such studies in reaching have proven valuable in understanding the relationship between the visual features of objects (e.g., intrinsic and extrinsic features) and the components of reaching movements which are sensitive to these features.

References

Arbib, M.A. (1985). Schemas for the temporal organization of behavior. *Human Neurobiology, 4*, 63-72.

Arbib, M.A. (1990). Programs, schemas and neural networks for control of hand movements: Beyond the RS framework. In M. Jeannerod (Ed.), *Attention and performance XIII* (pp. 111-138). Hillsdale, NJ: Erlbaum.

Cermak, S. (1985). Developmental dyspraxia. In E.A. Roy (Ed.), *Neuropsychological studies of apraxia and related disorders* (pp. 225-250). Amsterdam: North-Holland.

Charlton, J., Roy, E.A., Marteniuk, R.G., MacKenzie, C.L., & Square-Storer, P.A. (1988) Disruptions to reaching in apraxia. *Society for Neuroscience Abstracts, 14*, 1234.

Coltheart, M. (1984) Acquired dyslexias and normal reading. In R.N. Malatesha & H.A. Whitaker (Eds.), *Dyslexia: A global issue* (pp. 265-294). The Hague: Martinus Nijhoff.

Decety, J., & Ingvar, D.N. (1990). Brain structures participating in mental simulation of motor behavior: A neuropsychological interpretation. *Acta Psychologica, 73*, 13-34.

De Renzi, E. (1985). Methods of limb apraxia examination and their bearing on the interpretation of the disorder. In E.A. Roy (Ed.), *Neuropsychological studies of apraxia and related disorders* (pp. 45-64). Amsterdam: North-Holland.

De Renzi, E., Faglioni, P., & Sorgato, P. (1982). Modality-specific and supramodel mechanisms of apraxia. *Brain, 105*, 301-312.

De Renzi, E., & Luchelli, F. (1988). Ideational apraxia. *Brain, 111*, 1173-1185.

De Renzi, E., Motti, F., & Nichelli, P. (1980). Imitating gestures: A quantitative approach to ideomotor apraxia. *Archives of Neurology, 37*, 6-10.

Eddy, J.K., & Glass, A.L. (1981). Reading and listening to high and low imagery sentences. *Journal of Verbal Learning and Verbal Behavior, 20*, 333-345.

Elliott, D., & Weeks, D.J. (1990). Cerebral specialization and the control of oral and limb movements for individuals with Down's syndrome. *Journal of Motor Behavior, 22*, 6-18.

Farah, M.J. (1984). The neurological basis of mental imagery: A componential analysis. *Cognition, 18*, 245-272.

Farah, M.J., Gazzaniga, M.S., Holtzman, J.D., & Kosslyn, S.M. (1985). A left hemisphere basis for visual imagery? *Neuropsychologia, 23*, 115-118.

Fel'dman, A.G. (1966). Functional tuning of the nervous system with control of movement or maintenance of a steady posture. II. Controllable parameters of the muscle. *Biophysics, 11*, 565-578.

Frank, J.S., & Earl, M. (1990). Coordination of posture and movement. *Physical Therapy, 70*, 855-863.

Friesen, H., Roy, E.A., Square-Storer, P.A., & Adams, S. (1987). *Apraxia: Interrater reliability of a new error notation system for limb apraxia.* Paper presented at the meeting of the North American Society for the Psychology of Sport and Physical Activity, Vancouver, B.C.

Goodglass, H., & Baker, E. (1976). Semantic field, naming and auditory comprehension in aphasia. *Brain & Language, 3*, 359-374.

Goodglass, H., & Kaplan, E. (1972). *The assessment of aphasia and related disorders.* Philadelphia: Lea and Febiger.

Haaland, K.Y., & Flaherty, D. (1984). The different types of limb apraxia errors made by patients with left or right hemisphere damage. *Brain and Cognition, 3*, 370-384.

Hall, C., Buckolz, E., & Fishburne, G. (in press). Imagery and the acquisition of motor skills. *Canadian Journal of Sport Sciences.*

Heilman, K.M. (1973). Ideational apraxia - A redefinition. *Brain, 96*, 861-864.

Heilman, K., & Rothi, L.J.G. (1985). In K. Heilman & E. Valenstein (Eds.), *Clinical neuropsychology* (pp. 131-150). New York: Oxford University Press.

Heilman, K.M., Rothi, L.J., & Valenstein, E. (1982). Two forms of ideomotor apraxia. *Neurology, 32*, 342-346.

Hécaen, H., & Rondot, P. (1985) Apraxia as a disorder of a system of signs. In E.A. Roy (Ed.), *Neuropsychological studies of apraxia and related disorders* (pp. 45-64). Amsterdam: North-Holland.

James, W. (1890a). *Principles of psychology,* (Vol. 1). New York: Holt.

James, W. (1890b). *Principles of psychology,* (Vol. 2). New York: Holt.

Jason, G. (1983a). Hemispheric asymmetries in motor function: I. Left hemisphere specialization for memory but not performance. *Neuropsychologia, 21*, 35-46.

Jason, G. (1983b). Hemispheric asymmetries in motor function. II. Ordering does not contribute to left hemisphere specialization. *Neuropsychologia, 21*, 47-58.

Jason, G. (1985). Manual sequence learning after focal cortical lesions. *Neuropsychologia, 23*, 35-46.

Jason, G. (1986). Performance of manual copying tasks after focal cortical lesions. *Neuropsychologia, 23*, 41-78.

Jason, G. (1990). Disorders of motor function following cortical lesions: Review and theoretical considerations. In G. Hammond (Ed.), *Cerebral control of speech and limb movements* (pp. 141-168). Amsterdam: Elsevier Science Publishers.

Jeannerod, M. (1988) *The neural and behavioural organization of goal-directed movements.* Oxford: Clarendon Press.

Kertesz, A., & Hooper, P. (1982). Praxis and language: The extent and variety of apraxia in aphasia. *Neuropsychologia, 20*, 275-286.

Kertesz, A. (1985). Apraxia and aphasia: Anatomical and clinical relationship. In E.A. Roy (Ed.), *Neuropsychological Studies of Apraxia and Related Disorders* (pp. 163-178). Amsterdam: North-Holland

Kimura, D. (1982). Left-hemisphere control of oral and brachial movements and their relationship to communication. *Philosophical Transactions of the Royal Society of London, B298,* 135-149.

Kosslyn, S.M. (1980). *Image and mind.* Cambridge, MA: Harvard University Press.

Kosslyn, S.M., Holtzman, J.D., Farah, M.J., & Gazzaniga, M.S. (1985). A computational analysis of mental image generation: Evidence from functional dissociations in split-brain patients. *Journal of Experimental Psychology: General, 114,* 311-341.

Kosslyn, S.M., & Jolicoeur, P. (1980). A theory-based approach to the study of individual differences in mental imagery. In R.E. Snow, P.A. Federico, & W.E. Montague (Eds.), *Aptitude, learning and instruction: Cognitive processes analysis of aptitude, Volume 1* (pp. 139-175). Hillsdale, NJ: Erlbaum.

Lehmkuhl, G., Poeck, K., & Willmes, K. (1983) Ideomotor apraxia and aphasia: An examination of types and manifestations of apraxic syndromes. *Neuropsychologia, 21,* 199-212.

Liepmann, H. (1900). Das krankheitsbild der Apraxie (motorischen Asymbolie). *Monateschrift fur Psychiatrie und Neurologie, 8,* 1-44, 102-132, 182-197.

Miller, G.A., & Johnson-Laird, P.N. (1976). *Language and perception.* Cambridge, MA: Harvard University Press.

Norman, D.A. (1981). Categorization of action slips. *Psychological Review, 88,* 1-15.

Paivio, A. (1986). *Mental representations: A dual coding approach.* New York: Oxford.

Paivio, A., & te Linde, J. (1982). Imagery, memory, and the brain. *Canadian Journal of Psychology, 36*, 243-272.

Poeck, K. (1983) Ideational apraxia. *Journal of Neurology, 230*, 1-5.

Poeck, K. (1985). Clues to the nature of disruptions to limb apraxis. In E.A. Roy (Ed.), *Neuropsychological studies of apraxia and related disorders* (pp. 45-64). Amsterdam: North-Holland.

Poizner, H., Mack, L., Verfaellie, M., Rothi, L.J.G., & Heilman, M. (1990). Three-dimensional computergraphic analysis of apraxia: Neural representations of learned movement. *Brain, 113*, 85-101.

Roland, P.E., Larsen, B., Lassen, N.A., & Skinhoj, E. (1980). Supplementary motor area and other cortical areas in organization of voluntary movements in man. *Journal of Neurophysiology, 43*, 118-136.

Rothi, L.J.G., Heilman, K.M., & Watson, R.T. (1985). Pantomime comprehension and ideomotor apraxia. *Journal of Neurology, Neurosurgery and Psychiatry, 48*, 207-210.

Rothi, L.J.G., Mack, L., & Heilman, K.M. (1986). Pantomime agnosia. *Journal of Neurology, Neurosurgery and Psychiatry, 49*, 451-454.

Rothi, L.J.G., Mack, L., Verfaellie, M., Brown, P., & Heilman, K.M. (1988). Ideomotor apraxia: Error pattern analysis. *Aphasiology, 2*, 381-387.

Roy, E.A. (1978). Apraxia: A new look at an old syndrome. *Journal of Human Movement Studies, 4*, 191-210.

Roy, E.A. (1981). Action sequencing and lateralized cerebral damage: Evidence for asymmetries in control. In J. Long & A. Baddeley (Eds.), *Attention and performance IX* (pp. 487-498). Hillsdale, NJ: Erlbaum.

Roy, E.A. (1982). Action and performance. In A. Ellis (Ed.), *Normality and pathology in cognitive function* (pp. 265-298). New York: Academic Press.

Roy, E.A. (1983). Neuropsychological perspectives on apraxia and related action disorders. In R.A. Magill (Ed.), *Memory and control of action* (pp. 293-320). Amsterdam: North-Holland.

Roy, E.A. (Ed.). (1985). *Neuropsychological studies of apraxia and related disorders.* Amsterdam: North-Holland.

Roy, E.A., & Square, P.A. (1985). Common considerations in the study of limb, verbal and oral apraxia. In E.A. Roy (Ed.), *Neuropsychological studies of apraxia and related disorders* (pp. 111-159). Amsterdam: North-Holland.

Roy, E.A., & Square-Storer, P.A. (1990). Evidence for common expressions of apraxia. In G. Hammond (Ed.), *Cerebral control of speech and limb movements* (pp. 477-502). Amsterdam: Elsevier Science Publishers.

Roy, E.A., Square, P.A., Adams, S., & Friesen, H. (1985). Error/movement notation systems in apraxia. *Recherches Semiotiques/Semiotics Inquiry, 5,* 402-412.
Roy, E.A., Square-Storer, P.A., Adams, S., & Friesen, H. (1989). Disruptions to central programming of sequences. *Canadian Psychology, 30,* 423.
Roy, E.A., Square-Storer, P.A., Hogg, S., & Adams, S. (1991). Analysis of task demands in apraxia. *International Journal of Neuroscience, 56,* 177-186.
Saltzman, C., & Kelso, J.A.S. (1987). Skilled actions: A task-dynamic approach. *Psychological Review, 94,* 83-106.
Silva, J.M. (1983). Covert rehearsal strategies. In M.H. Williams (Ed.), *Ergogenic aids in sport* (pp. 140-152). Champaign, IL: Human Kinetics.
Sivak, B., & MacKenzie, C.L. (1990). Integration of visual information and motor output in reaching and grasping: The contributions of peripheral and central vision. *Neuropsychologia, 28,* 1095-1116.
Square-Storer, P.A., Roy, E.A., & Hogg, S. (1990). The dissociation of aphasia from apraxia of speech, ideomotor limb and bucchofacial apraxia. In G. Hammond (Ed.), *Cerebral control of speech and limb movements* (pp. 451-476). Amsterdam: Elsevier Science Publishers.

Acknowledgement
Preparation of this manuscript was partially supported by grants from the Parkinson Foundation of Canada and the Ontario Mental Health Foundation to Eric A. Roy.

Part 3:
Spatial-temporal anticipation

VISION AND MOTOR CONTROL
L. Proteau and D. Elliott (Editors)
© 1992 Elsevier Science Publishers B.V. All rights reserved.   285

CHAPTER 12

PREDICTIVE VISUAL INFORMATION SOURCES FOR THE REGULATION OF ACTION WITH SPECIAL EMPHASIS ON CATCHING AND HITTING

REINOUD J. BOOTSMA and C. (LIEKE) E. PEPER

*Department of Psychology, Faculty of Human Movement Sciences,
Free University, Van der Boechorststraat 9, 1981 BT Amsterdam,
The Netherlands*

In order to interact with an approaching ball, as in catching or hitting, an actor needs predictive information about *when* the ball will be *where*. The goal of the present chapter is to formally identify some of the available visual information sources, and to evaluate the experimental evidence available for the use of these information sources in the regulation of behavior, with special emphasis on catching and hitting.

Predictive Temporal Information
When a ball, or indeed any object, directly approaches an observer[1], the magnitude of the solid angle subtended by the object at the point of observation increases symmetrically. Importantly, this increase is non-linear (see Figure 12.1), because the magnitude of the solid angle is inversely related to the distance from the point of observation. While the solid angle increases only slowly during the first part of the approach, the last part shows a sudden 'explosion'. This 'explosion' or rapid increase has been coined *looming,* and is typical of approaching objects that are going to collide with the point of observation. As such, it provides information about imminent collision and could be useful in a behavioral setting. Indeed, Schiff (1965; Schiff, Caviness, & Gibson, 1962) demonstrated that a multitude of animal species (frogs, crabs,

---

[1] For reasons of convenience, an object approaching a stationary observer (or another object) will be considered as the paradigmatic case of relative approach.

chickens, monkeys) exhibited avoidant reactions when confronted with optically expanding shadow patterns. Importantly, such reactions did not occur when the shadows did not expand, but rather contracted. Schiff concluded that the animals were sensitive to the information contained in the expanding optical pattern. Bower, Broughton, and Moore (1970) reported similar results for ten-day old human babies.

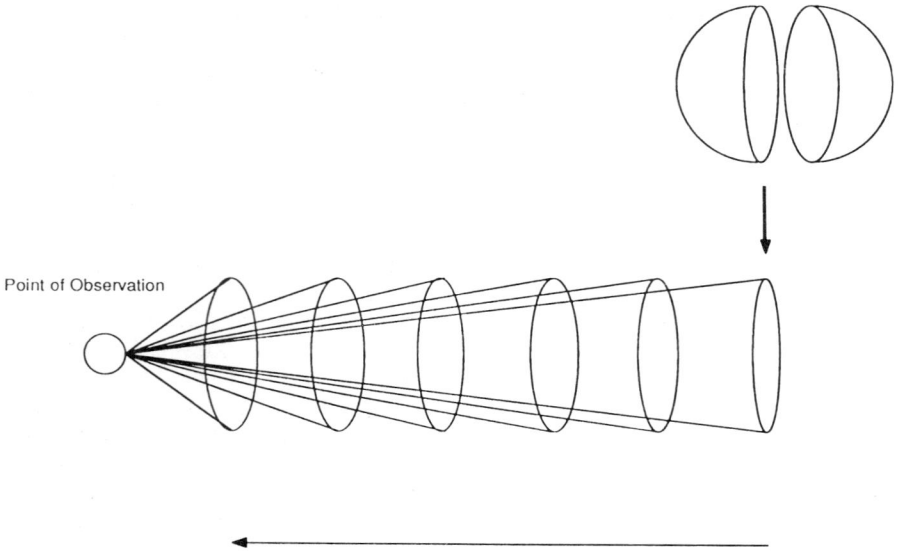

Figure 12.1. Geometrical representation of the non-linear increase in the solid angle subtended by an approaching ball at the point of observation.

Much more detailed information is available in the expanding optical pattern, however, than the mere fact that collision is imminent (see Schiff, 1965). Lee (1976)[2] formally demonstrated that the rate of dilation (i.e., the relative rate of expansion) of the optical contour of an approaching object is proportional to the inverse of the time remaining until collision, if speed of approach is constant. Moreover, he demonstrated that the rate of change over

---

[2] Before Lee (1976), others had also addressed the issue of time until collision (e.g., Carel, 1961; Hoyle, 1957; Purdy, 1958), but Lee has been the one who has popularised this source of information in tying it explicitly to the regulation of behavior.

time of this rate of dilation specifies whether a 'soft' or a 'hard' collision will occur. Taken together, these information parameters can be used to understand the regulation of, for instance, braking behavior in driving (Cavallo & Laurent, 1986; Lee, 1976). Put simply, a positive rate of dilation of the optical contour of the car in front specifies that one is going to collide with it if no action is taken. The driver should brake (i.e., decelerate the car) with an intensity that results in a rate of change of the rate of dilation that specifies that the upcoming collision will be, at most, soft. Preferably, the driver would brake a little more than that, which will result in the avoidance of a collision all together. Keeping the magnitude of the rate of change of the rate of dilation within a certain safety margin thus allows one to brake enough to avoid collision, without coming to an abrupt stop every time imminent collision is detected.

Braking

To demonstrate that humans are able to utilise the information contained in the rate of dilation of an optical contour, Schiff and Detwiler (1979) presented subjects with animated film clips of approaching objects. At particular stages of the approach, the film would go blank and subjects were asked to indicate when they thought the object would have reached them had it continued on the same trajectory. By using objects of different sizes, approaching the observer with different velocities, and blanking-out the film while the objects were at different distances from the observer, Schiff and Detwiler were able to demonstrate that their subjects did use the rate of dilation information in formulating their responses. For instance, responses for objects travelling at 25 cm s^{-1} that were blanked-out at a distance of 100 cm from the observer were found to be similar to responses for objects travelling at 50 cm s^{-1}, blanked-out at a distance of 200 cm from the observer. While both velocity and distance differed between these conditions, the times remaining until the object would have reached the observer were equal (4 s). In fact, the responses given were best described as a linear function of time remaining until contact, rather than as a function of velocity or distance to contact. Enhancing the visual information available, by adding a perspective grid over which the objects travelled, thus allowing subjects to use independent information about the rate of change of distance, did not lead to a better performance.

The performance in the experimental paradigm proposed by Schiff (Schiff & Detwiler, 1979; Schiff & Oldak, 1990) is however somewhat displeasing, as a consistent underestimation of time remaining until contact is normally found. In an alternative paradigm, Todd (1981) presented human observers with visual simulations of two simultaneously approaching objects. The objects differed in size, and starting position. Velocities were chosen such that the objects differed in their time of arrival at the point of observation by 10, 20,

50, 100, 200, or 300 ms. Subjects were asked to indicate which object would arrive first. Differences as small as 200 ms were judged with a discrimination accuracy of well over 90%, demonstrating that human observers are quite sensitive to the rate of dilation information.

Because the inverse of the rate of dilation of the optical contour of an approaching object specifies the time remaining until contact if speed of approach is constant, this variable (which has been denoted 'τ' by Lee, 1976) does not equal the real time to contact if the speed of approach is changing. Before we embark on a discussion of the effects hereof, let us first present a mathematical derivation of τ and $d\tau/dt$. For reasons of mathematical convenience we will only consider the simple situation depicted in Figure 12.2 (i.e., we will focus on the 2D angle, rather than on the solid [3D] angle subtended at the point of observation).

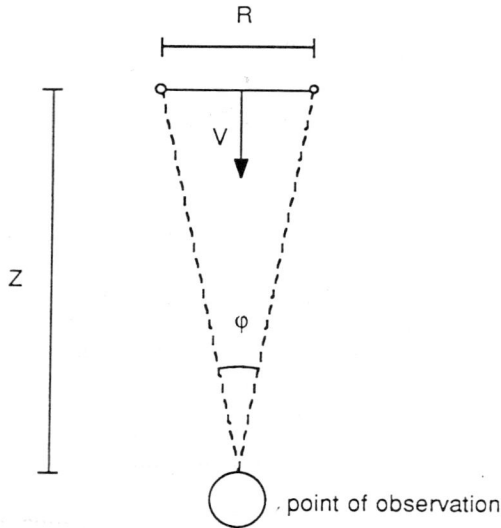

Figure 12.2. Geometrical relations between ball size (R), distance (Z), and optical angle (φ).

For an object of size R, at a current distance Z, travelling toward the point of observation with a velocity V (= dZ/dt), the optical angle φ subtended at the point of observation is given by

$$\tan \frac{1}{2} \varphi = \frac{R}{2*Z} \qquad (12.1)$$

For small φ, $\tan \frac{1}{2} \varphi = \frac{1}{2} \varphi$, hence

$$\varphi = \frac{R}{Z} \qquad (12.2)$$

The rate of change of φ over time is therefore

$$d\varphi/dt = - \frac{R*dZ/dt}{Z^2} \qquad (12.3)$$

The ratio of φ and $d\varphi/dt$ specifies the remaining time until contact[3], if dZ/dt is constant, as is evident from

$$\tau = \frac{\varphi}{d\varphi/dt} = \frac{R}{Z} * - \frac{Z^2}{R*dZ/dt} = - \frac{Z}{dZ/dt} = \text{time to contact} \qquad (12.4)$$

(If the velocity with which the distance is closed is considered to be negative, the velocity with which the angle φ grows is, per definition, positive; hence the minus sign). Thus, while physical size (R), physical distance (Z), and physical velocity (dZ/dt) are visually ambiguous (see equations 12.2 and 12.3)[4], the time remaining until contact (if speed of approach is constant) is directly available in the optic flow field, as it is completely specified in optical parameters in equation (12.4). In terms of the evolutionary pressures that have acted on almost all visually equipped species, it makes sense that an observer confronted with an unknown approaching object does not have to rely on estimates of distance and velocity based on ambiguous cues (or worse, have to wait until binocular information becomes available), but is able to directly perceive the time remaining until collision, and hence act accordingly. Recently, the availability of a similar source of information in the acoustic flow field has been demonstrated by Schiff and Oldak (1990), suggesting that

---

[3] The rate of dilation is defined as the ratio of speed of growth and instantaneous size, which in this case equals $[(d\varphi/dt)/\varphi]$. Therefore, τ equals the inverse of the rate of dilation.

[4] Obviously, this holds only for the monocular case. For the types of task of interest for the present contribution, the distance from the point of observation at which relevant information needs to be picked up will typically not allow the use of binocular information sources.

direct access to this important information quantity is not limited to visually equipped species.

An ongoing discussion in the literature concerns itself with the question whether τ is a unitary, higher order variable, or whether the remaining time to contact is computed on the basis of the separately assessed visual distance and velocity estimates (Abernethy & Burgess-Limerick, in press; Cavallo & Laurent, 1988; Laurent, 1991; Laurent, Dinh Phung, & Ripoll, 1989; see also Fleury, Bard, Gagnon, & Teasdale, this volume). In fact, even Lee (1976) might be interpreted to imply the latter, when he suggests that the threshold value for the perception of τ is dependent on the threshold for the perception of $d\varphi/dt$. It is important to realize that the inverse of τ, that is $(d\varphi/dt)/\varphi$, can simply be rewritten as $d(\ln \varphi)/dt$, and all that is needed therefore is a mechanism sensitive to the rate of change of the logarithm of the size of the angle subtended by the approaching object at the point of observation. Regan and Beverley (1978) have identified what they call 'looming detectors' in the human visual pathway, showing that a separate channel exists for the detection of change in angular size. While they did not ascertain whether this channel is sensitive to change in absolute angular size or its logarithm, the latter is certainly a very real option. Whether τ is the only source of information used in the tasks addressed by Abernethy and Burgess-Limerick (in press), Cavallo and Laurent (1988), and Laurent et al. (1989) is perhaps debatable, but the fact that it is a unitary, higher order variable seems less problematic (see also Tresilian, 1990).

What would happen if the speed of approach is not constant, as we have assumed above? Obviously, τ then no longer represent the real time to contact, because it does not take d^2Z/dt^2 into consideration. (For reason of conceptual clarity, Lee and Young [1985] propose to denote the physical variable that is the ratio of instantaneous distance [Z] and instantaneous approach speed [dZ/dt] the 'tau-margin'. In that terminology, the optical variable τ always specifies the tau-margin, which only equals the remaining time to contact if the speed of approach is constant.) Consider the case in which the speed of approach is decreasing at a constant rate by exactly the amount needed to bring the object to a stop right at the point of observation (i.e., dZ/dt = 0 when Z = 0). From first principles it follows that this is true when

$$d^2Z/dt^2 = \frac{(dZ/dt)^2}{2*Z}$$

(12.5)

The rate of change over time of τ (equation 12.4) can be seen to be

$$d\tau/dt = \frac{Z*d^2Z/dt^2}{(dZ/dt)^2} - 1$$

(12.6)

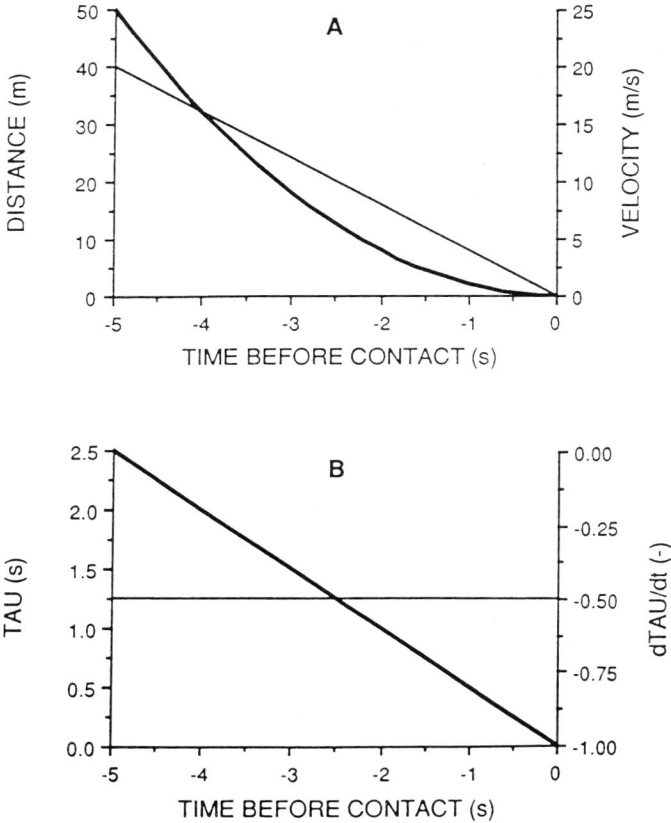

Figure 12.3. *(A) Physical distance (thick line) and velocity (thin line) as a function of time for a relative approach that will result in 'soft' contact with the point of observation, utilising a constant deceleration rate. Initial distance was set at 50 m, and initial velocity was set at 20 m s⁻¹. Deceleration was 4 m s⁻² throughout. (B) The optical variables τ (thick line) and dτ/dt (thin line) as a function of time for the situation depicted in A.*

Substituting equation (12.5) in equation (12.6) we find that in the case of a speed of approach decreasing at a constant rate so that the velocity at contact is exactly zero, $d\tau/dt = -0.5$ (see Figure 12.3). It is also evident that if the deceleration is not enough, $d\tau/dt < -0.5$, and that if the deceleration is large enough to stop before contact is made, $d\tau/dt > -0.5$. Thus, 'soft' contact is specified[5] by $d\tau/dt \geq -0.5$, and 'hard' contact is specified by $d\tau/dt < -0.5$. A driver, who wants to avoid colliding with a stationary car in front, thus simply needs to decelerate the car by an amount that leads to an optically specified value of $d\tau/dt > -0.5$. The reader is referred to Lee (1976) for a full treatment of the visual control of braking (see also Yilmaz [1991] for an experimental corroboration).

If an observer were able to use τ and $d\tau/dt$ simultaneously, the real remaining time to contact could in fact be arrived at for this particular condition (constant deceleration), using equations (12.4), (12.5), and (12.6), since

$$\text{time to contact} = \frac{\tau * (1 + \sqrt{1 - 2 * (d\tau/dt + 1)})}{d\tau/dt + 1} \qquad (12.7)$$

which is completely specified in optical variables in equation (12.7). However, biology might prefer to keep things simple. Rather than going through the hassle of ascertaining the real time to contact, with the chances of serious errors if, for instance, on occasion $d\tau/dt$ is not perceived accurately, biological perception-action systems might prefer to stick with τ. Note also that equation (12.7) only specifies the real time remaining until contact if the approaching object is, relatively to the observer, decelerating at a constant rate. If any change would occur in (i) the forces acting on the perceived object (which is quite likely if it were, for instance, another animal) or (ii) in the velocity of the observer, even more complicated relations would have to be worked out. Suppose the observer would not want to make any *a priori* guesses. What would happen if perceivers/actors would simply gear their actions to τ?

A first study addressing this issue was performed by Lee and Reddish (1981). They analysed films of gannets as they dived into the sea to catch fish. These birds typically do so from varying heights. As they start their dive, they go into a posture that allows them to experience little air resistance, but at the same time maintain some basic steering capabilities. If they were to hit the water in that configuration, however, the surface tension could impart serious

---

[5] Note that only the constant deceleration that will lead to $dZ/dt = 0$ when $Z = 0$ actually gives rise to a constant value of $d\tau/dt$ (namely -0.5) over time. In order to maintain a constant value of $d\tau/dt$ other than -0.5, the deceleration will have to be adjusted during the braking process.

injury in their outspread wings. Hence, just prior to entering the water the gannets streamline their wings, assuming an optimal aero- and aquadynamic posture. The question posed by Lee and Reddish concerns the timing of this final streamlining. The idea was that the birds, irrespective of the height from which they started their dive, would streamline when the tau-margin (optically available as τ) reached a certain critical value. As this critical tau-margin value is reached while the bird is farther away in time from the water surface for dives that are initiated from a larger altitude, the relation between the real time before contact at which the streamlining occurred and the duration of the dive (which is equal to the square root of two times the starting height divided by the gravitational acceleration) is non-linear. The moment of streamlining was found to follow the predicted function. Interestingly, following this strategy the bird has more time to streamline itself the longer the dive. Thus, the greater risk of injury associated with a longer dive (i.e., from a larger altitude, and therefore having a larger velocity at contact) is paired with a larger margin allowed for error.

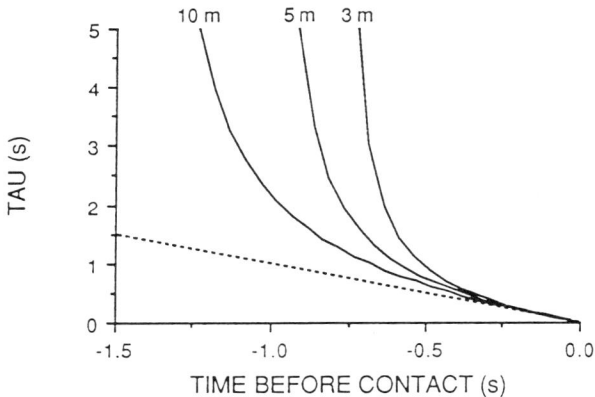

Figure 12.4. The tau-margin as a function of time before contact for balls dropped from different heights. The dotted line represents the identity condition.

A second study, by Lee, Young, Reddish, Lough, and Clayton (1983), examined the way in which human subjects jump up to hit a falling ball. Here, again, there is a difference between the tau-margin and the real time to contact, because a falling ball accelerates due to the gravitational force. Subjects were required to first crouch down, and then jump up, attempting to contact the ball with their fist at the highest possible point. By monitoring the angles at the elbow and the knee, Lee et al. were able to evaluate the temporal unfolding of the act as a function of the height from which the ball was dropped. A plot of the tau-margin as a function of time to contact for different starting heights (see Figure 12.4) reveals that differences between the tau-margin and the time to contact are large early on in the drop but grow progressively smaller as the ball approaches the point of observation.

The temporal unfolding of the act for balls dropped from different heights could be understood as a single function of the tau-margin, while different functions were needed for each height to describe the behavior as a function of the time to contact. The invariance found for the tau-margin strategy (and *not* for time to contact) provides strong evidence for the argument that the action was geared to information about the tau-margin (accessible via τ), rather than information about the real time to contact. Thus, biology indeed seems to prefer to keep things simple!

Since these original experiments by Lee and co-workers, quite a number of studies have been reported, testifying to the usefulness of tau-margin information in the regulation of action (Bootsma, 1989; Bootsma & Van Wieringen, 1988, 1990; Fitch & Turvey, 1978; Laurent et al., 1989; Lee, Lishman, & Thomson, 1982; Savelsbergh, Whiting, & Bootsma, 1991; Van der Horst, 1990; Warren, Young, & Lee, 1986). With the focus of the present chapter being on catching and hitting, we will only discuss studies relating to these skills specifically.

Timing in Hitting

The classical work on batting a pitched baseball by Hubbard and Seng (1954) has been re-interpreted by Fitch and Turvey (1978) as exemplary of the use of τ. Examination of Hubbard and Seng's data revealed that the duration of the swing was essentially constant, irrespective of the speed of the pitch.[6] If the batter would use a τ-strategy, this is exactly what would be predicted. Suppose that the batter would initiate the swing when the tau-margin reached a certain critical value. Ignoring the marginal deceleration due to air resistance, this implies that the same amount of time would always remain before the ball crossed the plate. In order to make contact with the ball, the duration of the swing would, therefore, necessarily have to be kept constant, irrespective of the speed of the ball. Along very similar lines, but from a different

---

[6] Examination of the data reported in Table 2 of Hubbard and Seng (1954) reveals that the duration of the swing is, in fact, not completely independent of the ball speed, $r(70) = -.257$, $p < 0.05$. However, the best fit linear relationship is MT (ms) = 250.0 - 2.42 * Ball Speed (m/s), and it is evident that the slope is very small indeed.

perspective, Tyldesley and Whiting (1975) proposed an 'operational timing hypothesis' to account for the consistency that they observed in the execution of an attacking forehand drive by an expert table tennis player. Tyldesley and Whiting argued that, over hundreds of thousands of trials, an expert player learns to produce a consistent movement pattern (for a particular drive) in both the spatial and the temporal domain. The ability to reproduce this consistent movement pattern would enable the expert player to deal only with the problem of when to initiate the drive. Although Tyldesley and Whiting were, at the time, unaware of the τ variable, it does not take a great leap to suggest that this would provide exactly the information needed by the expert player.

In a recent study, Bootsma and Van Wieringen (1988) presented an expert table tennis player with balls approaching at different velocities. As expected, the duration of the forward swing phase was found to be fairly constant, and independent of the approach velocity. This led them to conclude that the initiation of the drive is indeed regulated on the basis of tau-margin information, as the use of some type of distance information would have necessarily affected the duration of the drive. In a subsequent study (Bootsma & Van Wieringen, 1990), the 'operational timing hypothesis' was directly addressed. While all five expert players studied were found to produce quite consistent movement patterns (e.g., standard deviations of movement time were between 5 and 21 ms), it was also found that all five revealed a better temporal accuracy at the moment of ball/bat contact (2-5 ms!) than at the moment of movement initiation. This clearly contradicts the operational timing hypothesis, because its starting point is that the player would control the moment of initiation, after which the movement pattern would run off ballistically. If anything, between-trial variability should increase from the moment of initiation until the moment of ball/bat contact, due to inherent 'noise' in the neuromuscular system (Anderson & Pitcairn, 1986). The finding of decreasing between-trial variability therefore dictated a different interpretation.

Bootsma and Van Wieringen (1990) argue for a continuous perception-action coupling perspective. A funnel-like type of control (Bootsma, Houbiers, Whiting, & Van Wieringen, in press) would have the player initiate the drive within a certain spatio-temporal bandwidth, rather than at a specific, precisely defined point in space and time. According to the perceptual information available, the drive may then be adjusted so that the act becomes appropriately constrained for the all-important moment of ball/bat contact. Such a strategy would naturally lead to the progressive increase in accuracy as the act unfolds, as reported by Bootsma and Van Wieringen (1990). Additional experimental evidence comes from the finding of negative intra-individual correlations between the magnitude of the optically available time remaining until contact (i.e., τ) and the average acceleration of the bat that were found for all five players studied. Having started a little earlier than average (i.e., τ at initiation is somewhat larger than average, but still within the allowable bandwidth), and therefore having a little more time, was found to be associated with an acceleration of the bat somewhat less than average. Vice versa, having started a little later than average, and therefore having a little less time than average, was found to be associated with an acceleration of the bat that was somewhat larger than average. In this way, each individual drive is optimally attuned to the situation at hand. Importantly, the continuous perception-action coupling perspective allows one to understand the invariance (e.g., MT is independent of ball velocity) and variance (e.g., variations in MT reflect functional adaptations) within the same framework. Various variants of the generalised motor program concept (Gentner, 1987; Heuer, 1988; Schmidt, 1975) typically invoke different levels for the explanation of variance and invariance. Moreover, more often than not, variance is regarded to be the result of error, or noise, while the results of Bootsma and Van Wieringen (1990) strongly suggest that it constitutes functional (or compensatory) variability (see also Beek, in press, for a concurrent explanation of variance and invariance in juggling).

Timing in Catching

While catching[7] is similar to hitting in a number of ways, there are some important differences. In hitting, the main objective is to make contact with the ball, at a particular place and time (and often with the end effector moving in a particular direction). In catching this is to be complemented with the task of closing the fingers around the ball. Alderson, Sully, and Sully (1974) have demonstrated that in order to catch a tennis ball approaching at a speed of some 10 m s^{-1}, the catcher has a time window of 50 ms in which the hand needs to be closed. Closing too early will cause the ball to be deflected off the fingers, while closing too late will result in the ball bouncing off the palm of the hand/fingers. Thus, the 'transport' component of a hitting action needs to be coordinated with a 'grasp' component in catching. To date, not much work has been reported on the way in which these two components are coordinated in catching (but see Bootsma and Van Wieringen [in press]; Jeannerod [1984]; Marteniuk, Leavitt, MacKenzie, and Athenes [1990] for a discussion on possible modes of coordination of transport and grasp components in prehension). The distinction drawn in some of the literature on one-handed catching between *position* and *grasp* errors (e.g., Fischman & Schneider, 1985; Smyth & Marriott, 1982; Whiting, 1986; see also Smyth, 1986; Fischman,

---

7 The focus here will be on one-handed catching.

1986) might be considered to be related to these different components. Typically (e.g., Smyth & Marriott, 1982, p. 146) a trial in which the ball does not make contact with the outstretched hand is identified as a position error, while a trial in which the ball does contact the hand (sometimes refined to contacting the metacarpophalangeal pads of the fingers or the palm of the hand), but is not caught, is identified as a grasp error. The terminology chosen (i.e., position and grasp errors) is derived from the Alderson, Sully, and Sully (1974) description of the catching act in which a distinction is made between a positioning component (that can be fractionated into gross and fine positioning components [Alderson et al., 1974; Whiting, 1986]) and a grasping component. The suggestion that the two types of error described above are related to errors in the two components of the catching action is however not warranted. A failure to catch a ball can only be designated to be due to a grasp error if the hand was positioned correctly in the first place, because it is defined as a ball that did contact the hand. Obviously, grasp component errors can occur also when this is not the case. The terminology adopted by Savelsbergh and Whiting (1988; Savelsbergh, 1990; Whiting, Savelsbergh, & Faber, 1988), in which position errors are equated with spatial errors and grasp errors are equated with temporal errors, is also somewhat misleading. Catching is often not performed with the hand held stationary at the anticipated location of contact. Positioning of the hand is therefore a spatio-temporal rather than a mere spatial problem. Hence, being at the wrong place at a certain time can be the result of (i) a spatial error, or (ii) a temporal error, which then amounts to the same thing as being at the right place at the wrong time. Both the positioning (or transport) component and the grasping component are liable to suffer spatial and temporal mistakes. In accordance with the proposition of Rosengren, Pick, and Von Hofsten (1988), we suggest to only distinguish between *catch, hand contact,* or *miss* in studies that do not take direct measures of the kinematics of the catching movement.

The finding reported by Rosengren et al. (1988) and Savelsbergh and Whiting (1988) that subjects are actually quite able to catch luminescent balls in an otherwise completely darkened room is of importance for the present chapter, as it testifies to the availability of useful temporal and spatial information sources that can be attributed solely to the perceived ball motion (see also Proteau, this volume). The facilitating effect on catching afforded by a minimal visual frame of reference (Rosengren et al., 1988; Savelsbergh, 1990) will be of interest in the sections on generalised tau-margin information and predictive spatial information sources.

The timing of the grasp component of a catching action has recently been addressed by Savelsbergh, Whiting, and Bootsma (1991). As for hitting, it is hypothesised that the timing of the grasp component is regulated on the basis of the temporal information specified by τ. Whereas the behavioral experiments of Lee and Reddish (1981), and Lee et al. (1983) discussed earlier demonstrated that the temporal unfolding of a number of acts is geared to the tau-margin, rather than to the real time to contact, this provides strong but strictly speaking nevertheless only circumstantial evidence for the use of τ. Savelsbergh et al. (1991) decoupled the information provided by the rate of dilation of the optical contour of the approaching ball and its instantaneous position and velocity by using a ball that changed size as it approached the subject. As control conditions, a ball of constant size, equal to the size of the 'deflating' ball at the start of its trajectory, and a ball of constant size, equal to the size of the 'deflated' ball at the end of its trajectory, were used. If the information used for the timing of the closing action of the hand was indeed τ, it can be predicted that subjects should perform the closing action later for the deflating ball, relative to the closing action for the two balls of constant size, because the rate of dilation would specify a longer time to contact. In fact, this was exactly what was found. Under both binocular and monocular viewing conditions, the timing of the closing of the hand, as characterised by the time at which peak closing velocity of the hand was reached, was found to occur significantly later for the deflating ball than for the balls of constant size (Savelsbergh et al., 1991). The conclusion can be drawn that the timing of the grasping component is indeed based on the perceived rate of dilation of the optical contour generated in the optic array by the approaching ball.

Tau-margin Information for Approach to a Position Other than the Point of Observation

In the previous sections we have assumed implicitly that the ball was directly approaching the point of observation. Typically, however, the actor will not position the point of observation directly in line with the ball's trajectory. In one-handed catching the ball is normally contacted at eyeheight, but at some distance from the head. A baseball is typically hit at a location both lower than and to the side of the head. Thus, the catcher and the hitter are in need of predictive information that specifies how long it will take until the ball reaches a designated position, rather than the point of observation. Consider the situation depicted in Figure 12.5. If the ball is to be contacted at point p, the tau-margin information needed is the time until the ball reaches p (see Bootsma, 1988; Bootsma & Oudejans, 1991; Tresilian, 1990).

Let φ denote, as before, the angle subtended at the point of observation by the ball, and let θ denote the angle formed by the current position of the ball, the point of observation, and the future point of contact p. From the sine rule it follows that

$$D = \frac{\sin \theta * Z}{\sin \beta}$$

(12.8)

Because the angle β does not change over time, the rate of change of D is

$$dD/dt = \frac{dZ/dt * \sin \theta + Z * \cos \theta * d\theta/dt}{\sin \beta}$$

(12.9)

The inverse of the tau margin is therefore given by

$$- \frac{dD/dt}{D} = - \frac{dZ/dt}{Z} - \frac{d\theta/dt}{\tan \theta}$$

(12.10)

For small angles φ and θ, $- \frac{dZ/dt}{Z} = \frac{d\varphi/dt}{\varphi}$, and $\tan \theta = \theta$,

and equation (12.10) becomes

$$- \frac{dD/dt}{D} = \frac{d\varphi/dt}{\varphi} - \frac{d\theta/dt}{\theta} = d(\ln \varphi)/dt - d(\ln \theta)/dt$$

(12.11)

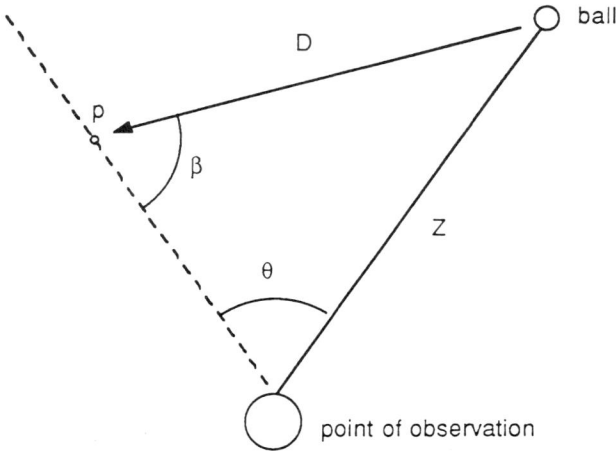

Figure 12.5. Geometrical relations between distance to be covered (D), distance from the point of observation (Z), and optical angle (θ).

The required tau-margin (the inverse of equation [12.11]) is described completely in optical terms, and could therefore potentially be used to regulate behavior. Using a forced-choice discrimination paradigm similar to that of Todd (1981), Bootsma and Oudejans (1991) experimentally addressed the question whether human observers were sensitive to the information described in equation (12.11). Note that if the object approaches the point of observation directly ($d\theta/dt = 0$), equation (12.11) reduces to equation (12.4). Also note that if the object does not approach the point of observation ($d\varphi/dt = 0$), the tau margin is specified simply by the inverse of $d(\ln \theta)/dt$. This latter situation was considered first. Subjects were presented with simulations of two objects approaching a common wall from the left and right side. For each of the two objects, object size and starting position were varied. On the basis of the starting positions selected for a particular trial, approach velocities were chosen such that the difference in the arrival times at the common wall was 10, 20, 50, 100, or 200 ms. Similar to Todd's procedure, subjects were required to respond before either of the objects reached the common wall. It was found that subjects were able to correctly discriminate which of the objects would arrive first well over 90% in the 200 ms arrival time difference condition. Performance was significantly better than chance already for the 50 ms arrival time difference condition. In a subsequent series of experiments, the objects were simulated to move in depth ($d\varphi/dt \neq 0$) as well as sideward ($d\theta/dt \neq 0$). Subjects were found to perform quite accurately in these conditions as well, although movement in depth was found to affect performance. Bootsma and Oudejans suggest that this could be due to the fact that rate of constriction of the optical gap (the term of equation [12.11] relating to θ) is given somewhat more weight than the rate of dilation of the optical contour (the term of equation [12.11] relating to φ). This might however have been the result of the simulations not conveying a real sense of 3D, and any definite conclusions about differential weighting in real movement situations must await further experimentation. Nevertheless, given the overall accuracy of performance, Bootsma and Oudejans conclude that human observers are sensitive to the information described in equation (12.11).

The availability of a visual information source specifying the tau-margin for situations in which one object approaches another object within the field of view allows for a consideration of this source of information for the regulation of a number of actions. For instance, the movement of the hand as it approaches a target, as in an aiming movement, or a target object, as in a prehensile movement, might be regulated on the basis of the perceived tau-margin (Bootsma & Van Wieringen, in press). MacKenzie, Marteniuk, Dugas, Liske, and Eickmeier (1987), Marteniuk, MacKenzie, Jeannerod, Athenes, and Dugas (1987), and Marteniuk et al. (1990) have demonstrated that changing the spatial accuracy demands in aiming and prehension tasks leads to differential effects in the deceleration phase of the transport component movement, while the acceleration phase is not affected. An analogy with the regulation of braking in avoiding hard collisions while driving a car (Lee, 1976), as discussed earlier, can be suggested.

Tresilian (1990) notes that the finding of Rosengren et al. (1988) and Savelsbergh (1990) of a facilitating effect of a minimal visual frame of reference might be understood on the basis of the frame allowing a better assessment of θ, and hence a better assessment of the temporal information described in equation (12.11). As we shall see, a similar beneficial effect might be expected for the pick up of spatial information.

Predictive Spatial Information

The issues addressed in this last section raise the point of how an actor can determine *where* an approaching ball will go. Already in 1965, Schiff demonstrated that fiddler crabs presented with asymmetrically expanding shadow patterns tended to move at approximately right angles to the apparent path of approach, and he suggested that the animals may pick up information concerning the path of approach as specified in the degree of skew in the magnification. Lee and Young (1985) have pointed to the fact that the displacement of the center of expansion that occurs when the ball will not make contact with the point of observation could be used to obtain the direction of motion. Fitch and Turvey (1978) had in fact formulated the same point, but from the perspective of occlusion of background. They pointed out that an asymmetrical deletion/accretion of background texture can be used to obtain the direction of motion. Regan, Beverley, and Cynader (1979) also address the specification of the direction of motion, but do so from a binocular perspective, arguing that the ratio of velocities on the left and right retinae provides such information for movement in the horizontal plane. It is important to note however that *direction of motion* is of itself not enough to allow a prediction of *future passing distance,* as it needs to be combined with a current distance estimate.

Todd (1981) presents an interesting solution for what he denotes the 'outfielder problem'. When a batter hits a fly ball toward the outfield, the outfielder needs to position himself near the future place of landing of the ball in order to catch it. Interestingly, outfielders seem to be able to determine almost instantaneously where the ball will land relative to their current position. Todd proposes that information about the future landing position relative to the outfielder's current position is available in the ratio of two times, relating to different planes, namely the horizontal and the vertical plane. Let T_Y denote the time until the ball returns to eyeheight (vertical plane of motion) and let T_Z denote the time until the ball reaches the fronto-parallel plane (horizontal plane of motion). The ratio T_Y/T_Z then specifies whether the outfielder should move forward, backward, or stay where he is. If T_Y/T_Z = 1, the ball will land exactly at the current position of the point of observation. If $T_Y/T_Z > 1$, the ball will still be above eyeheight when it reaches the fronto-parallel plane, and it will therefore land behind the player. Conversely, if $T_Y/T_Z < 1$, the ball will reach eyeheight before it reaches the fronto-parallel plane and the ball will land in front of the player.[8] In order for this strategy to be viable, T_Y and T_Z need to be optically available. Let us therefore consider the situation depicted in Figure 12.6 in the two planes of interest.[9]

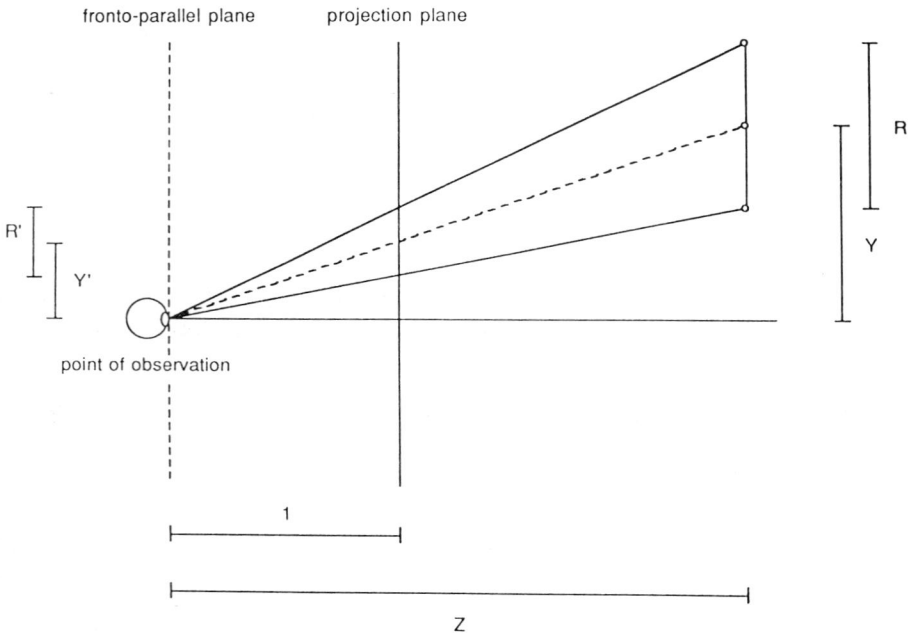

Figure 12.6. Geometrical relations between a moving object (R) and a projection plane.

---

[8] Landing position is taken here with reference to the horizontal plane located at eyeheight.

[9] For reasons of mathematical convenience, the present analysis is performed relative to a plane, at a distance of unity from the point of observation. As Lee (1974) has demonstrated that optic variables associated with one surface can be uniquely transformed into optic variables associated with any other surface, this convenient choice is warranted.

In the situation under examination we will neglect the decelerative influence of air resistance and assume that $d^2Z/dt^2 = 0$.

$$Z = \frac{R}{R'}$$ (12.12)

$$dZ/dt = \frac{-R*dR'/dt}{R'^2}$$ (12.13)

$$Y = Z*Y'$$ (12.14)

$$dY/dt = Z*dY'/dt + dZ/dt*Y'$$ (12.15)

$$d^2Y/dt^2 = Z*d^2Y'/dt^2 + 2*dZ/dt*dY'/dt$$ (12.16)

Combining equations (12.12) and (12.13) it can easily be shown that the time (T_Z) until the ball reaches the fronto-parallel plane is given by (see equation [12.4])

$$T_Z = \frac{R'}{dR'/dt}$$ (12.17)

For the vertical plane the situation is a little more complicated, because we need to incorporate the acceleration due to gravity. The time until the ball returns to eyeheight is given by

$$0 = Y + dY/dt * T_Y + 0.5 * d^2Y/dt^2 * T_Y^2$$ (12.18)

Division by Y yields

$$0 = 1 + \frac{dY/dt}{Y} * T_Y + 0.5 * \frac{d^2Y/dt}{Y} * T_Y^2$$ (12.19)

The individual terms of equation (12.19) can expressed completely in optical terms, using equations (12.12) to (12.16):

$$\frac{dY/dt}{Y} = \frac{dY'/dt}{Y'} - \frac{dR'/dt}{R'}$$ (12.20)

$$\frac{d^2Y/dt^2}{Y} = \frac{d^2Y'/dt^2}{Y'} - \frac{2*dY'/dt * dR'/dt}{Y' * R'}$$ (12.21)

and T_Y can therefore also be expressed completely in optical terms using the quadratic formula. With T_Y and T_Z both being optically specified, the T_Y/T_Z ratio can, potentially, be used.

Using a forced-choice paradigm, Todd (1981) found that subjects could accurately discriminate between balls that would eventually land in front of them and balls that would land at the point of observation. More than 70% correct responses were obtained for a difference in future landing position as small as 3 m, and more than 90% correct responses for a difference of 9.1 m. Importantly, this surprising accuracy was achieved on the basis of the information available during the initial phase of ball flight (i.e., before it reached the top of its trajectory), when the distance from the observer was between 15 and 30 m.[10]

Note however that the T_Y/T_Z ratio does not specify where the ball is going to break the fronto-parallel plane of the observer, which is the kind of predictive spatial information needed if one is going to attempt to catch or hit the ball from the current point of observation. It is possible to demonstrate that this future *place of contact* is specified optically as well (Bootsma, in press; Bootsma & Carello, 1991). For reasons of clarity we will first consider the situation in which the approaching ball does not have a downward acceleration (i.e., $d^2Y/dt^2 = 0$). While this situation is unrealistic in our normal environment, it should be borne in mind that it does apply quite often when we consider the future place of contact in the horizontal plane. The future place of contact in the fronto-parallel plane (Y_c) is then given by

$$Y_c = Y + dY/dt * T \qquad\qquad (12.22)$$

where T stands for the time remaining until the ball reaches the fronto-parallel plane (i.e., $Z = 0$). Using equation (12.17) if follows that

$$Y_c = Y + dY/dt * \frac{R'}{dR'/dt} \qquad\qquad (12.23)$$

---

[10] With the parameters used by Todd (1981; see p. 802) in his simulations, the optical expansion velocity was always greater than 0.2 deg/s. This is well above the threshold of 0.08 deg/s mentioned by Lee (1976), and suggests that subjects could indeed use information based (in part) in optical expansion. For a real outfielder, standing at a distance of 75 m from home plate, a baseball (diameter 7.5 cm) hit with a horizontal velocity component of 50 m/s will produce a rate of expansion of approximately 0.04 deg/s. This is a factor two below the threshold described by Lee (1976) and would, at first sight, seem to implicate that expansion information could not be used. Additional experimentation will have to address this point in order to decide whether higher order variables incorporating expansion velocity are subject to the same threshold values.

Appropriate substitution, using equations (12.12) to (12.15), enables equation (12.23) to be written as

$$\frac{Yc}{R} = \frac{dY'/dt}{dR'/dt} \qquad (12.24)$$

Thus, the ratio of the optical displacement velocity and the optical expansion velocity specifies the future passing height (relative to eyeheight, where $Y = 0$) in units ball size. In almost every ball game, the ball size is regulated to be constant, and for a player to perceive distance in terms of ball size is therefore appropriate.

Including an acceleration d^2Y/dt^2, Y_c is given by

$$Y_c = Y + dY/dt * T + 0.5 * d^2Y/dt^2 * T^2 \qquad (12.25)$$

which can be rewritten to become

$$\frac{Yc}{R} = \frac{d^2Y'/dt^2 * R'}{2 * (dR'/dt)^2} \qquad (12.26)$$

Equation (12.26) suggests that observers would need to be sensitive to optical acceleration, in order to arrive at an estimate of future place-of-contact. A number of studies have suggested that observers can only crudely assess such second order information (Bootsma & Oudejans, 1991; Todd, 1981; Todd & Bressan, 1990). Interestingly, the right part of equation (12.26) can also be expressed as

$$\frac{Yc}{R} = \frac{dY'/dt}{dR'/dt} + 0.5 * d\left(\frac{dY'/dt}{dR'/dt}\right)/dt * \frac{R'}{dR'/dt} \qquad (12.27)$$

Obviously, equation (12.27) reduces to equation (12.24) if no acceleration were to occur in Y-direction. As in equation (12.24), the future place of contact is specified in units of ball size in equation (12.27). Notice that the optical specification of Y_c/R also allows the type of judgement to be made that was required by Todd (1981): If $Y_c/R > 0$ the ball will land behind the point of observation, and if $Y_c/R < 0$ the ball will land in front of the point of observation.

The feasibility of the use of the information described in equations (12.24) and (12.27) rests on the claim that future passing distance is perceived in units ball size. The experiment to be described shortly was set up to test this claim.

As it has not yet been reported elsewhere, we will provide a little more detail than in the foregoing.

We have previously reported that human observers are able to judge quite accurately whether an approaching ball, that is viewed for only a brief period of time during its approach, will be reachable when it passes the fronto-parallel plane (Bootsma, in press; Bootsma, Bakker, Van Snippenberg, & Tdlohreg, 1991). Note that these judgements required predictive spatial information in the horizontal rather than the vertical plane, as balls would pass the observers at constant height, but at different lateral distances from their midpoint. The information described in equations (12.24) and (12.27) can be transformed to the required direction (which would be X-direction for the reference frame defined in Figure 12.6, perpendicular to the page plane) by simply substituting X for Y. Future passing distance in the horizontal plane, X_c, is specified by the ratio of dX'/dt and dR'/dt if no acceleration occurs in X-direction. Having established that observers can actually make accurate perceptual judgements of reachability in this plane, we decided to use this task to evaluate the effect of unexpectedly and infrequently introducing a ball of a different size. Based on the hypothesis that subjects use information similar to that described in equation (12.24) or (12.27) to arrive at such a judgement, it can be predicted that introducing a ball that is smaller than the standard ball will lead to an underestimation of the judged critical passing distance[11], and introducing a ball that is larger than the standard ball will lead to an overestimation of the judged critical passing distance, as long as subjects are unaware of the fact that ball size has been changed and are therefore still scaling distance to the standard ball size.

Eight subjects participated in the experiment, all of whom reported normal or corrected to normal vision. From a rail attached to the ceiling (6.25 m above the floor) of a large experimentation hall 10 balls were suspended using fishing line (4.90 m in length), with 10 cm horizontal distance between adjacent suspension points. After having been pulled back and up to a fixation beam, 3.20 m in length and 5.05 m above the floor, the balls were held in place by means of 10 solenoids.

During a block of trials, all 10 balls were released one by one in a randomized order. Upon release a ball would fall out of the solenoid and swing across the room, reaching its lowest point of 1.35 m at a distance of 1.30 m in front of the observer. After passing through the observer's fronto-parallel plane, the ball would be caught on a 4 m long rod, positioned 0.7 m

---

[11] Judged critical passing distance is defined as the distance at which subjects judge 50% of the balls to be reachable.

behind the observer at a height of 1.70 m. After completion of a block of trials (10 balls), the balls could be pulled back up to the fixation beam using this rod. Balls passed through the subject's fronto-parallel plane at a distance of 0.5 m to 1.4 m to the right of the subject's midpoint, with 10 cm increments between adjacent balls. Three types of ball, differing in diameter only, were used: 4.9 cm (smaller), 5.75 cm (standard), and 6.9 cm (larger). All balls were painted mat black.

Subjects wore Liquid Crystal Display spectacles that could be made to change from opaque to transparent and vice versa, with a rise time of 3-5 ms. Balls were seen against a plain white background. The computer that regulated release of the balls from the solenoids also controlled the opening (700 ms after release) and closing (1300 ms after release) of the spectacles, with only the glass of the preferred eye ever becoming transparent (i.e., monocular viewing only). Total flight time of the balls, from release until passage through the observer's fronto-parallel plane was 1500 ms. During the task the subject stood on a thin (2 cm) wooden platform with the feet strapped to this platform, providing a constant, slightly opened stance.

In every block, all ten balls passed the subject on the right hand side. After closing of the spectacles the subject indicated verbally whether the ball was judged to be reachable or not. Actual reaching movements were not allowed. Two test blocks were presented in which only standard sized balls (diameter 5.75 cm) were used. In each of the following experimental blocks, one alternative ball was mixed with the remaining nine standard sized balls. The position of this odd ball was changed over the blocks, following a staircasing procedure (or up-and-down method, Dixon & Massey, 1969). If the odd ball was judged to be reachable, it would be presented one position further to the right in the next block. If it was judged to be not-reachable, it would be presented one position closer to the subject in the next block of trials. Thus, the positions occupied by the different-sized ball concentrated around the critical distance. The number of staircasing trials varied between 18 and 20, resulting in a fixed nominal sample size of 15 (Dixon & Massey, 1969).

Rest periods of two minutes were administered between blocks, during which time the balls were pulled up and repositioned. The subject would simply sit in a chair, still wearing the opaque glasses. Because we wanted to test two alternative ball sizes (smaller and larger than standard), each subject was tested in two sessions, on consecutive days. After the second session had been completed, the subject was asked whether he or she had noticed any differences in ball size. None of them reported having noticed this.

To obtain the critical passing distance, defined as the point at which 50% of the balls were judged to be no longer reachable, two different analyses were required. The staircasing method was used to find the judged critical distance for the alternative balls (Dixon & Massey, 1969). The judged critical distance for the standard ball was calculated for both sessions separately, and was derived by fitting the data with the logistic function:

$$\text{percent judged reachable} = \frac{100}{1 + e^{-k * (c-x)}} \tag{12.28}$$

where 100 is the maximum percentage, x is the passing distance, c is the 50% point, and k is a measure of the slope at this point (see Bootsma et al., 1991). R-squared for the fit obtained ranged from 0.96 to 1.00. Expressed as a percentage of the real maximum reaching distance (measured independently), Table 12.1 presents the judged critical passing distances found for each of the eight subjects.

Table 12.1.
Judged critical passing distances, expressed as a percentage of actual maximum reaching distance, for the standard (5.75 cm diameter), smaller (4.9 cm), and larger (6.9 cm) balls for each of the eight subjects.

	Smaller	Standard	Larger	Standard
Subject				
1	95.68	108.95	121.81	102.37
2	109.84	121.05	120.25	113.33
3	100.37	112.92	128.53	105.14
4	88.08	109.32	122.65	107.29
5	114.30	130.40	157.95	131.23
6	90.87	106.36	115.29	96.51
7	82.79	93.68	108.20	90.34
8	114.36	123.24	145.87	134.21
mean	99.54	113.24	127.57	110.05
sd	12.23	11.46	16.46	15.61
	$t(7) = 10.02, p < 0.001$		$t(7) = 7.89, p < 0.001$	

Paired t-tests revealed that the judged critical passing distance was significantly smaller for the smaller ball than for the standard ball, and that the judged critical passing distance was significantly larger for the larger ball than for the standard ball. No statistically significant difference was found to exist between the two standard ball sessions ($p > 0.10$). The magnitude of the overestimation of the critical point for the larger ball and the magnitude of the underestimation of the critical point for the smaller ball could be predicted for each individual subject to be respectively 6.9/5.75 and 4.9/5.75 times the critical point for the standard ball size. No statistically significant difference was found between the predicted and the actual data for the larger ball (t [7] = 1.77, $p > 0.05$). The effect of using the smaller ball on the other hand was a little less than predicted (t [7]= 2.01, $p < 0.05$, one-tailed).

Nevertheless, both the direction and the magnitude of the changes in judged critical distance were found to closely approximate the predictions made. We conclude that future passing distance indeed seems to be perceived in units balls size, as we had hypothesized on the basis of our analysis of available optical information (equations [12.24] and [12.27]). It is important to realise that in ball games, for instance, the size of the ball is always constant (in any case throughout the game, if not throughout the season), and that the use of ball size as the metric for perceiving future passing distance is therefore viable in these situations at least.

Finally, let us return to the finding of Rosengren et al. (1988) and Savelsbergh (1990) that provision of a (minimal) frame of reference facilitates the catching of luminescent balls in an otherwise completely darkened environment. As noted earlier, Tresilian (1990) rightly argues that one factor of importance can be the more accurate perception of the angle θ (see Figure 12.5), allowing a better perception of the time remaining until the ball reaches the position in which it will be caught (see equation [12.11]). On the basis of the analysis provided for the perception of future place of contact, however, it can also be argued that the provision of a (minimal) frame of reference allows a better assessment of displacement velocity, dY'/dt, and hence a better perception of the future place of contact, as specified (at least partially) by the ratio of displacement and expansion velocities (see equations [12.24] and [12.27]).

Concluding Remarks

The analyses presented above have shown that predictive information about *when* an approaching ball will be *where* is indeed available in the transformations of optic array sampled at the point of observation of the actor. Moreover, the experiments reviewed have demonstrated that humans (and an interesting range of other animals) are able to utilise these information sources. The power of *predictive information* lies in the possibilities it offers for *prospective control*. Rather than relying on feedback about the current state of affairs and having to regulate behavior on the basis thereof, predictive information allows an assessment of what will happen in the nearby future. Success in actions like catching and hitting, as well as many others (e.g., Lee, 1976; Warren, Morris, & Kalish, 1988; Warren, Mestre, Blackwell, & Morris, 1991), would be almost impossible without such predictive information.

References

Abernethy, B., & Burgess-Limerick, R. (in press). Visual support for the timing of skilled movements: A review. In J.J. Summers (Ed.), *Approaches to the study of motor control and learning.* Amsterdam: North-Holland.

Alderson, G.J.K., Sully, D.J., & Sully, H.G. (1974). An operational analysis of a one-handed catching task using high-speed photography. *Journal of Motor Behavior, 6,* 217-226.

Anderson, M., & Pitcairn, T. (1986). Motor control in dart throwing. *Human Movement Science, 5,* 1-18.

Beek, P.J. (in press). Inadequacies of the proportional duration model: Perspectives from a dynamical analysis of timing in juggling. *Human Movement Science.*

Bootsma, R.J. (1988). *The timing of rapid interceptive actions: Perception-action coupling in the control and acquisition of skill.* Amsterdam: Free University Press.

Bootsma, R.J. (1989). The accuracy of perceptual processes subserving different perception-action systems. *Quarterly Journal of Experimental Psychology, 41A,* 489-500.

Bootsma, R.J. (in press). Predictive information and the control of action: What you see is what you get. *International Journal of Sport Psychology.*

Bootsma, R.J., Bakker, F.C., Van Snippenberg, F.J., & Tdlohreg, C.W. (1991). *The effects of anxiety on perceiving the reachability of passing objects.* Manuscript submitted for publication.

Bootsma, R.J., & Carello, C. (1991). *Optical information for perceiving the future position of a pitched baseball: A reaction to McBeath (1990).* Manuscript submitted for publication.

Bootsma, R.J., Houbiers, M.H.J., Van Wieringen, P.C.W., & Whiting, H.T.A. (in press). Acquiring an attacking forehand drive: The effects of static and dynamic environmental information. *Research Quarterly for Exercise and Sport.*

Bootsma, R.J., & Oudejans, R.R.D. (1991). *Visual information about time to collision between two objects.* Manuscript submitted for publication.

Bootsma, R.J., & Van Wieringen, P.C.W. (1988). Visual control of an attacking forehand drive in table tennis. In O.G. Meijer & K. Roth (Eds.), *Complex movement behaviour: 'The' motor-action controversy* (pp. 189-199). Amsterdam: North-Holland.

Bootsma, R.J., & Van Wieringen, P.C.W. (1990). Timing an attacking forehand drive in table tennis. *Journal of Experimental Psychology: Human Perception and Performance, 16,* 21-29.

Bootsma, R.J., & Van Wieringen, P.C.W. (in press). Spatio-temporal organization in natural prehension. *Human Movement Science.*

Bower, T.G.R., Broughton, J.M., & Moore, M.K. (1970). The coordination of visual and tactile input in infants. *Perception & Psychophysics, 8,* 51-53.

Carel, W.L. (1961). Visual factors in the contact analog. G.E. Advanced Electronics Center Publication No. R6ELC60. Ithaca, NY: Cornell University Press.

Cavallo, V., & Laurent, M. (1988). Visual information and skill level in time to collision estimation. *Perception, 17,* 623-632.

Dixon, W.J., & Massey, F.J. (1969). *Introduction to statistical analysis.* New York: McGraw-Hill.

Fischman, M.G. (1986). A note on issues in recent work on one-handed catching. *Journal of Motor Behavior, 18,* 497-501.

Fischman, M.G., & Schneider, T. (1985). Skill level, vision and proprioception in simple one-handed catching. *Journal of Motor Behavior, 17,* 219-229.

Fitch, H.L., & Turvey, M.T. (1978). On the control of activity: Some remarks form an ecological point of view. In D. Landers & R. Christina (Eds.), *Psychology of motor behavior and sport* (pp. 3-35). Champaign, IL: Human Kinetics.

Gentner, D.R. (1987). Timing of skilled motor performance: Tests of the proportional duration model. *Psychological Review, 94,* 255-276.

Heuer, H. (1988). Testing the invariance of relative timing: Comment on Gentner (1987). *Psychological Review, 95,* 552-557.

Hoyle, F. (1957). *The black cloud.* London: Penguin.

Hubbard, A.W., & Seng, C.N. (1954). Visual movements of batters. *Research Quarterly, 25,* 42-57.

Jeannerod, M. (1984). The timing of natural prehension movements. *Journal of Motor Behavior, 16,* 235-254.

Laurent, M. (1991). Visual cues and processes involved in goal-directed locomotion. In A.E. Patla (Ed.), *Adaptability of human gait: Implications for the control of locomotion* (pp. 99-123). Amsterdam: North-Holland.

Laurent, M., Dinh Phung, R., & Ripoll, H. (1989). What visual information is used by riders in jumping? *Human Movement Science, 8,* 481-501.

Lee, D.N. (1974). Visual information during locomotion. In R.B. McLeod & H. Pick (Eds.), *Perception: Essays in Honor of J.J. Gibson* (pp. 250-267). Ithaca, NY: Cornell University Press.

Lee, D.N. (1976). A theory of visual control of braking based on information about time-to-collision. *Perception, 5,* 437-459.

Lee, D.N., Lishman, J.R., & Thomson, J.A. (1982). Visual regulation of gait in long jumping. *Journal of Experimental Psychology: Human Perception and Performance, 8,* 448-459.

Lee, D.N., & Reddish, P.E. (1981). Plummeting gannets: A paradigm of ecological optics. *Nature, 293,* 293-294.

Lee, D.N., & Young, D.S. (1985). Visual timing in interceptive actions. In D.J. Ingle, M. Jeannerod, & D.N. Lee (Eds.), *Brain mechanisms and spatial vision* (pp. 1-30). Dordrecht: Martinus Nijhoff.

Lee, D.N., Young, D.S., Reddish, P.E., Lough, S., & Clayton, T.M.H. (1983). Visual timing in hitting an accelerating ball. *Quarterly Journal of Experimental Psychology, 35A,* 333-346.

Marteniuk, R.G., Leavitt, J.L., MacKenzie, C.L., & Athenes, S. (1990). Functional relationships between grasp and transport components in a prehension task. *Human Movement Science, 9,* 149-176.

Marteniuk, R.G., MacKenzie, C.L., Jeannerod, M., Athenes, S., & Dugas, C. (1987). Constraints on human arm movement trajectories. *Canadian Journal of Psychology, 41,* 365-378.

MacKenzie, C.L., Marteniuk, R.G., Dugas, C., Liske, D., & Eickmeier, B. (1987). Three-dimensional movement trajectories in Fitts' task: Implications for control. *Quarterly Journal of Experimental Psychology, 39A,* 629-647.

Purdy, W.C. (1958). *The hypothesis of psychophysical correspondence in space perception.* Unpublished doctoral dissertation. Cornell University, Ithaca, NY.

Regan, D., & Beverley, K.I. (1978). Looming detectors in the human visual pathway. *Vision Research, 18,* 415-421.

Regan, D., Beverley, K.I., & Cynader, M. (1979). The visual perception of motion in depth. *Scientific American, 241,* 136-151.

Rosengren, K.S., Pick, H.L., & Von Hofsten, C. (1988). Role of visual information in ball catching. *Journal of Motor Behavior, 20,* 150-164.

Savelsbergh, G.J.P. (1990). *Catching behaviour.* Amsterdam: Free University Press.

Savelsbergh, G.J.P., & Whiting, H.T.A. (1988). The effect of skill level, external frame of reference and environmental changes on one handed catching. *Ergonomics, 31,* 1655-1663.

Savelsbergh, G.J.P., Whiting, H.T.A., & Bootsma, R.J. (1991). 'Grasping' tau! *Journal of Experimental Psychology: Human Perception and Performance, 17,* 315-322.

Schiff, W. (1965). Perception of impending collision: A study of visually directed avoidant behavior. *Psychological Monographs: General and Applied, 79* (11 Whole No. 604).

Schiff, W., Caviness, J.A., & Gibson, J.J. (1962). Persistent fear responses in rhesus monkey to the optical stimulus of 'looming'. *Science, 136,* 982-983.

Schiff, W., & Detwiler, M.L. (1979). Information used in judging impending collision. *Perception, 8,* 647-658.

Schiff, W., & Oldak, R. (1990). Accuracy of judging time-to-arrival: Effects of modality, trajectory and gender. *Journal of Experimental Psychology: Human Perception and Performance, 16,* 303-316.

Schmidt, R.A. (1975). A schema theory of discrete motor skill learning. *Psychological Review, 82,* 225-260.

Smyth, M.M. (1986). A note: Is it a catch or a fumble? *Journal of Motor Behavior, 18,* 492-496.

Smyth, M.M., & Marriott, A.M. (1982). Vision and proprioception in simple catching. *Journal of Motor Behavior, 14,* 143-152.

Todd, J.T. (1981). Visual information about moving objects. *Journal of Experimental Psychology: Human Perception and Performance, 7,* 795-810.

Todd, J.T., & Bressan, P. (1990). The perception of 3-dimensional affine structure from minimal apparent motion sequences. *Perception & Psychophysics, 48,* 419-430.

Tresilian, J.R. (1990). Perceptual information for the timing of interceptive action. *Perception, 19,* 223-239.

Tyldesley, D.A., & Whiting, H.T.A. (1975). Operational timing. *Journal of Human Movement Studies, 1,* 172-177.

Van der Horst, A.R.A. (1990). *A time-based analysis of road user behaviour in normal and critical encounters.* Soest: Practicum Drukkerij.

Warren, W.H., Mestre, D.R., Blackwell, A.W., & Morris, M.W. (1991). Perception of circular heading from optical flow. *Journal of Experimental Psychology: Human Perception and Performance, 17,* 28-43.

Warren, W.H., Morris, M.W., & Kalish, M. (1988). Perception of translational heading from optical flow. *Journal of Experimental Psychology: Human Perception and Performance, 14,* 646-660.

Warren, W.H., Young, D.S., & Lee, D.N. (1986). Visual control of step length during running over irregular terrain. *Journal of Experimental Psychology: Human Perception and Performance, 12,* 259-266.

Whiting, H.T.A. (1986). Isn't there a catch in it somewhere? *Journal of Motor Behavior, 18,* 486-491.

Whiting, H.T.A., Savelsbergh, G.J.P., & Faber, C.M. (1988). Catch questions and incomplete answers. In A.M. Colley & J.R. Beech (Eds.), *Cognition and action in skilled behaviour* (pp. 257-272). Amsterdam: North-Holland.

Yilmaz, E.H. (1991). *The use of visual expansion in braking regulation: An experimental study.* Unpublished manuscript, Brown University, Department of Psychology, Providence, RI.

Acknowledgement

This paper was written while the first author held a fellowship from the Royal Netherlands Academy of Arts and Sciences and was at leave at the Department of Kinesiology, University of Waterloo, Canada. Current e-mail address is: R_BOOTSMA@SARA.NL.

VISION AND MOTOR CONTROL
L. Proteau and D. Elliott (Editors)

CHAPTER 13

COINCIDENCE-ANTICIPATION TIMING: THE PERCEPTUAL-MOTOR INTERFACE

MICHELLE FLEURY, CHANTAL BARD MARIE GAGNON and NORMAND TEASDALE

Laboratoire de Performance Motrice Humaine
Université Laval, Sainte-Foy, Québec, G1K 7P4

The Nature of the Coincidence-Anticipation Task

In the early fifties, Poulton (1950, 1952) proposed an explanation of the motor interception process. He suggested that interceptive actions involve two temporal predictions: first, the time at which an object will arrive at the point of contact, based on its velocity and original distance (receptor anticipation), and second, the moment to initiate an action, based on knowledge about movement duration (effector anticipation). More recently, Konzag and Konzag (1980; see Buekers, Pauwels, & Meugens, 1988 for more details on this conceptual framework) also made a distinction between "external" and "internal" anticipation. In a broad sense, external anticipation can be defined as the prediction of environmental events, whereas internal anticipation describes the prediction of the performer's own movements. Both types of anticipation have three components: event anticipation (what will happen?), spatial anticipation (where will it happen?), and temporal anticipation or timing (when will it happen?).

Interceptive actions take place in a dynamic environment. The performer, therefore, must relate directly with the environment to be able to adapt responses to the punctual characteristics of a temporally unstable environment. Such a task involves both the perceptual and the motor control domains. The subject's displacement in space, or the presence of a moving object in the visual field, may create environmental temporal uncertainty. In the present chapter, we shall deal mainly with tasks in which the subject is static and the object is moving (but see Mestre, this volume).

While performing a coincidence-anticipation task, the subject must visually detect the presence of the moving stimulus and respond by intercepting the stimulus. Processes involved from detection to response are numerous and complex: movement detection, judgment on and prediction of stimulus speed, integration of perception and action for coincidence production. These processes can also be influenced by numerous determinants (stimulus speed, response complexity, practice). We shall make an integrative attempt at presenting the processes and determinants involved in coincidence anticipation performance.

Movement Detection
The study of the perceptual aspect of the coincidence-anticipation task relates to the psychology of movement perception. The human visual system can detect movement with great accuracy. Lateral displacements as small as 5 to 10 minutes, much inferior to the inter-cones interval, may be detected under certain conditions (Legge & Campbell, 1981; McKee & Nakayama, 1984; Nakayama, 1981, 1985; Nakayama & Tyler, 1978, 1981; Tyler & Torres, 1972). Considering the astonishing efficiency of the visual system in detecting movement, it is worth studying the underlying biological mechanisms responsible for movement perception.

Most authors interested in movement perception generally introduce their work with questions on basic structures. Take for example, the first few sentences of Nakayama's (1985) review of literature on movement perception:

> Physics provides no special status for visual motion, striking the issue whether it is fundamental or whether it is just the displacement of visual image over time. Introspection is no more decisive. Is motion a basic phenomenological dimension like color and stereopsis, or does it derive from sensory processes, like space and time? (p. 626)

Single-cell recording studies (Albus, 1980; Barlow, 1953; Barlow, Hill, & Levick, 1964; Hubel & Wiesel, 1962, 1965) have provided some answers to these questions. Specifically, the identification of cells sensitive to displacement indicates that movement perception has an underlying biological dimension. It is now widely accepted that motion is initially evaluated in parallel by two-dimensional arrays of local motion detectors (retinotopically organized), operating, in the simplest case, directly on the local light intensity (Borst & Egelhaaf, 1989). Regardless of the cellular mechanisms responsible for movement detection, all species rely on common network principles for information processing (Borst & Egelhaaf, 1989). According to these authors: "Motion detection may, therefore be one of the first examples in computational neurosciences where common principles can be found not only at the cellular level, but also at the level of the computation performed by small neural networks" (Borst & Egelhaaf, 1989, p. 297). Of course, interceptive actions involve more than mere detection, and the way judgments and predictions originate from stimulus speed evaluation must be taken into consideration.

Judgment and Prediction of Speed

Whenever a coincidence motor response has to be made to a moving object, an evaluation of the metric value of the displacement (i.e., the speed) must occur. Speed has proven to be a major determinant of performance in interceptive tasks (Ball & Glencross, 1985; Bard, Fleury, Carrière, & Bellec, 1981; Dunham, 1977; Haywood, 1977; Isaacs, 1983). For a given displacement, the total response time, as well as the time-sharing between information processing and response execution are modified by speed (Gagnon, Fleury, & Bard, 1990). Moreover, stimulus speed also interacts with factors such as age, practice and knowledge of results; these variables will be discussed in a later section.

Shea, Krampitz, Tolson, Ashby, Howard, and Husak (1981), had adult subjects displace a barrier in coincidence with the arrival of a moving stimulus at a target location. One of six speeds (67.1, 89.4, 111.8, 134.1, 156.5, and 178.8 cm/s) was presented to the subjects. The task consisted of displacing a barrier coincident with the arrival of the moving stimulus at a *target* location. When stimulus speed was unknown to subjects (random presentation), temporal accuracy was better when the stimulus was moving at 134.1 cm/s than when it was moving at a lower speed. Yet, the temporal accuracy was better when the three slowest speeds were presented, than when the fastest speeds (156.5 and 178.8 cm/s) were presented. On the other hand, when stimulus speed was known, the temporal absolute error decreased monotonically with increased speed (up to 134.1 cm/s). Prior knowledge of the speed improves temporal performance.

Haywood, Greenwald, and Lewis (1981) submitted subjects to three different stimulus speeds under each of three speed contexts: 44.7, 134.1, and 223.5 cm/s; 134.1, 223.5, and 312.9 cm/s; 223.5, 312.9, and 402.3 cm/s. The results of interest are those obtained at 223.5 cm/s. Constant temporal error was smaller when the *target* speed was presented in the first condition than when it was presented within the third context. Thus, the same single speed used in different contexts induces different timing performance, reflecting a range effect. Weeks and Shea (1984) showed that test speeds of 67 cm/s or 201 cm/s brought no sequential effect of speed, regardless of the absolute value of the preceding speed (67, 89, 134, 156, 179, or 201 cm/s). The authors noted, however, a tendency toward error reduction as the absolute value of the preceding speed got close to the test speed. Therefore, the range effect seems more important under extreme speed values. Gagnon, Bard, Fleury, and Michaud (1991) found an invariant bias (constant temporal error) according to the speed context. The obtained biases (early or late) clearly reflect a *response compromise* strategy specific to the stimulus context; they are not a function of the absolute speed values *per se*. Thus, the above experiment demonstrates that, from a purely perceptual viewpoint, it is difficult to compare results of different experiments when the number of speeds used and the absolute speed values differ.

In other respects, is the resolution capacity of the perceptual system affected by speed? Shea and Northam (1982) demonstrated that discrimination of linear speeds follows a continuum. In this study, the just noticeable differences (JNDs) were computed for criterion speeds of 134.11, 268.22, and 402.33 cm/s. Increments of 9.94 cm/s were added to variable speeds to evaluate the JND of criteria speeds, their absolute value being distributed over a range of 47.7 cm/s. JND increased proportionally with an increased criterion speed. The length of the apparent movement as well as the stimulus direction (toward, across, away), however, did not affect the JND.

When subjects estimate speed and distance from films illustrating real driving situations, speed is usually underestimated (Hakkinen, 1963). The discrepancy between objective and estimated speeds further increases with absolute speed. Brown (1931) already reported that perceived speed could correspond to many actual speeds, thus depending on the ambient visual field. Therefore, subjects may perceive two contexts having similar inter-speed intervals as very different according to the absolute speed values.

In brief, stimulus speed is an overwhelming factor in movement prediction. Its effect is very context specific, that is, not attached to absolute speed values. Moreover, unlike stimulus length and direction, speed affects the perceptual threshold (perceived JND).

Other Factors Related to Stimulus Characteristics

Additional factors, related to stimulus characteristics are also liable to influence performance in coincidence anticipation tasks. Gagnon, Bard, Fleury, and Michaud (1988), using an apparent movement perpendicular to the subject, showed that subjects displayed better temporal performances when the stimulus came from the left rather than from the right. Payne (1986) demonstrated that when a stimulus moves at a speed of 268 cm/s, an increased temporal length of the apparent movement results in an improvement in timing performance. In this study, however, duration of stimulus presentation and length of apparent movement were confounded. The results may also be attributed to the fact that the presentation duration of the stimulus increases as the length of apparent movement increases. Shea et al. (1981), and Ball and Glencross (1985) reported that, for identical speeds, an increased presentation duration also results in an increased positive constant error bias.

Role of Retinal and Eye/Head Systems in Movement Perception

Visual perception of a moving object results from the comparison of a retinal signal, coding image movements, and of an efferent signal, coding retinal movements in space (De Graaf & Wertheim, 1988). In order for the movement of the object to be perceived, the amplitude of both signals must be greater than one JND, otherwise the object is perceived as stationary (Wertheim, 1981).

According to Raymond and Shapiro (1984), an object speed may be determined by different mechanisms depending upon whether the image of the object is or is not moving relative to the retina. When the object is moving, its speed is signaled by neural units responding to movements of the image on the retina. However, during visual pursuit, if the eye moves at the same speed as the object, there is no movement of the retinal image and, therefore, perception of the moving object comes from an efference copy (Von Holst & Mitelstaedt, 1950) or a corollary discharge (Teuber, 1960). However, Raymond and Shapiro (1984) observed that speed perception may be modified when the background consists of large moving contours, similar to team sport situations. They suggested that visual pursuit in a complex environment is the result of two combined ocular mechanisms: (a) the voluntary pursuit system and (b) the optokinetic (vestibular) reflex system. Voluntary ocular movements produce the efferent copies necessary for visual perception during pursuit, whereas the optokinetic reflexive eye movement system does not produce such copies. However, stimulus parameters related to the background influence the optokinetic gain. Therefore, during pursuit tasks, this optokinetic gain may influence the ratio between reflexive and voluntary activities and thus create an incorrect estimation of the stimulus speed.

In baseball, the batter intercepts balls moving at speeds of 150 km/h and over (500 degrees/s). Bahill and LaRitz (1984) anlayzed the pursuit performance of the retinal system. They measured three basic ocular movements: saccadic eye movement, vestibulo-ocular eye movement and, vergence eye movement. In addition they measured smooth-pursuit eye movement. These authors justified the use of these types of ocular movements on the basis of their dynamic properties (latency, speed, and high frequency cutoff values), which differ and are not similarly affected by fatigue, drugs and illness. The analysis of the batter's task in baseball revealed that a slight vestibulo-occular displacement toward the left, accompanied by a head displacement toward the right, occurs during the first portion of the ball trajectory. Such a vestibulo-ocular movement must be inhibited rapidly since it induces a compensatory eye movement in the direction opposite to the ball. Comparative results between intermediate and professional players showed that professionals can inhibit this nonfunctional reflex, thus releasing the head for ball pursuit. A few subjects exhibited a pursuit strategy during the first third of the trajectory, followed by an anticipatory saccade ahead of the ball. This strategy allows the view of the ball/bat impact. In brief, better performance in expert players can be explained by a faster visual pursuit, a better ability to suppress vestibulo-ocular reflex, and the occasional use of an anticipatory saccade.

Shanks and Haywood (1987) also studied eye movements in baseball batters. They demonstrated that pursuit eye displacements begin only 150 ms after ball release by the pitcher and pursuit is not continuous. Information regarding ball speed is therefore obtained via expansion of the retinal image. Pursuit accuracy, during a coincidence-anticipation task presented on a computer screen, is poor (Haywood, 1977). The relatively proximal location of the screen places subjects in a situation such that information is probably collected via displacement of the retinal image rather than by visual pursuit. The pursuit operation is therefore neither accurate *per se* nor does it explain much of the variance of the cognitive-motor performance.

Bard and Fleury (1986) showed that, in children aged 9, and 11 years, pointing movements to a perpendicular apparent movement were more accurate when the head was fixed then when it was free to move. This was not the case for children aged 6.

Overall, these experiments demonstrate that, at least in non-experts, it is easier to adopt a strategy of head stabilization. This strategy does not confront subjects with errors induced by the optokinetic reflex gain during head movements.

The Effect of Occlusion on Movement Perception and its Role in Coincidence-Anticipation Tasks

In the early seventies, several authors evaluated the effects of trajectory occlusion on catching performance (Sharp & Whiting, 1974; Whiting, Gill, & Stephenson, 1970; Whiting & Sharp, 1974). More recently, Smyth and Marriott (1982), Rosengren, Pick, and Von Hofsten (1988), and Whiting (1986) refined the initial paradigm by further controlling for different ball vision and handling conditions. A major conclusion of these studies is that catching performance significantly improves whenever the ball can be viewed even for a brief period. Thus, the perceptual system is equipped with mechanisms allowing the detection and judgement of a moving stimulus. Some of these mechanisms are generic and can be observed across phylogenetic development. Others require the elaboration of a more complex computational network in the perceptual system.

Perceptual and Motor Interface

The work reported above was predominantly based on a classical *information processing* framework. A number of authors have also approached the study of interceptive actions using an ecological framework (Gibson, E.J., 1950, Gibson, J.J., 1979; Johansson, 1950; Reed, 1982; Von Hofsten & Lee, 1982; Whiting, 1986; Bootsma, 1988). We shall now review this portion of the literature.

Gibson, E.J. (1950) and Johansson (1950) were among the first to consider movement as an integral component of perception. More recently, Von Hofsten and Lee (1982) also emphasized the tight link existing between perception and action. They suggested that the motor response is modulated by changes occurring in the environment. Although the theory of perception/action was chosen as a theoretical basis by many researchers interested by coincidence-anticipation, it is still somewhat difficult to uncover a leading thread from their work. The utilization of interception tasks has often been based on purely ecological justification, since they best simulate sport activities of an open nature. As we shall see, ecology alone does not seem to provide a sufficient interpretative tool for the different findings (but see also Bootsma & Peper, this volume).

Recently, researchers (Bootsma & Van Wieringen, 1988; Reed, 1988; Von Hofsten & Lee, 1982; Von Hofsten, 1986, 1987; Whiting, 1986) applied coincidence-anticipation findings to a more global theory that has been called an "ecological theory of perception-action." In a developmental context, Von Hofsten (1987) emphasized the simultaneous evolution of perception and action. The question mainly relates to the way different perception/action systems are organized. The perception/action approach allows us to reconcile results from existing studies, because different stimulus characteristics, coupled with different motor task complexities, facilitate the understanding of the operational mode of the visuomotor interface in the timing of interceptive actions. From this viewpoint, several factors have heen investigated as potential determiners for the stability of the interface.

Tasks

Interception has also been labeled transit reaction, timing, anticipatory timing, coincidence timing, anticipation, coincidence-anticipation, and interceptive actions. These labels were often a function of the types of task used, the disciplinary background of the researcher, and the relative recency of the experiments being described.

The various tasks exhibit similarities and differences. The aim of this section is to determine whether there is a general capacity for coincidence-anticipation or several specific abilities, each having their perceptual or motor particularities.

Von Hofsten (1986), in a review of manual abilities, elegantly presented the specificity of the perception/action interface:

> Describing the information that drives the motor system becomes as important as the description of the motor system itself. In the past, problems of motor skills have often been studied without considering their specific ties to perception. It is not that the motor system has been studied in isolation of any controlling stimuli but rather in ignorance of the possibility that different stimuli might drive the system in qualitatively different ways (p. 167).

To support his argument, Von Hofsten (1987) compared results from two different coincidence anticipation tasks. In one experiment (Dorfman, 1977), subjects displaced a cursor dot shown on an oscilloscope screen, with a manual slide control, in order to intercept a moving target dot. In another experiment (Dunham, 1977), subjects had to lift a foot from a spring switch in coincidence with the arrival of a rolling ball at a pre-determined target. Performance was better in the latter task. Temporal errors in both studies were about three times as large as when subjects are required to catch a small ball (Alderson, Sully, & Sully, 1974). Von Hofsten judiciously emphasized the specificity of the perception/action systems, and questioned the generalizability of data analyzed in isolation, even though the authors in each of these experiments pretend to a better understanding of *skilled behavior*.

We (Bard, Fleury, & Gagnon, 1990) often observed that timing biases of coincidence-anticipation (temporal constant errors) increase with increased complexity of the motor response. For instance, in a simple button-press task, subjects and particularly children are late with the slowest stimulus speed and early with the fastest speed. On the other hand, whenever the task involved limb displacements (throwing task), children's reponses are always too early with the slowest stimulus speed and too late with the fastest speed (Fleury & Bard, 1985). These findings show that a modification in the complexity of the response results in a modification in timing performance.

Von Hofsten (1986) suggests that the relation between action and perception is organized around specific perception-action systems. Each action/perception system is specifically related to the characteristics of the interface between the specific perceptual characteristics of the mobile stimulus and the specific characteristics of the movement to be produced, in relation to a precise goal-direction (here, interception of the moving object). Moreover, the strength of the link between the mobile stimulus and the motor response increases with the experience of the coupling of specific perceptual-motor characteristics of the task used during practice (Gagnon, Bard, Fleury, & Michaud, 1989). Repetition establishes a new nervous structure relatively specific to information prevaling during practice (Marteniuk, MacKenzie, & Leavitt, 1988). Such a phenomenom has been called "the functional specificity of motor learning" (Proteau, Marteniuk, Girouard, & Dugas, 1987). Yet, it is difficult to discriminate whether time-to-collision refers to the same action-perception system when driving a car as when catching a ball. In this regard, Von Hofsten (1983) states that the systems responsible for catching a ball moving perpendicularly or parallel to the viewer are based on different visual cues. Beek (1986) rightly wonders if this implies the existence of two perception-action systems for catching. This question is still being debated.

Time-to-Collision

It has been argued (Gibson, E.J., 1950; Gibson, J.J., 1958) that the environment comprises material substances, each delimited by a specific surface. The light reflected by these surfaces makes visual perception possible. The texture and pigmentation of each surface reflects light in a non-uniform way. The light reflected at a precise location of the environment forms a densely structured "optic array." For each observation point, a unique optic array is captured. For example, as a ball moving at a constant speed approaches an observer, the image of the coming ball in the optic array dilates. According to Lee (1976), the inverse of the dilation rate ($1/\tau$) of such an image will then equal "time to contact" (Tc). Therefore, Tc calculation results from automatic information processing produced by radial expansion or dilation of the retinal image. Thus, when a ball approaches at a constant speed, perception of speed and distance is not necessary to perceive time to contact. This automatic processing has been documented in many approach and escape locomotor behaviors (Laurent, 1987). This has also been examined in the development of ball-grasping (Lee, Young, Reddish, Lough, & Clayton, 1983), and in ball-hitting behaviors (Bootsma, 1988; Bootsma & Van Wieringen, 1990; McLeod, McLaughlin, & Nimmo-Smith, 1986).

Simplicity and efficiency make the "optic flow procedure" very useful, since it allows direct access to Tc. There is no need for computational effort, looking at the target is sufficient. More recently, Cavallo and Laurent (1988) hypothesized that other sources of information are taken into consideration when calculating time-to-contact. Cavallo and Laurent (1988) conducted an *in vivo* experiment (car driving), in which amplitude of visual field (normal and restricted), distance evaluation mode (binocular and monocular), approach speed (30 and 90 km/h), and time-to-contact (3 and 6 s) were manipulated. Tc accuracy increases with a normal visual field, binocular vision, greater speed, and a high level of expertise, suggesting that information on speed and distance are also used for Tc computation. These results do not exclude the use of τ to evaluate Tc, they simply show that, in certain situations, τ is not the unique variable used to compute Tc. In catching tasks, Von Hofsten (1987) suggested that retinal dilation is particularly useful when the ball is coming perpendicularly to the receiver. However, there are situations in which the target does not approach but circles or passes in front of the subject (thus yelding no retinal expansion), yet catching is still possible. In these cases, timing rests on other kinds of visual information.

In summary, detection and evaluation of moving stimuli in coincidence-anticipation tasks are complex and rely on different processes, depending on the duration of movement in relation to the subject, and also on the size and duration of stimulus exposure. Moreover, the nature of the motor task also alters the evaluation of some characteristics of the moving simulus.

Factors Influencing Automatization of a Coincidence-Anticipation Task

Besides characteristics of the moving stimulus, other factors influence coincidence-anticipation performance. We shall briefly review the main motor determinants.

The Effects of Practice

Practice has been consistently found to be the most important variable affecting learning and performance in coincidence anticipation tasks in children (Dorfman, 1977; Gagnon, 1986; Gagnon et al., 1991; Hoffmann, Imwold, & Koller, 1983; Stadulis, 1971; Wade, 1980; Wrisberg & Mead, 1983), but not in adults (Del Rey, 1982; Del Rey, Wughalter, & Whitehurst, 1982; Haywood, 1983; Shea, 1980).

Considering task complexity, minimal practice effects have been observed in coincidence tasks involving a simple motor response (button-press); conversely, with complex motor responses (e.g., a ball throw), larger effects of practice have been reported (Bard et al., 1981; Grose, 1967; Hoffmann et al., 1983). The lack of sensitivity to practice for simple responses, due to the infant's well established capacity to pursue a moving trajectory, is a well supported result (Von Hofsten, 1979, 1980). When performing simple tasks, the motor component is minimal, resulting in a ceiling effect in performance (cf., Schmidt, 1982). It is thus imperative to control for stimulus and/or response complexity when investigating practice effects. Gagnon et al. (1989) reported interactions of practice (7 sessions, 20 trials/session) with stimulus speeds (randomly presented) when performing a sliding-throw. With practice, subjects accelerated their response to rapid stimulus velocity, and slowed their response at low speed. Such a modification in the timing strategy leads to a decreased bias between the slow (early) and the fast speeds (late). Overall, studies have show that repetition affects both the perceptual (reaction time) and motor (movement time) components of the coincidence anticipation task.

The practice problem has also been approached by comparing different levels of expertise. Tyldesley and Whiting (1975) suggested that interception errors in novice table tennis players result from an inability to (a) spatially predict the initiation point, (b) temporally predict the initiation point, (c) select the required motor program, and (d) produce the correct motor program consistently. In contrast, experts judged the spatial location of movement initiation more accurately than intermediate and novice players. Intermediate and expert table tennis players also showed similar speed and displacement patterns and identical movement durations. Whiting (1986) reviewed two additional studies conducted on table tennis players. Experts produced strokes with similar duration, trajectory, and spatial initiation point. Beginners demonstrated a longer and more variable movement pattern. With practice novices rapidly reduced the degrees of freedom originating from spatial orientation by establishing a more stable spatial initiation point with respect of the table edge.

Bootsma (1988) trained beginners in table tennis. Even after 1,600 practice trials (5 days), there were still important variations in the acceleration profiles of the bat. The improvement in performance originated from consistency of the direction of travel of the bat at the moment of contact (a parameter describing a relation between actor and environment) rather than through the exact location where contact occurred. Movement constancy observed in experts (Franks, Weicker, & Robertson, 1985; Hubbard & Seng, 1954; Tyldesley & Whiting, 1975) appears only very late in the learning process, suggesting that it is a subproduct of performance in experts rather than a determinant (Bootsma, 1988).

Young (1988) had subjects produce a forward hitting movement requiring a reversal component to coincide with the arrival of an apparent movement of 134 cm/s. Kinematic study of the proficient performers demonstrated that the reversing location of the movement (approximately (165°), its variability, and the moment of acceleration (temporal onset of acceleration toward the target) were good predictors of performance. As suggested by Tyldesley and Whiting (1975), the initiation point of the forward hitting movement is an important determinant of expertise.

In conclusion, for different hitting tasks there are several invariant parameters associated with high-level performance. Moreover, improvement in coincidence anticipation performance with practice is linked to motor response improvement (in timing and pattern) rather than to modifications in the perceptual component of the task. Alternatively, during the early acquisition phase of the task, improvement in performance through repetition is related to the perceptual and action components of the coincidence-anticipation task. Thus, when looking at practice, it is important to make a distinction between beginners and experts.

The Effects of Knowledge of Results

Knowledge of results (KR) constitutes an important variable for the learning of coincidence anticipation tasks. It is possible to provide spatial and/or temporal KR (Gagnon et al., 1991), or knowledge of performance (KP) on certain kinematic variables (Young, 1988). One can also vary the accuracy level of KR (Jensen, Picado, & Morenz, 1981), and the presentation schedule of KR (Winstein & Schmidt, 1990). For these studies, it is important to discriminate between temporary (performance) and permanent (learning) effects of KR (Salmoni, Schmidt, & Walter, 1984; see also Schmidt, 1988, pp. 345-363, for a detailed discussion of the importance of using transfer design for studying learning).

Haywood (1975), Magill and Chamberlin (1988), and Green, Chamberlin, and Magill (1988) demonstrated that, for adults, the coincidence-anticipation task provides enough information in itself to produce adequate performance. They suggested that KR on temporal error becomes redundant with visual feedback. However, this is not always the case. Gagnon, Fleury, and Bard (1988) manipulated the informational content of KR in children aged 6, 8 and 10 years. During practice sessions, one experimental group received spatial KR, one group temporal KR and a third group was given no KR. KR affected significantly the parameter(s) directly linked to the information conveyed. Temporal KR significantly reduced constant temporal error, whereas spatial KR improved spatial accuracy (absolute and variable spatial errors). Moreover, when transferred to a task involving both temporal and spatial constraints, children receiving temporal KR exhibited the greatest improvement in temporal timing with respect to stimulus speed. Compared to spatial KR subjects' performance, subjects receiving temporal KR performed efficiently when fast stimuli were presented (150 cm/s) and erratically when a slower stimulus (100 cm/s) was used. On the other hand, subjects in the spatial KR groups showed similar temporal performance at slow and fast speeds indicating that the information provided was used. The spatial KR group showed a decreased temporal variability compared to the no-KR and temporal KR groups. Therefore, whenever the perception/action interface changed (i.e., when different speeds were presented), the temporal performance improved when temporal KR was given. In adults, Young (1988) also observed differential effects of supplementary information according to its informational content (KR or KP). The different types of KP and KR yielded improvements which were specific to the type of feedback provided, both in performance and learning.

In Young's (1988) and Gagnon's et al. (1990) studies, information on spatial location, was more readily usable than temporal information. Thus, it seems that spatial information rests on a more concrete basis than temporal information and is easier to interpret (Gagnon, 1986). Indeed, spatial referents can be directly perceived, whereas there is no specific sense to support temporal referents. Because of the importance of the informational content conveyed by KR or KP, one must proceed to a cautious identification of the task components before choosing the type of information to provide to the performer.

Conclusion

As Newell (1989) said:

> One of the significant limitations of the motor control and skill acquisition domain is that the theories, models, and hypotheses are, in most cases, task specific. Many lines of theorizing fail to hold up under even small changes in tasks constraints ... (p. 92)

In future studies, it will be imperative to identify different perception/action systems to verify how perception modifies action, utilizing perceptual-motor tasks characterized by their ecological significance. If the control of a manual slide to intercept a dot on an oscilloscope constitutes a perception/action system worth understanding, then we should study it. Conversely, if the interest *per se* of understanding the control modes that govern such activity is trivial, then we should orient our efforts toward the study of what Whiting (1986) calls "culture specific skills" such as catching (Whiting, 1986), car driving (Cavallo & Laurent, 1988), or plumetting gannets' dive in the sea (Lee & Reddish, 1981). The marvelous adaptability of the nervous system to execute so differently tasks that seem very similar is quite disarming. Nevertheless, concentration on the study of the system's adaptation and modulation rather than lingering on rigid invariance seems to have stronger experiential and experimental support.

The solution might rest in systematically varying the perceptual and motor components of specific perception/action systems (catching, throwing, grasping, walking) to verify how perception modulates action and vice-versa.

References

Albus, K. (1980). The detection of movement direction and effects of contrast reversal in the cat's striated cortex. *Vision Research, 20*, 289-293.

Alderson, G.J.K., Sully, D.J., & Sully, H.G. (1974). An operational analysis of one-handed catching task using high speed photography. *Journal of Motor Behavior, 6*, 217-226.

Bahill, T.A., & LaRitz, T. (1984). Why can't batters keep their eyes on the ball? *American Scientist, 72*, 249-253.

Ball, C.T., & Glencross, D. (1985). Developmental differences in a coincident timing task under speed and time constraints. *Human Movement Science, 4*, 1-15.

Bard, C., & Fleury, M. (1986). Contribution of head movement to the accuracy of directional aiming and coincidence-timing tasks. In M.G. Wade & H.T.A. Whiting (Eds.), *Motor development in children: Aspects of coordination and control* (pp. 309-319). Dordrecht: Martinius Nijhoff.

Bard, C., Fleury, M., Carrière, L., & Bellec, J. (1981). Components of the coincidence-anticipation behavior of children aged from 6 to 11 years. *Perceptual and Motor Skills, 52*, 547-556.

Bard, C., Fleury, M., & Gagnon, M. (1990). Coincidence anticipation timing: An age-related perspective. In C. Bard, M. Fleury, & L. Hay (Eds.), *Development of eye-hand coordination across life span* (pp. 283-305). Columbia: South Carolina University Press.

Barlow, H.B. (1953). Summation and inhibition in the frog's retina. *Journal of Physiology, 119*, 69-88.

Barlow, H.B., Hill, R.M., & Levick, W.R. (1964). Retinal ganglion cells responding selectively to direction and speed of image motion in the rabbit. *Journal of Physiology, 173*, 377-407.

Beek, P.J. (1986). Perception-action coupling in the young infant: An appraisal of Von Hofsten's research programme. In M.G. Wade & H.T.A. Whiting (Eds.), *Motor development in children: Aspects of coordination and control* (pp. 187-206). Dordrecht: Martinius Nijhoff.

Bootsma, R.J. (1988). *The timing of rapid interceptive actions: Perception-action coupling in the control and acquisition of skill*. Amsterdam: Free University Press.

Bootsma, R.J., & Van Wieringen, P.C.W. (1988). Visual control of an attacking forehand drive. In O.G. Meijer & K. Roth (Eds.), *Complex movement behaviour: 'The' motor-action controversy* (pp. 189-199). Amsterdam: North-Holland.

Borst, A., & Egelhaaf, M. (1989). Principles of visual motion detection. *Trends in Neurosciences, 12*, 297-306.

Brown, J.F. (1931). The visual perception of velocity. *Psychologische Forschung, 14*, 199-232.

Buekers, M.J.A., Pauwels, J., & Meugens, P. (1988). Temporal and spatial anticipation in twelve-year-old boys and girls. In A.M. Colley & J.R. Beech (Eds.), *Cognition and action in skilled behaviour* (pp. 283-292). Amsterdam: North-Holland.

Cavallo, V., & Laurent, M. (1988). Visual information and skill level in time-to-collision estimation. *Perception, 17*, 623-632.

De Graff, B., & Wertheim, A.H. (1988). The perception of object motion during smooth pursuit eye movements: Adjacency is not a factor contributing to the filehne illusion. *Vision Research, 28*, 497-502.

Del Rey, P. (1982). Effects of contextual interference and retention on transfer. *Perceptual and Motor Skills, 54*, 467-476.

Del Rey, P., Wughalter, E.H., & Whitehurst, M. (1982). The effect of contextual interference on females with varied experience in open sport skills. *Research Quarterly, 53*, 108-115.

Dorfman, P.W. (1977). Timing and anticipation: A developmental perspective. *Journal of Motor Behavior, 9*, 67-79.

Dunham, P. (1977). Age, sex, speed and practice in coincidence-anticipation performance of children. *Perceptual and Motor Skills, 45*, 187-193.

Fleury, M., & Bard, C. (1985). Age, stimulus velocity and task complexity as determiners of coincident timing behavior. *Journal of Human Movement Studies, 11*, 305-317.

Franks, I.M., Weicker, D., & Robertson, D.G.E. (1985). The kinematics, movement phasing and timing of a skilled action in response to varying conditions of uncertainty. *Human Movement Science, 4*, 91-105.

Gagnon, M. (1986). *Connaissance des résultats et évolution du contrôle temporel et spatial dans une tâche d'anticipation-coïncidence.* [Knowledge of results and development of temporal and spatial control in a coincidence-anticipation task]. Unpublished master's thesis, Université Laval, Québec.

Gagnon, M., Bard, C., Fleury, M., & Michaud, D. (1988, June). *Age and stimulus velocity as determinants of time-sharing between information processing and motor response in a coincidence anticipation task.* Paper presented at the annual meeting of the North American Society for the Psychology of Sports and Physical Activity, Knoxville, TN.

Gagnon, M., Bard, C., Fleury, M., & Michaud, D. (1989, March). *Modification de la stratégie de partage temporel avec la pratique lors d'une tâche d'anticipation-coïncidence chez des enfants de 6 et 10 ans.* [Modifications to the time sharing strategy resulting from practice in a coincidence-anticipation task by children aged 6 and 10 years]. Paper presented at the annual meeting of L'Association Québécoise des Sciences de l'Activité Physique, Montréal, Québec, Canada.

Gagnon, M., Bard, C., Fleury, M., & Michaud, D. (1991). *Influence de la vitesse du stimulus sur l'organisation temporelle de la réponse motrice lors d' une tâche d' anticipation-coïncidence chez des enfants de 6 et 10 ans.* [Effect of the stimulus speed on the temporal organization of a motor response in a coincidence-anticipation task by children aged 6 to 10 years]. Manuscript submitted for publication.

Gagnon, M., Fleury, M., & Bard, C. (1988). Connaissance des résultats et contrôle spatiotemporel dans une tâche d'anticipation-coïncidence chez des enfants de 6 à 10 ans. [Knowledge of results and spatio-temporal control in a coincidence-anticipation task with children aged from 6 to 12 years old]. *Revue Canadienne de Psychologie, 42,* 347-363.

Gagnon, M., Bard, C., & Fleury, M. (1990). Age, knowledge of results, practice and stimulus characteristics as determiners of performance in a coincidence anticipation task with spatial and temporal requirements. In J.E. Clark & J.H. Humphrey (Eds.), *Advances in Motor Development Research* (pp. 56-79). New York: AMS Press.

Gibson, E.J. (1950). *The perception of the visual world.* Boston: Houghton-Mifflin.

Gibson, J.J. (1958). Visually controlled locomotion and visual orientation in animals. *British Journal of Psychology, 49,* 182-194.

Gibson, J.J. (1979). *The ecological approach to visual perception.* Boston: Houghton-Mifflin.

Green, K.J., Chamberlin, C.J., & Magill, R.A. (1988, June). *Verbal KR and novel transfer for an anticipation timing task : A case for information redundancy.* Paper presented at the annual meeting of the North American Society for the Psychology of Sports and Physical Activity, Knoxville, TN.

Grose, E.J. (1967). Timing control and finger, arm and whole body movements. *Research Quarterly, 38,* 10-21.

Hakkinen, S. (1963). *Estimation of distance and velocity in trafic situations.* Reports from the Institute of Occupational Health, No. 3. Helsinki.

Haywood, K.M. (1975). Relative effects of three knowledge of results treatments on coincidence-anticipation performance. *Journal of Motor Behavior, 7,* 271-274.

Haywood, K.M. (1977). Eye movements during coincidence-anticipation performance. *Journal of Motor Behavior, 9,* 313-318.

Haywood, K.M. (1983). Response to speed changes after extended practice in coincidence-anticipation judgments. *Research Quarterly, 54,* 28-32.

Haywood, K.M., Greenwald, G., & Lewis, C. (1981). Contextual factors and age group differences in coincidence-anticipation performance. *Research Quarterly, 52,* 458-464.

Hoffman, J.S., Imwold, C.H., & Koller, J.A. (1983). Accuracy and prediction in throwing: A taxonomic analysis of children's performance. *Research Quarterly, 54,* 33-40.

Hubbard, A.W., & Seng, C.N. (1954). Visual movements of batters. *Research Quarterly, 25*, 42-57.

Hubel, D.H., & Wiesel, T.N. (1962). Receptive fields, binocular interaction and functional architecture in the cat's visual cortex. *Journal of Physiology, (London), 160*, 106-154.

Hubel, D.H., & Wiesel, T.N. (1965). Receptive fields and functional architecture in two non-striate visiual areas (18 and 19) of the cat. *Journal of Neurophysiology, 28*, 229-289.

Isaacs, L.D. (1983). Coincidence anticipation in simple catching. *Journal of Human Movement Studies, 9*, 195-201.

Jensen, B.E., Picado, M.E., & Morenz, C. (1981). Effects of precision of knowledge of results on performance of a gross motor coincidence-anticipation task. *Journal of Motor Behavior, 13*, 9-17.

Johansson, G. (1950). *Configuration in event perception.* Uppsala: Almkvist & Wiksell.

Konzag, I., & Konzag, G. (1980). Anforderungen an die kognititiven functionen in der psychischen regulation sportlicher spielhandlungen. [Requirements associated to the cognitive functions in the psychoregulation of play elements in sport]. *Theorie und Praxis der Korperkultur, 29*, 20-31.

Laurent, M. (1987). *Les coordinations visuo-locomotrices: étude comportementale chez l'homme.* [Visuo-locomotor coordination: Behavioral study in man]. Thèse de doctorat d'état. Université d'Aix-Marseille.

Lee, D.N. (1976). A theory of visual control of braking based on information about time-to-collision. *Perception, 5*, 437-459.

Lee, D.N., & Reddish, P.E. (1981). Plummeting gannets: A paradigm of ecological optics. *Nature, 293*, 293-294.

Lee, D.N., Young, D.S., Reddish, P.E., Lough, S., & Clayton, T. M. H. (1983). Visual timing in hitting an accelerating ball. *Quarterly Journal of Experimental Psychology, 35A*, 333-346.

Legge, G.E., & Campbell, F.W. (1981). Displacement detection in human vision. *Vision Research, 21*, 205-214.

Magill, R.A., & Chamberlin, C.J. (1988, June). *Verbal KR can be redundant information in motor skill learning.* Paper presented at the annual meeting of the North American Society for the Psychology of Sports and Physical Activity, Knoxville, TN.

Marteniuk, R.G., MacKenzie, C.L., & Leavitt, J.L. (1988). Representational and physical accounts of motor control and learning: Can they account for the data? In A.M. Colley & J.R. Beech (Eds.), *Cognition and action in skilled behaviour* (pp. 173-190). Amsterdam: North-Holland.

McKee, S.P., & Nakayama, K. (1984). The detection of motion in the peripheral field. *Vision Research, 28*, 25-32.

McLeod, P., McLaughlin, C., & Nimmo-Smith, I. (1986). Information encapsulation and automaticity: Evidence from the visual control of finely timed actions. In M.I. Posner & O.S.M. Malin (Eds.), *Attention and performance XI* (pp. 391-406). Hillsdale, NJ: Erlbaum.

Nakayama, K. (1981). Differential motion hyperacuity under conditions of common image motion. *Vision Research, 21*, 1475-1482.

Nakayama, K. (1985). Biological image motion processing: A review. *Vision Research, 5*, 625-660.

Nakayama, K., & Tyler, C.W. (1978). Relative motion induced between stationary lines. *Vision Research, 18*, 1663-1668.

Nakayama, K., & Tyler, C.W. (1981). Psychophysical isolation of movement sensitivity by removal of familiar position cues. *Vision Research, 21*, 427-433.

Newell, K.M. (1989). On task theory specificity. *Journal of Motor Behavior, 21*, 92-96.

Payne, V.G. (1986). The effects of stimulus runway length on coincidence-anticipation timing performance. *Journal of Human Movement Studies, 12*, 289-295.

Poulton, E.C. (1950). Perceptual anticipation and reaction time. *Quarterly Journal of Experimental Psychology, 2*, 99-112.

Poulton, E.C. (1952). The basis of perceptual anticipation in tracking. *British Journal of Psychology, 43*, 295-302.

Proteau, L., Marteniuk, R.G., Girouard, Y., & Dugas, C. (1987). On the type of information used to control and learn an aiming movement after moderate and extensive training. *Human Movement Science, 6*, 181-199.

Raymond, J.E., & Shapiro, K.L. (1984). Optokinetic backgrounds affect perceived velocity during occular tracking. *Perception & Psychophysics, 36*, 221-224.

Reed, E.S. (1982). An outline of a theory of action systems. *Journal of Motor Behavior, 14*, 98-134.

Reed, E.S. (1988). Applying the theory of action systems to the study of motor skills. In O.G. Meijer & K. Roth (Eds.), *Complex movement behaviour: 'The' motor-action controversy* (pp. 45-86). Amsterdam: North-Holland.

Rosengren, K.S., Pick, H.L., & Von Hofsten, C. (1988). Role of visual information in ball catching. *Journal of Motor Behavior, 20*, 150-164.

Salmoni, A.W., Schmidt, R.A., & Walter, C.B. (1984). Knowledge of results and motor learning: A review and critical reappraisal. *Psychological Bulletin, 95*, 355-386.

Schmidt, R.A. (1982). *Motor control and learning: A behavioral emphasis.* Champaign, IL: Human Kinetics.

Schmidt, R.A. (1988). *Motor control and learning: A behavioral emphasis (2nd ed.).* Champaign, IL: Human Kinetics.

Shanks, M.D., & Haywood, K.M. (1987). Eye movements while viewing a baseball pitch. *Perceptual and Motor Skills, 64*, 1191-1197.

Sharp, R.H., & Whiting, H.T.A. (1974). Exposure and occluded duration effects in ball-catching skills. *Journal of Motor Behavior, 6*, 139-147.

Shea, C.H. (1980). Effects of extended practice and movement time on motor control of a coincident timing task. *Research Quarterly, 51*, 369-381.

Shea, C.H., Krampitz, J.B., Tolson, H., Ashby, A.A., Howard, R.M.,& Husak, W.S. (1981). Stimulus velocity, duration and uncertainty as determiners of response structure and timing accuracy. *Research Quarterly, 52*, 86-99.

Shea, C.H., & Northam, C. (1982). Discrimination of visual linear velocities. *Research Quarterly, 53*, 222-225.

Smyth, M.M., & Marriott, A.M. (1982). Vision and proprioception in simple catching. *Journal of Motor Behavior, 14*, 143-152.

Stadulis, R.E. (1971). *Coincidence-anticipation behavior of children.* Unpublished doctoral dissertation, Teachers College, Columbia University, New York.

Teuber, H.L. (1960). Perception. In H. Magoun (Ed.), *Handbook of physiology III: Section on neurophysiology* (pp. 1595-1668). Washington: American Physiological Society.

Tyldesley, D.A., & Whiting, H.T.A. (1975). Operational timing. *Journal of Human Movement Studies, 1*, 172-177.

Tyler, C.W., & Torres, J. (1972). Frequency response characteristics for sinusoidal movement in the fovea and periphery. *Perception & Psychophysics, 12*, 232-236.

Von Hofsten, C. (1979). Development of visually directed reaching: The approach phase. *Journal of Human Movement Studies, 5*, 160-178.

Von Hofsten, C. (1980). Predictive reaching for moving objects by human infants. *Journal of Experimental Child Psychology, 30*, 369-382.

Von Hofsten, C. (1983). Catching skills in infancy. *Journal of Experimental Psychology: Human Perception and Performance, 9*, 75-85.

Von Hofsten, C. (1986). The emergence of manual skills. In M.G. Wade & H. T. A. Whiting (Eds.), *Motor development in children: Aspects of coordination and control* (pp. 167-185). Dordrecht: Martinius Nijhoff.

Von Hofsten, C. (1987). Catching. In H. Heuer & A.F. Sanders (Eds.), *Perspectives on perception and action* (pp. 33-46). Hillsdale, NJ: Erlbaum.

Von Hofsten, C., & Lee, D.N. (1982). Dialogue on perception and action *Human Movement Science, 1*, 125-138.

Von Holst, E., & Mittelstaedt, M. (1950). Das reafferenz prinzip, Wechselwirkungen zwischen Zentralnervensystem und peripherie [The principle of reaference, mutual exchanges between the central nervous system and the periphery]. *Naturwissenschaften, 37*, 464-476.

Wade, M.G. (1980). Coincidence-anticipation of young normal and handicapped children. *Journal of Motor Behavior, 12*, 103-112.

Weeks, D.L., & Shea, C.H. (1984). Assimilation effects in coincident timing responses. *Research Quarterly, 55*, 89-92.

Wertheim, A.H. (1981). On the relativity of perceived motion. *Acta Psychologica, 8*, 97-110.

Whiting, H.T.A. (1986). Movement invariances in culture-specific skills. In M.G. Wade & H.T.A. Whiting (Eds.), *Motor Development in children: Aspects of coordination and control* (pp. 147-166). Dorcrecht: Martinus Nijhoff.

Whiting, H.T.A., Gill, E.B., & Stephenson, J.M. (1970). Critical time intervals for taking in flight information in a ball-catching task. *Ergonomics, 13*, 265-272.

Whiting, H.T.A., & Sharp, R.H. (1974). Visual occlusion factors in a discrete ball-catching task. *Journal of Motor Behavior, 6*, 11-16.

Winstein, C.J., & Schmidt, R.A. (1990). Reduced frequency of knowledge of results enhances motor skill learning. *Journal of Experimental Psychology: Learning, Memory, and Cognition 16*, 677-691.

Wrisberg, C.A., & Mead, B.J. (1983). Developing coincident timing skill in children: A comparison of training methods. *Research Quarterly, 54*, 67-74.

Young, D.E. (1988). *Knowledge of performance and motor learning.* Unpublished doctoral dissertation. University of California, Los Angeles.

Acknowledgment

The preparation of this chapter was supported by grants awarded to M. Fleury, C. Bard, and N. Teasdale by the Natural Sciences and Engineering Research Council of Canada and by the Fonds F.C.A.R., gouvernement du Québec.

VISION AND MOTOR CONTROL
L. Proteau and D. Elliott (Editors)

CHAPTER 14

TIME TO CONTACT AS A DETERMINER OF ACTION: VISION AND MOTOR CONTROL

DAVID GOODMAN* and DARIO G. LIEBERMANN**

*School of Kinesiology, Simon Fraser University,
Burnaby, British Columbia, V5A 1S6*

**Center for Research and Sports Medicine Sciences,
Wingate Institute, Netanya, 42902, Israel*

In this chapter we examine the role vision, and in particular the "time to contact" variable, plays in determining subsequent action. We adopt a perspective following that of Gibson's (1950) notion of direct perception. Although Gibson's ecological approach provided an alternative to the traditional views of perception it was left to Lee (1974, 1976, 1980a, b) to operationalize many of the conjectures. Lee (Lee, 1980a, 1990; Lee & Young, 1986) presented a number of examples of how action is geared to the environment. The basic premise was simply that to survive, an organism must gear its actions to the environment. The role of vision in mediating information in such tasks as stabilizing the head and eyes, maintaining balance, steering, and locomotion was clearly of major significance. The importance Gibson (1950) placed on movement in perception is something that simply can no longer be ignored (see Bootsma & Peper, this volume). Of course, he was not alone in this, as suggested by Trevarthen (1968): "visual perception and plans for voluntary action are so intimately bound together that they may be considered products of one cerebral function" (p. 39). Direct perceptionists would argue about the 'richness of light', and how light carries information about the environment through which it has travelled and from which it has been reflected. As Gordon (1989) summarizes the view:

If we examine light arriving at the eye in real situations we find that it is structured. It is highly complex and potentially rich in information. A single momentary retinal image may be impoverished, but this is not true of the nested solid visual angles through which the head and eyes sweep in normal perceiving. As we come to understand more and more about these arrays and the potential information contained in their structure, less frequently will we need to invoke supplementary, indirect processes in explanations of seeing (p. 153).

The Direct Perception of Dynamic Events

In the motor domain actions have been too often studied separately from both perception and the environment in which the actions are carried out. This is in spite of the fact that they are clearly interrelated (cf., Michaels & Carello, 1981; Turvey & Kugler, 1984). Turvey (1977) described this relationship as a coalition, whereby information about environmental changes constrains adaptation. This is accomplished by means of exteroceptive perceptual inputs under the conditions that: (a) the actor resonates or is sensitive to them (i.e., tuned to the stimulus) and, (b) that such environmental changes are meaningful to the specific actor (i.e, they afford a change in the state of the organism) (Michaels & Carello, 1981). The implication of this approach is that the study of one component of the animal-environment continuum concerns all levels of analysis (ecological, perceptual, and motor). By ignoring this interrelationship investigators may well be making conclusions based on only a partial view of the space-time event being investigated.

Ecological psychology advocates concepts and tentative mechanisms by which this interactive process may be carried out at least at the perceptual and environmental levels. Gibson (1950, 1966, 1979) proposed that visual perception is direct and continuous, and may be used to regulate motion. Perception, occurs by means of 'invariants' (i.e., features that remain constant under environmental transformations). Movement produced invariants carry lawful kinematic information (Solomon, 1988) which may only be perceived by special purpose perceptual devices (Runeson, 1977). A cornerstone of the ecological approach is the concept that there is information in the environment ready to be used, which becomes manifested only when relative motion occurs. In these cases, invariants play a major role as units of information. Body-scaled relationships such as the stride frequency, the body mass, and the length of the hindlimbs which may underly the adoption of a metabolically least costly pattern found in animal gait (Hoyt & Taylor, 1981) may be considered invariants. Similarly, the perceived relationship between body proportions and the height of a stair (the so called Pi or π numbers) (Warren, 1984) which determine whether climbing is possible for the specific actor, or whether a stair is perceived as a sitting chair (Mark & Vogele, 1987) may be considered invariants. In landings from a free fall or jumps this relationship would be expressed as ratio between body mass and momentum gained at contact with the ground, which may guide how much muscular tension is needed to dissipate hard impacts at collision (cf., Warren & Kelso, 1985). In the latter example, however, rescaling information in intrinsic actor terms is not enough to overcome inertia. Timing may be critical, and therefore, the individual may also need information about when to initiate action.

Action Modulation and Visual Information in Free Fall Landing

Following this line of thought, the present chapter focuses on the timing of muscle activation while undergoing free fall conditions. We present results from two laboratories in which individuals had to perform self-initiated free falls. The free fall task allows for direct manipulation of height of fall (and hence time of fall), and ultimately the role of vision in preparation for landing. Although this task has been used in a number of studies, only a few have specifically examined the role of vision.

Initial studies on the reaction to vertical motion demonstrated that sudden transitions in acceleration caused a reflexive limb extension in babies (Schaltenbrand, 1925, cited in Greenwood & Hopkins, 1980, p. 295). Using a rather rudimentary experimental design, Gurnee (1934) demonstrated that sensitivity to downward movements was greater than to upward movements. The task consisted of pressing down a key with the finger on the right hand side when the blindfolded subject detected a downward movement, and to press a second key on the left side when the movement was perceived as being upwards.

This early interest on reactions to falls was later followed up by Matthews (1956) and Matthews and Whiteside (1960) among others. These investigators were interested in the reflexive reactions of human subjects during sudden transitions from 0 to 1 g in the vertical direction. The specific focus however, was on the vestibular system as opposed to the visual-action system. For example, Melvill Jones and Watt (1971) studied the latencies of onset and the pattern of muscle electrical activity (EMG) related to the otolith-vestibular structures during unexpected falls. A design was adopted whereby subjects were suddenly dropped while griping a suspended bar with their hands. Their findings provided initial evidence regarding the role of long-latency reflexive responses during brief falls. The stereotyped EMG pattern observed in the gastrocnemius muscle indicated an initial burst of activity at a mean onset latency of 74.2 ms after release. The finding was initially labelled by Melvill Jones and Watt (1971) as a functional stretch reflex (FSR) of otolith-spinal origin. The general pattern of EMG activity found in unexpected falls lasting over 200 ms showed, however, that the actual response was bi-phasic

(Greenwood & Hopkins, 1976a). The typical pattern of EMG activity is illustrated in Figure 14.1.

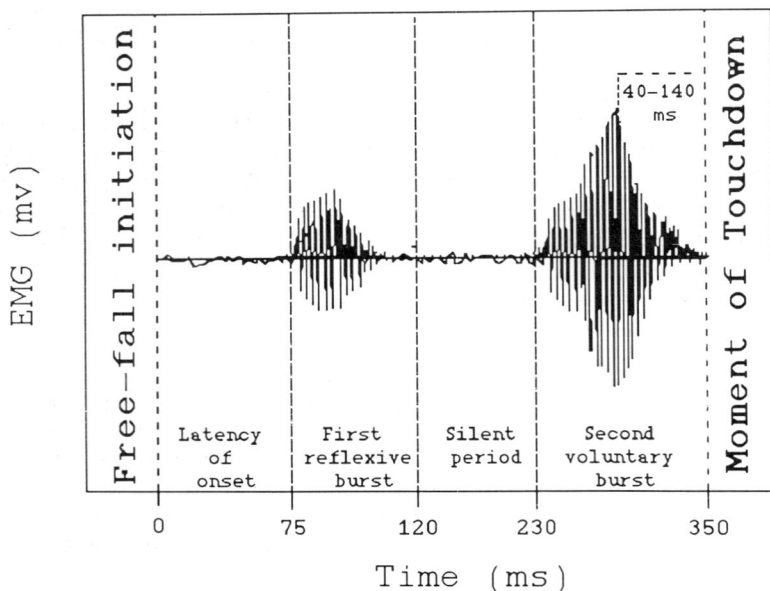

Figure 14.1. Typical pattern of EMG activation in the preparatory muscles during unexpected falls lasting over 200 ms.

Typically one notes a latency period of 30-80 ms following release before the first burst of activity, and if the flight time is longer than 200 ms, a second burst is consistently observed. Greenwood and Hopkins suggested that the initial burst of EMG activity is primarily a response to release rather than to the moment of touchdown. They attributed this response to a startle reflex due to sudden unexpected falls. Their task also controlled for acceleration by a system of counterbalancing weights while the subjects were suspended in a parachute harness. They showed that accelerations slightly lower than 1 g enhanced activity in the soleus muscle at a mean latency of 80 ms after release. However, the amplitude and the latency of onset of this initial response remained unchanged for unexpected falls lasting up to 520 ms, whether subjects were blindfolded or in a visual condition. When acceleration was reduced this activity was reduced, and at accelerations lower than 2 m·s⁻² it disappeared. Their results provided evidence that the reflexive pattern is initiated by otolith-vestibular stimulation rather than by startle responses as first proposed by Greenwood and Hopkins (1976a).

The second burst of EMG activity usually appeared between 40-140 ms prior to landing. It was this activity that led Greenwood and Hopkins (1976a, b) to conclude that a preprogrammed neural message was activated and completed prior to contact with the ground in free fall landings. That is, the voluntary muscular response (indexed by the second burst of activity) was clearly evident prior to contact with the ground, and its onset latency depended somewhat on the height of fall (Greenwood & Hopkins, 1976a). It is this result that we find most interesting. Indeed in our view self-release falls are of some ecological significance, and may well demonstrate the role of the visual system in preparation for action. Moreover, the voluntary activity represented by the second burst appears to be the candidate for visual information.

Time to Contact and the Tau (τ) Heuristic

The difficulty in interpreting the landing preparatory actions previously discussed may be a consequence of the approach taken. The computational view, discussed at length in Ullman (1980) and elsewhere (cf., Gregory, 1980) appears somewhat deficient in accounting for how coordinated acts occur in response to rapidly changing environmental cues. Thus, in an attempt to reconcile evidence from free-fall experiments we turn to an approach from direct perception, and focus on the time to contact variable as described by Lee.

According to Lee the information available from the optic flow field as one approaches a surface allows for the immediate perception of the time to contact with that surface without going through a priori mental computations through another important invariant the so called Tau ratio (τ) (Lee, 1976; 1980a, b). Time to contact is assumed to be a low order parameter that facilitates behavioral pattern formations when the flight time (the control parameter) is manipulated. The perceived time to contact is generally considered an important though unconscious variable in many real life situations (see Bootsma & Peper's chapter in this volume).

Evidence for the use of information about time to contact while abundant is primarily based on observational studies. Lee and Young (1985) summarized several examples in which this variable is obtained using the τ ratio. According to Lee (1980a, b), τ may be regarded as a relative quantity, perceivable by any tuned biological eye that 'resonates' to time to contact information τ arises from the relationship between the instantaneous optical angle specified by the layouts of the approaching environmental surface (at any given moment of the event), and its respective rate of angular change (i.e., its angular velocity).   At constant approach velocity τ is equal to the time to contact. As Lee (1980a) acknowledges, under accelerative conditions the value of the τ ratio provides only an estimate of the time to contact. This estimate can be used for the adoption of a strategy based on a critical value of τ termed τ_m.

Tau and the Timing of the Landing Response

In the free-fall task, the optic elements are assumed, for convenience, to be parallel to the direction of the fall, and orthogonal to the ground. Figure 14.2 presents the optic variables on the time continuum of the free-fall event along with the EMG activity obtained from a sample trial (Liebermann, 1988).

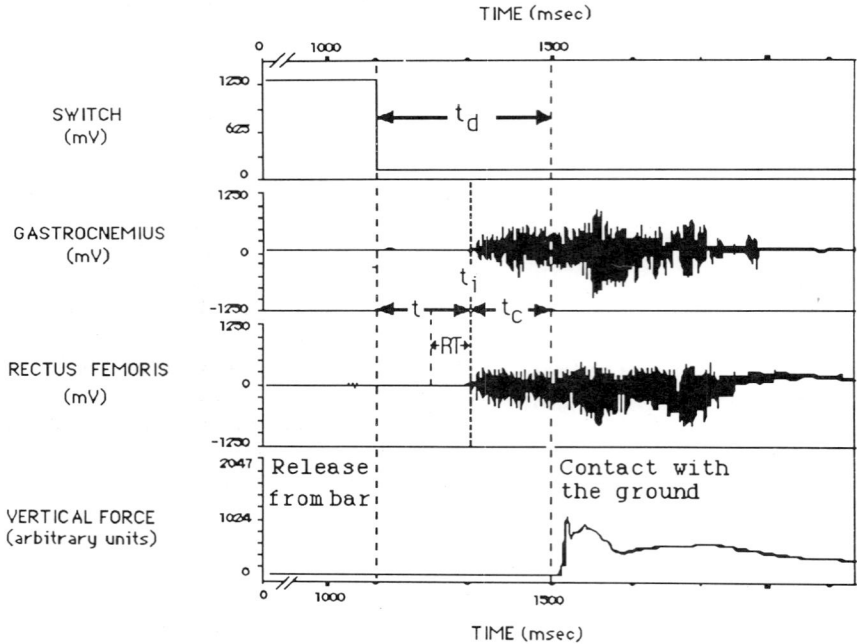

Figure 14.2. Electromyographic activity from a sample trial illustrating onset of EMG from gastrocnemius and rectus femoris. Also indicated are the optic variables τ and τ_c.

The time continuum of the event from initiation of the fall until ground contact (i.e., the total flight time t_d) may be divided in two distinctive periods defined by the moment of muscle activation (t_i). The first elapsed time before muscle activity is initiated (t), is dependent upon when the individual perceived that $\tau_{(t)}$ has reached a critical value τ_m. The second, the perceived time to contact (t_c), is simply the time from t_i until the moment of touchdown. Note that the time to contact is underestimated since it does not incorporate a reaction time period from perception of t_m until muscle activation. Thus, a reaction time unit RT should also be taken into account.

When relative movement occurs it turns out that τ represents the inverse of the rate of dilation of the layouts of any environmental surface projected on the retina of the eye. Under conditions in which the approach velocity is constant, $\tau_{(t)}$ equals time to contact at any time t of an event of total duration t_d. Time differentiation of the optic variables allows calculation of $\tau_{(t)}$, as

$$\tau_{(t)} = Z(t) / V(t) \tag{14.1}$$

The equivalent expressions in angular form are provided by Bootsma and Peper in this volume.

The use of the available information obtained through the τ model has been initially suggested by Lee (1974) in walking towards static or moving objects on a collision course. In athletic performances, the last steps of the run-up to the take-off board in long jumping appear to be regulated by τ (Lee, Lishman, & Thomson, 1982), as is the vertical thrust of the running steps during running over irregular terrains (Warren, Young, & Lee, 1986). Similar conclusions have been reported in the performance of a forehand in table tennis (Bootsma & Van Wieringen, 1990), in regulating the take-off moment in ski-jumping performances (Lee & Young, 1985, 1986), in catching (Savelsbergh, Whiting, & Bootsma, 1991), and in adjusting the running steps in clearing an obstacle during horse-riding (Laurent, Dinh Phung, & Ripoll, 1989).

Evidence for the use of the Tau margin strategy, due to accelerative conditions, has been suggested by Lee and Reddish (1981) in the timing of the wing folding action of gannets diving down in the direction of the water. Wagner (1982) also demonstrated that information contained in the optical flow field triggers the pre-landing deceleration and leg extension of houseflies. Wagner suggested that the relevant modulating cue conveyed by this optic flow was time, and not distance. In humans, the principles of the τ strategy are apparently used to time the braking response of drivers in situations of imminent collision (Lee, 1976). As well adjustment of actions may be visually guided by the τ_m strategy in jumping to hit a ball dropped from different heights (Lee, Young, Reddish, Lough, & Clayton, 1983).

Both Liebermann (1988) and Sidaway, McNitt-Gray, and Davis (1989) independently conducted experiments using the free fall task in order to investigate how subjects time muscle preactivation for landings. Sidaway et al. (1989) examined the timing of preactivation in the *rectus femoris* muscle in preparation for landings. They had six subjects step off three different height platforms (72, 104, and 159 cm). Throughout the drop subjects directed their gaze at a force plate that also served as a landing surface. For each drop Sidaway et al. (1989) determined the interval from onset of the preparatory muscle activity to foot contact with the ground (t_C). Similarly, Liebermann (1988) used four different heights (5, 20, 60, 90 cm). However, rather than have subjects step off a platform, Liebermann's subjects were suspended from an overhead bar, and initiated the free-fall by simply releasing their handgrip on the bar. Subjects in this study also directed their gaze to the force platform on which they were to land. Subjects were instrumented so as to enable the determination of activation time of both rectus femoris and gastrocnemius muscles. In addition, on two separate days, subjects performed under vision and no-vision conditions.

Table 14.1a and b presents the results of the Sidaway et al. (1989) and Liebermann (1988) experiments.

Note that flight times for Sidaway et al. (1989) are predicted flight times under free fall conditions. In this case, flight time is simply predicted as FT = $(2 \cdot \text{height}/g)^{1/2}$. Liebermann, on the other hand, lists actual mean flight time as recorded from bar release to first impact on the force platform. The predicted flight times (101, 202, 350, 428 ms, respectively for the 5, 20, 60 and 90 cm heights) overestimate the actual flight time values. This was due to a combination of factors, including release from the overhead bar (subjects tended to let the hand slide along the bar as they started their decent) and extension of the foot on landing. What is clearly apparent from both studies is the monotonically increasing function between height of release (or flight time) and elapsed time to activation of preparatory muscle action. For Sidaway et al. this increased from 323 ms at the 72 cm height to 429 ms at the 159 cm height. Liebermann indicates a change from 37 ms at the lowest height (5 cm) to 271 ms at the highest height (90 cm). Thus, it is clear that in both studies subjects did not adopt a strategy of initiating muscular action at a constant time after release. These findings contrast quite sharply with earlier reported studies on free fall (Greenwood & Hopkins, 1974; Melvill Jones & Watt, 1971). Indeed the majority of studies postulated that an otolith-vestibular reflex triggered muscle activation at a constant time after release. On the same vein, this finding contrast with those reported in animal studies which showed that preparation to land (from 60 to 120 cm height) takes place at relatively constant times from the moment of touchdown (McKinley, Smith, & Gregor,

1983). The major difference in the two studies reported here and the earlier studies was the use of self-initiated free falls (either by stepping off a platform or releasing a bar) rather than the unexpected falls.

Table 14.1a.
Data from Sidaway et al. (1989).

	Drop height (cm)		
Dependent variables	72	104	159
Flight time (ms)	383	470	569
Mean time of EMG (*rectus femoris*) onset prior to ground contact (ms)	60	73	149
Tau margin (ms)	592	521	583

Table 14.1b.
Data from Liebermann (1988)

	Drop height (cm)			
Dependent variables	5	20	60	90
Flight time (ms)	100	202	350	428
Mean actual flight time (ms)	60	140	291	377
Mean time of EMG (*rectus femoris*) onset prior to ground contact (ms) with vision available	23	27	79	109
without vision available	32	60	86	149
Tau margin (ms)	2234	999	717	661

Interestingly, the data do not completely conform to expectations from Lee's model of τ. That is, under conditions of constant velocity (not the case here), time to contact (t_c) should remain constant. When the person is undergoing acceleration, τ may constitute the basis for the strategy used to time the preparatory action. Thus, in light of the urgency of the situation in the landing task, individuals may start the voluntary action during the flight path when the optic variable τ reaches a margin value τ_m. Working backwards from the data one can determine τ_m at the time of muscle preactivation, which should be a constant value.

Sidaway et al. (1989) reports t_m values of 592, 521, and 583 ms for the low, medium and high heights respectively, which, based on an ANOVA, were not significantly different from each other. Liebermann's data produced τ_m values of 2234, 999, 717 and 661 ms for the four heights. Only at the two higher heights did τ_m stabilize. Note that in both Sidaway et al.'s (1989) study and Liebermann's (1988) study τ_m values are clearly not the actual values used to initiate action as they exceed, in every case, the total flight time. One must remember, however, that when considering an accelerative approach, the τ_m provides only an estimate of the time to contact.

We conclude, as did Sidaway et al. (1989) that "the evidence contradicts the hypothesis that subjects could compute time to contact, and therefore, initiate preactivation at some constant time to contact for all heights" (p. 260). Note, however, that for heights 60 cm and greater, the τ_m values (in their respective studies) were quite stable despite large differences in the mean preactivation intervals at the different heights. In neither the Liebermann (1988) study or the Sidaway et al. (1989) study was there evidence for cognitive processes that could compute and then trigger preactivation at some constant height above the ground or at some constant time after initiating the free fall. Although Sidaway et al. (1989) suggest that "lawfully generated properties of the optic array appeared to be the basis on which prospective control was regulated" (p. 222), we remain skeptical and are not quite ready to entirely dismiss cognitive processes as playing a role. Indeed, Liebermann (1988) presents rather compelling evidence for an alternate strategy.

Cognitive Alternatives in the Landing Preparation
First, it is clear that at the lowest heights, subjects could not have waited until τ_m values were exceeded in order to initiate action, as flight time was simply not long enough. This is consistent with earlier reported experiments of Melvill Jones and Watt (1971) who argued that during unexpected falls from heights lower than 5 cm (flight time < 100 ms) did not allow for sufficient time for even reflexive muscle activity. Consistent with the foregoing, Liebermann (1988) found that in over 60% of the trials from the lowest height

subjects exhibited EMG activity in the *rectus femoris* prior to release from the overhead bar. In fairness to Lee, however, he did not point out that the τ_m strategy may only be effective when total time of activity is greater than 300 ms.

The second aspect of Liebermann's study which is difficult to explain by incorporating the τ_m strategy are the findings from the no-vision condition. In this condition subjects were blindfolded prior to self-release from the bar. Thus, while they had visual information as to the height from the landing surface prior to initiating the free fall, no visual information was available during the actual decent. Nevertheless, as indicated in Table 14.1b, results were quite similar in the no-vision and vision conditions.

Adoption of a landing strategy based solely on time to contact obtained through τ appears unlikely. While under conditions of constant acceleration the τ strategy is modified (Lee & Reddish, 1981; Lee et al., 1983), it is still based on the premise of expansion of the optical array, or the rate of dilation of the image reflected on the retina. At this point we are not sure of how one could account for differing circumstances under which an act is carried out.

By way of example, consider the free fall task as described earlier. Had the jump been made head first, rather than foot first, the rate of dilation of the image on the retina would be similar. But, the distance of the eyes from the landing surface is quite different. Hence, the value of τ_m would surely be different. The τ_m assumes that the vantage point during the activity is directed orthogonally to the surface or the object. In this manner the rate of dilation of the retinal image is maximized. This, however, is not the way many activities are performed. For instance, in the free fall task, one could use a reference point quite distant from the actual landing point. Lastly, subjects in our task (Liebermann, 1988) could clearly make pre-contact adjustments even in the absence of vision (Liebermann & Goodman, in press).

Our data suggests that subjects might be ready to initiate the landing response prior to release regardless of the availability of visual cues during the performance. That is, there may have been some short-term visual representation of the environment with an appropriate time code to guide the preprogrammed voluntary response in preparation to land. A plausible account is that performers used information based on a visual representation of the height of fall. A similar suggestion has been forwarded by Dyhre-Poulsen and Laursen (1984) in the case of primates.

The presence of an 'intermittent' visual mode of control may be appropriate to interpret our data (Elliott in this volume, Elliott & Madalena, 1987; Laurent & Thomson, 1988; Thomson, 1983). Evidence suggests that visual cues are utilized at specific times rather than continuously within time limits of approximately 2 s (Elliott & Madalena, 1987). In the Liebermann (1988) experiment the period from the moment the subjects vision was occluded, until the end of the landing was somewhat longer (5-15 s, approximately). Yet, the delay had no deteriorous effects on performance. In addition to the environmental information, subjects in our study may have estimated an approximate duration of the flight for each height of fall. In general, subjects are capable of estimating event durations even without intention to do so (Liebermann, Raz, & Dickinson, 1988). Such an estimation may be of help in determining when to initiate action, particularly, when visual cues are not available. Other cognitive factors, such as knowledge about the type of landing place, may also modify the responses in reaction to the free falls. For instance, stiff *vs.* softer landing surfaces cause a change in the amplitude and the onset time of EMG activity even at low landing heights (Fukuda, Miyashita, & Fukuoka, 1987; Gollhofer, 1987).

In conclusion, the approach we have taken in this chapter was to examine what we considered functional for the perceiving organism. We were guided primarily by the work of Gibson (1950) and Lee (1980a, b) and remain convinced that the visual and motor systems can only be understood by examining them together. What remains to be determined is how that interaction takes place, or how vision guides action.

References

Bootsma, R.J., & Van Wieringen, P.C.W. (1990). Timing an attacking forehand drive in table tennis. *Journal of Experimental Psychology: Human Perception and Performance, 16,* 21-29.

Dyhre-Poulsen, P., & Laursen, A.M. (1984). Programmed electromyographic activity and negative incremental stiffness in monkeys jumping downward. *Journal of Physiology, 350,* 121-136.

Elliott, D., & Madalena, J. (1987). The influence of premovement visual information on manual aiming. *Quarterly Journal of Experimental Psychology, 39A,* 541-559.

Fukuda, H., Miyashita, M., & Fukuoka, M. (1987). Unconscious control of impact force during landing. In B. Jonsson (Ed.), *Biomechanics X-A* (pp. 301-305). Champaign, IL: Human Kinetics.

Gibson, J.J. (1950). *The perception of the visual world.* Boston: Houghton Mifflin.

Gibson, J.J. (1966). *The senses considered as perceptual systems.* Boston: Houghton Mifflin.

Gibson, J. J. (1979). *The ecological approach to visual perception.* Boston: Houghton Mifflin.

Gollhofer, A. (1987). Innervation characteristics of the m. gastrocnemius during landing on different surfaces. In B. Jonsson (Ed.), *Biomechanics X-B* (pp. 701-706). Champaign, IL: Human Kinetics.

Gordon, I.E. (1989). *Theories of visual perception.* New York: John Wiley & Sons.

Greenwood, R., & Hopkins, A. (1974). Muscle activity in the falling man. *Journal of Physiology (London), 241,* 26-27.

Greenwood, R., & Hopkins, A. (1976a). Landing from an unexpected fall and a voluntary step. *Brain, 99,* 375-386.

Greenwood, R., & Hopkins, A. (1976b). Muscle responses during sudden falls in man. *Journal of Physiology (London), 254,* 507-518.

Greenwood, R., & Hopkins, A. (1980). Motor control during stepping and falling in man. In J.E. Desmedt (Ed.), *Spinal and supraspinal mechanisms of voluntary motor control and locomotion. Progress in clinical neurophysiology, volume 8* (pp. 294-309). Basel: Karger.

Gregory, R.L. (1970). *The intelligent eye.* New York: McGraw-Hill.

Gurnee, H. (1934). Threshold of vertical movement of the body. *Journal of Experimental Psychology, 17,* 271-285.

Hoyt, D.F., & Taylor, C.F. (1981). Gait and the energetics of locomotion in horses. *Nature, 292,* 239-240.

Laurent, M., Dinh Phung, R., & Ripoll, H. (1989). What visual information is used by riders in jumping? *Human Movement Science, 8,* 481-501.

Laurent, M., & Thomson, J.A. (1988). The role of visual information in control of a constrained locomotor task. *Journal of Motor Behavior, 20,* 17-37.

Lee, D.N. (1974). Visual information during locomotion. In R.B. MacLeod & H.L. Pick (Eds.), *Perception: Essays in honor of James Gibson* (pp. 250-267). Ithaca, NY: Cornell University Press.

Lee, D.N. (1976). A theory of visual control of braking based on information about time-to-collision. *Perception, 5,* 437-459.

Lee, D.N. (1980a). Visuo-motor coordination in space-time. In G.E. Stelmach & J. Requin (Eds.), *Tutorials in Motor Behavior* (pp. 281-296). Amsterdam: North-Holland.

Lee, D.N. (1980b). The optic flow field: The foundation of vision. *Philosophical Transactions of the Royal Society of London, 290,* 169-179.

Lee, D.N. (1990). Getting around with light. In R. Warren & A.H. Wertheim (Eds.), *Perception and control of self-motion* (pp. 487-505). Hillsdale, NJ: Erlbaum.

Lee, D.N., Lishman, J.R., & Thomson, J.A. (1982). Regulation of gait in long jumping. *Journal of Experimental Psychology: Human Perception and Performance, 8,* 448-459.

Lee, D.N., & Reddish, P.E. (1981). Plummeting gannets: A paradigm of ecological optics. *Nature, 293,* 293-294.

Lee, D.N., & Young, D.S. (1986). Gearing action to the environment. *Experimental Brain Research Series, volume 15,* Heidelberg: Springer-Verlag.

Lee, D.N., Young, D.S., Reddish, P.E., Lough, S., & Clayton, T.M.H. (1983). Visual timing in hitting an accelerating ball. *Quarterly Journal of Experimental Psychology, 35A,* 333-346.

Liebermann, D.G. (1988). *A direct approach to landing in humans: Implications of the time to contact variable as a modulator of the voluntary timing response during free-falls.* Unpublished master's thesis, Simon Fraser University, Burnaby, British Columbia.

Liebermann, D.G., & Goodman, D. (in press). Effects of visual guidance on the reduction of impacts during landings. *Ergonomics.*

Liebermann, D.G., Raz, T., & Dickinson, J. (1988). On intentional and incidental learning and estimation of temporal and spatial information. *Journal of Human Movement Studies, 15,* 191-204.

Mark, L.S., & Vogele, D. (1987). A biodynamic basis for perceived categories of action: A study of sitting and stair climbing. *Journal of Motor Behavior, 19,* 367-384.

Matthews, B.H.C. (1956). Tendon reflexes in free fall. *Proceedings of the Physiological Society, 20,* 31P-32P.

Matthews, B.H.C., & Whiteside, T.C.D. (1960). Tendon reflexes in free fall. *Proceedings of the Royal Society of London, 153,* 195-204.

McKinley, P.A., Smith, J.L., & Gregor, R.J. (1983). Responses of elbow extensors to landing forces during jump downs in cats. *Experimental Brain Research, 49,* 218-228.

Melvill Jones, G., & Watt, D.G.D. (1971). Muscle control of landing from unexpected falls in man. *Journal of Physiology (London), 219,* 729-737.

Michaels, C.F., & Carello, C. (1981). *Direct perception.* Englewood Cliffs, NJ: Prentice-Hall.

Runeson, S. (1977). On the possibility of "smart" perceptual mechanisms. *Scandinavian Journal of Psychology, 18,* 172-179.

Savelsbergh, G.J.P., Whiting, H.T.A., & Bootsma, R.J. (1991). Grasping tau. *Journal of Experimental Psychology: Human Perception and Performance, 17,* 315-322.

Sidaway, B., McNitt-Gray, J., & Davis, G. (1989). Visual timing of muscle preactivation in preparation for landing. *Ecological Psychology, 1,* 253-264.

Solomon, H.Y. (1988). Movement-produced invariants in haptic explorations: An example of self-organizing, information driven, intentional system. *Human Movement Science, 7,* 201-233.

Thomson, J.A. (1983). Is continuous visual monitoring necessary in visually guided locomotion? *Journal of Experimental Psychology: Human Perception and Performance, 9,* 427-443.

Trevarthen, C.B. (1968). Two mechanisms of vision in primates. *Psychologische Forschung, 31*, 299-337.

Turvey, M.T. (1977). Preliminaries to a theory of action with reference to vision. In R. Shaw & J. Bransford (Eds.), *Perceiving, acting, and knowing. Toward an ecological psychology* (pp. 211-265). Hillsdale, NJ: Earlbaum.

Turvey, M.T., & Kugler, P.N. (1984). An ecological approach to perception and action. In H.T.A. Whiting (Ed.), *Human motor actions. Bernstein reassessed* (pp. 373-412). Amsterdam: North-Holland.

Ullman, S. (1980). Against direct perception. *The Behavioral and Brain Sciences, 3*, 373-415.

Wagner, H. (1982). Flow-field variables trigger landing in flies. *Nature, 297*, 14-15.

Warren, W.H. (1984). Perceiving affordances: Visual guidance of stair climbing. *Journal of Experimental Psychology: Human Perception and Performance, 10*, 683-703.

Warren, W.H., & Kelso, J.A.S. (1985). Work group on perception and action. In W.H. Warren & R.E. Shaw (Eds.), *Persistence and change. Proceedings of the first international conference on event perception* (pp. 269-281). Hillsdale, NJ: Erlbaum.

Warren, W.H., Young, D.S., & Lee, D.N. (1986). Visual control of step length during running over irregular terrain. *Journal of Experimental Psychology: Human Perception and Performance, 12*, 259-266.

Acknowledgment

The authors would like to thank Richard Carson, Winston Byblow, and Dan Weeks for their helpful comments on earlier drafts.

Part 4: Posture and locomotion

VISION AND MOTOR CONTROL
L. Proteau and D. Elliott (Editors)
© 1992 Elsevier Science Publishers B.V. All rights reserved.

CHAPTER 15

THE EFFECT OF EYE CLOSURE ON POSTURAL SWAY: CONVERGING EVIDENCE FROM CHILDREN AND A PARKINSON PATIENT

JANET STARKES*, CINDY RIACH**, and
BEVERLEY CLARKE***

* *Department of Physical Education, McMaster University
Hamilton, Ontario, L8S 4K1
and
College of Liberal Arts and Science, Ibaraki University
Bunkyo 2-2-1, Mito, Ibraki, Japan*

** *Department of Physical Education, McMaster University
Hamilton, Ontario, L8S 4K1*

*** *Department of Neurology, McMaster University Medical Centre
Hamilton, Ontario, L8N 3Z5*

This chapter will concentrate on postural sway and what happens to performance when one attempts to stand with eyes closed versus eyes open. When standing quietly, "sway" is the phenomenon that results from continual small deviations from the vertical and subsequent attempts to correct these deviations. When standing quietly, all human beings sway and this has formed the basis of many experimental tests and clinical assessments for over a century.

When normal adults are asked to stand as still as possible in a comfortable stance, performance with eyes closed is slightly worse than performance with eyes open. This is probably the most robust finding in posturography and has been the basis of the "Romberg test" in clinical medicine, since the 1800's.[1] Parkinson patients are substantially more destabilized with eye closure. There is some debate whether very young children benefit as much from visual information as their older peers. The evidence on the influence of normal vision in the postural stability of young children is conflicting. Riach and Hayes (1987) concluded that the Romberg quotients of young children (2 to 15 years) were generally lower than adult values. This finding indicates that young children are not as destabilized by eye closure as adults. However, more recent evidence by Starkes and Riach (1990) suggests that children between the ages of 8-12 years may be more destabilized by loss of vision than are adults. The results are thus inconclusive.

Parkinson patients routinely show degraded performance when they attempt to stand with eyes closed. There are conditions when performance with eyes closed may be offset in normal adults and children but remain problematic in Parkinson patients.

When subjects are asked to voluntarily lean sideways and back and forward as far as possible (i.e., to the outer limits of their stability margins), eye closure affects performance very little for normal adults and children. This is not the case for Parkinson patients. These patients have reduced stability margins with eyes open which persist and are usually worsened by eye closure.

Likewise sway velocity may be interesting from a behavioral perspective. In reaching, movement velocity has been developmentally linked to the changing strategies for ballistic or sensory guided control (Hay, 1979). In this chapter we suggest that this may also be true of postural control in children and Parkinson patients.

In order to examine these issues we will first describe the typical methodology of postural sway studies and examine the nature of the dependent measures routinely used to assess postural sway.

Measures of Postural Stability

One of the most common measures of postural control, both clinically and behaviorly is the Romberg test of quiet standing postural sway (Njiokiktjien & de Rijke, 1972). Historically, the Romberg test was an assessment of how much the patient swayed. This subjective measure was and is used clinically in neurological assessments with comparisons of postural sway being made between eyes open and eyes closed conditions. Romberg comparison data are

---

[1]The Romberg quotient is an expression of eyes closed sway value as a percentage of the eyes open value (Njiokiktjien & Van Parys, 1976).

available on patients with neurological conditions as diverse as cerebellar ataxia (Dichgans, Mauritz, Allum, & Brandt, 1976; Lucy & Hayes, 1985; Nashner & Grimm, 1978; Njiokiktjien & de Rijke, 1972; Njiokiktjien, de Rijke, Kieker-Van Ophen, & Voorhoeve-Coebergh, 1978), vestibular deficits (Nashner, Black, & Wall, 1982), and Parkinson's disease (Folkerts & Njiokiktjien, 1972; Njiokiktjien & de Rijke, 1972).

Recent technological developments have enabled automated and refined assessments such that postural sway parameters can now be easily quantified. With these developments postural control has become a more valuable measure in clinical assessment and a popular aspect of research in sensori-motor control. One of the quantifiable parameters of postural sway is the recording of changes in position of the centre of pressure of ground reaction forces beneath the feet. The subject is asked to stand as quietly as possible. A force plate, on which the subject stands records the position of centre of pressure as it fluctuates with postural sway.

Biomechanically the degree of physical stability during quiet standing or "static" equilibrium is proportional to the size of the base of support and stability is maximized in any direction when the vertical line of the centre of gravity is furthest from the edge of the base of support. Put succinctly, the bigger the base of support and the closer one is to the vertical centre, the more stable one is. Koozekanani, Stockwell, McGhee, and Firoozmand (1980) have devised an index of stability to describe this realtionship (equation 15.1).

$$s(t) = \min s_i (t) \quad (i=1,2) \tag{15.1}$$

The stability margin over time ($s(t)$) is the shortest distance (min) from the centre of pressure to either the front (S_1) or back (S_2) of the supporting foot. A person loses balance when the stability margin goes to zero. During quiet standing the vertical line of the centre of gravity is almost directly aligned with centre of pressure beneath the feet. By monitoring the moment-to-moment fluctuations in centre of pressure an assessment of postural stability can be obtained.

In our studies of postural control, basic changes in centre of pressure position were determined using a strain gauge force plate. Throughout the standing test conditions the subject's ground reaction force in the vertical direction (F_z) and moments of force (Mx and My) about the lateral and antero-posterior axes were recorded from the force plate. The signals were processed and estimates of centre of pressure position determined (see Riach & Starkes, 1989 for details). Assessment of centre of pressure fluctuations can be evaluated in the form of position (moment-to-moment stability), range of movement (stability over a period of time), standard deviation from mean position (sway magnitude over a time period) and velocity. While the behavioral interpretations of sway velocity are as yet unclear, it may be a reflection of postural control strategy (Riach & Starkes, 1991a).

The range of centre of pressure movements during quiet stance gives an indication of how stable the person is over a period of time. It shows how much of the base of support the subject uses over the test duration (Riach & Starkes, 1991b). A measure of maximum usable range or "stability limits" gives information about how much of the anatomical base of support can be used to sustain balance. While the anatomical base of support is the area between and beneath the feet, "stability limits" outline the functional base of support. As such the size of the functional base of support reflects many control features beyond size of the feet. Outer limits of stability have been studied in adults (McCollum & Leen, 1989; Murray, Seireg, & Sepic, 1975; Riach & Starkes, 1991b), children (Riach & Starkes, 1991b) and in the elderly (Lee & Deming, 1987; Murray et al., 1975).

The standard deviation about the mean position (or root mean square) of centre of pressure movements provides a measure of sway magnitude over a time period. This measure has been used to compare spontaneous sway magnitude changes with age in children (Odenrick & Sandstedt, 1984; Riach & Hayes, 1987), in the elderly (Hayes, Spencer, Riach, Lucy, & Kirshen, 1985), and under different visual conditions (Riach & Hayes, 1987; Riach & Starkes, 1989; Starkes & Riach, 1990).

Age Related Changes in Postural Control
Postural Sway

Postural control of children improves as they mature. Studies of sway magnitude (standard deviation of centre of pressure changes about the mean position) show that spontaneous postural sway during quiet standing decreases with increasing age in children. Also, young boys sway more than young girls but boys' performance improves such that there is no difference between the genders by 10 years (Odenrick & Sandstedt, 1984; Riach & Hayes, 1987). In regression analyses, age alone explains a small but significant amount of the between subject variance in sway. Physical stature (height and weight) also account for some of the variance. There remains, however a large amount of between subject variability not readily explained by age, sex or physical stature (Odenrick & Sandstedt, 1984; Riach & Hayes, 1987). Individual differences in maturation of elements within the postural control processes may well contribute to variability in sway. Maturation of postural mechanisms include: short and long loop proprioceptive reflexes (Bawa, 1981), vestibular function (Ornitz, 1983), cortical, and visual processing (Butterworth & Hicks, 1977; Lee & Aronson, 1974) and finally postural synergy organization (Forssberg & Nashner, 1982; Shumway-Cook & Woollacott, 1985). Most studies demonstrate large variability in maturation rate between individuals. Within children, there are also differences in the maturation rate of various processes of postural control. Understanding the basis of different sources of variance in performance is the foremost objective in studying postural stability in children.

Stability Limits

In quiet standing postural control, the degree of stability is proportional to the size of the base of support. Not surprisingly, children with smaller feet than adults, have a smaller base of support. As mentioned previously, they also have greater spontaneous postural sway than adults. Both of these factors contribute to poor postural stability in children. Furthermore, children may lack the strength and/or postural control to use as much of their anatomical base of support as do adults. Their stability limits may be a smaller ratio of their anatomical base of support. That is, they may not be able, through limited strength and/or neural control, to maintain stability when the vertical line of centre of gravity approaches the limits of the anatomical base of support. As outlined previously the anatomical base of support is the area within and between the feet. The "stability limits" define the area beneath the feet that the subject can use to contain and support the vertical line of centre of gravity.

In an initial study (Riach & Starkes, 1991b), we measured stability limits of children and adults. We also investigated maturational changes and the influence of vision. The question was whether closing the eyes affects the stability limits of children and/or adults. It was hypothesized that vision may contribute necessary sensory feedback for the control of stability and/or vision may contribute to the confidence required in venturing out to the limits of stability.

In our study, 70 children aged 4 to 14 years and 17 adults stood on a force plate with shoes off and feet together for 2 trials of 30 seconds, once with eyes open and once with eyes closed. During each trial the subject was asked to sway as far as possible forward, backward, left and right. They swayed first forward through the sequence and then backward, thus going to each extreme position twice in a trial. Each extreme position was held for 2-3 seconds. The centre of pressure of ground reaction forces was calculated using the vertical force and appropriate moment of force (cf. Riach & Starkes, 1989). The x-y movements of centre of pressure over the 30 second trial were recorded and the maximum ranges of antero-posterior and lateral sway were measured (i.e., stability limits). Figure 15.1 illustrates a "spaghetti" plot or recording of centre of pressure of a 4 year old and an adult who leaned in each direction twice.

J. Starkes, C. Riach, and B. Clarke

An outline of the feet together position was traced on paper as a measure of the length and width of the anatomical base of support. Each subject was also measured for maximum plantar flexion torque at the ankle joint.

At this point an attempt was made to determine the relative contributions of various factors in the prediction of the stability limits in both the antero-posterior and lateral directions. Regression analyses were employed and the predictor variables used were: age, gender, height, weight, ankle torque strength, foot length and width of the two feet together. The dependent measures were the maximum range of centre of pressure movements lengthwise (antero-posterior) and widthwise (lateral) (i.e., the stability limits of centre of pressure).

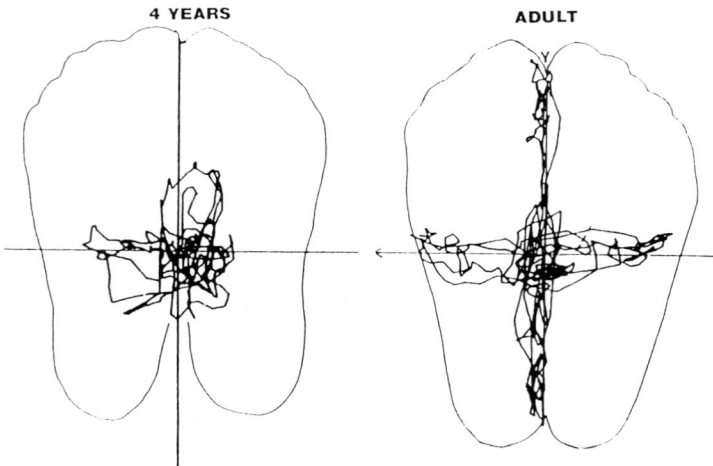

Figure 15.1. Stability limits of centre of pressure for a 4 year old and an adult (normalized foot size).

The first stage of analyses of the childrens' data was a correlation matrix of the variables. Sex did not correlate significantly with any of the other variables and so was not considered further. Not surprisingly, high positive correlations were found ($p < .01$) among age and the physical characteristics of height, weight, and foot length. Significant positive correlations ($p < .01$) were also found between maximum ranges (stability limits). This indicated that those subjects with greater stability limits in one direction also have wide stability limits in the other direction, again not a surprising finding. Figure 15.2 shows the mean maximum range of sway (how far subjects can lean on average) for children and adults.

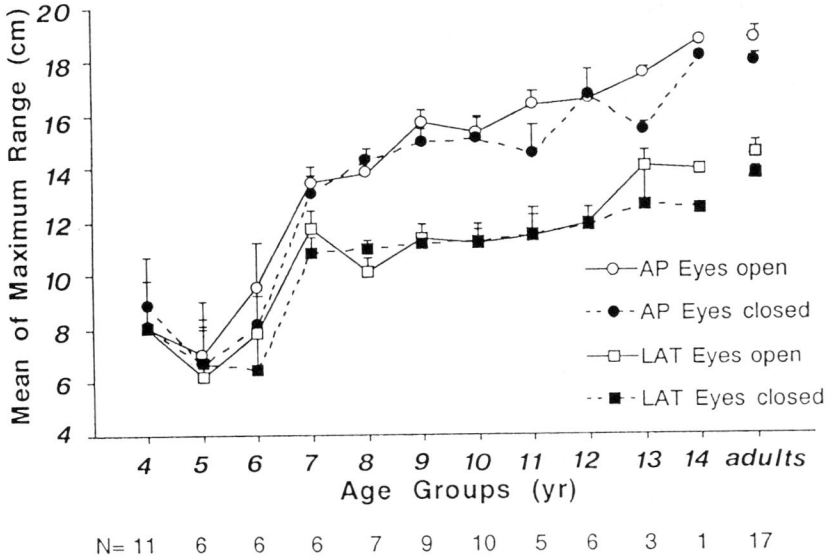

Figure 15.2. Mean of maximum range (cm) and standard errors for stability limits in adults and children of varying ages. Data are presented for antero-posterior and lateral directions with both eyes open and eyes closed.

From 4 to 6 years children show a great deal of inter subject variance about the mean stability limit, regardless of direction or visual condition (Figure 15.2). From 7 years to adulthood there is a gradual increase in stability limits and decreasing variability in performance. What was unexpected was a sudden jump in stability limits and decrease in inter subject variance between 6 and 7 years. Performance is relatively equal with eyes open or closed, and subjects gradually are able to lean further forward and backward than sideways.

Figure 15.3 indicates how much of the anatomical base is actually used at different ages. Foot size is normalized across ages by using the percentage of the total base size. Young children, 4-6 years, have stability limits between 6 and 9 cm. This constitutes approximately 38 to 57% of their anatomical base (as measured from feet tracings). There is a pronounced increase of 4 to 5 cm in the sway limits at 7 years in both visual conditions. At the same time there is a decrease in between-subject variance. When stability limits are taken as a percentage of the anatomical base of support there remains an increase in percentage (increasing by 19% to 26%) at this age. Also evident is the finding that children 7 years and above appear to have adult levels of sway limits when normalized to footsize.

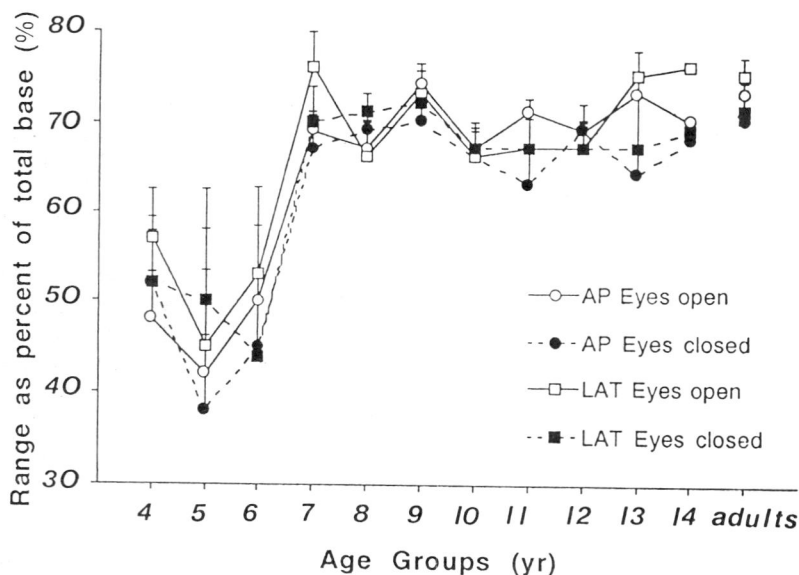

Figure 15.3. Range of sway as a percent of total base of support for adults and children of varying ages. Data are presented for antero-posterior and lateral directions with eyes open and eyes closed.

The adult data are quite comparable with data reported by Whitney (1962). Whitney reported that effective foot-base for adults when leaning was approximately 66% of foot length. In the present study we found stability limits of 70% (eye closed) and 73% (eyes open) antero-posterior.

Strategies in the developmental years

As mentioned previously, the control of quiet standing postural sway improves throughout childhood. Improvement is seen with maturational changes in sensorimotor control and continues at a fairly regular rate of development. The apparent consistency in improvement in postural control is no doubt, however, an oversimplification of the developmental process. In reaching skills, for example, it has been demonstrated that there is a change in control strategy at about 7 years of age (Hay, 1979). Subjects were tested wearing visual prisms which laterally displaced the visual field. Children younger than 7 years used a predominantly ballistic, open loop strategy of arm movements seen as fast movements with later feedback corrections. Children at 7 years used slower more accurate reaching movements with earlier correction which has been attributed to the utilization of a visually guided strategy.

There is evidence in postural control of a similar shift from fast open loop postural adjustments to a slower sensory guided (feedback based) control strategy. When young children (1-1/2 to 3 years) are tested with postural perturbations they show adjustments in leg and trunk muscles (EMG) similar to adults (Shumway-Cook & Woollacott, 1985). However, their EMG responses are larger in amplitude and longer in duration than those seen in older children and adults. The postural responses often overcompensate for platform induced sway and thus produce greater body oscillation (Shumway-Cook & Woollacott, 1985). These relatively large postural adjustments that overcompensate are characteristic of ballistic movements. Older children (7 to 10 years) and adults make slower more accurate corrections causing very little if any oscillation. This improvement may be indicative of a change in postural control strategy (i.e., toward a sensory guided control). It may, however, be argued that the improved control is a reflection of improved timing, magnitude and accuracy of the ballistic corrections.

Similar findings are seen in studies of static postural control (Shambes, 1976; Williams, Fisher, & Tritschler, 1983). Young children (4 and 6 years) show a greater frequency of "bursts" in muscle activity when stability is threatened than do older children (8 years).

One way of investigating control strategy is to look at the speed of postural adjustments. The purpose of the second study was to test the hypothesis that postural control strategy changes from fast ballistic control to slower more accurate sensory guidance. Children (n=81) aged 4 to 13 years and adults (n=26) were tested during quiet stance for 20 seconds on a force plate. The vertical force (F_z) and moments of force about the antero-posterior and lateral axes were recorded. From these signals the position and movements of the centre of pressure of ground reaction forces were estimated.

Sway velocity was derived by dividing the total sway distance by time (Riach & Starkes, 1991a).

The first stage of analyses was to determine the relationship between age and sway velocity. Scatter plots showed that sway velocity decreased with age. It was considered possible that sway velocity may be more related to physical development (height and weight) than to the non-physical developmental changes with age. A regression analysis determined that when the variance due to height and weight was removed there remained a significant influence of age in determining sway velocity. This suggests that while the physical parameters of height and weight influence sway velocity there remains a significant effect of non-physical changes related to age that influence sway velocity. Also, the significant reduction in sway velocity at 8 years is not parallelled by concomitant changes in physical characteristics.

In order to determine at what ages there were significant changes in sway velocity a two way analysis of variance was conducted for age (10 groups of children) x vision (eyes open, eyes closed) with repeated measures on the second variable. In essence the results showed two clusters with 4,5, and 7 years olds showing sway at a higher velocity than 8,9,11,12, and 13 year old children. Figure 15.4 illustrates this significant decrease in sway velocity at approximately 8 years of age.

Vision was also an important factor with velocities significantly higher in the eyes closed condition (mean = 2.63 cm/s) compared to eyes open condition (mean = 1.81 cm/s). There was no significant age by vision interaction indicating that the effect of eye closure was the same across all ages regardless of any changes in sway velocity. Because the change in sway velocity at 8 years seen in the eyes open condition was also evident in the eyes closed condition, it suggests that the change is toward sensory guidance (i.e., visual, vestibular, and proprioceptive) not just visual guidance (see also Proteau, this volume).

From the developmental work we are left with the following conclusions. First, in quiet standing, children sway less with increasing age in both antero-posterior and lateral directions. Second, stability limits are small and quite variable in children 4 to 6 years of age. There is a sudden increase in stability limits at 7 years up to adult levels when normalized to foot size. The age of 7 years appears to be an important maturational stage in the postural control of children. Third, stability limits of children do not appear to be influenced by eye closure. It appears that they can rely on feedback from vestibular and proprioceptive sensors in the absence of vision. A closer investigation of individual ages, however, may reveal age related changes in the importance of vision. Fourh, changes in sway velocity may reveal developmental changes in

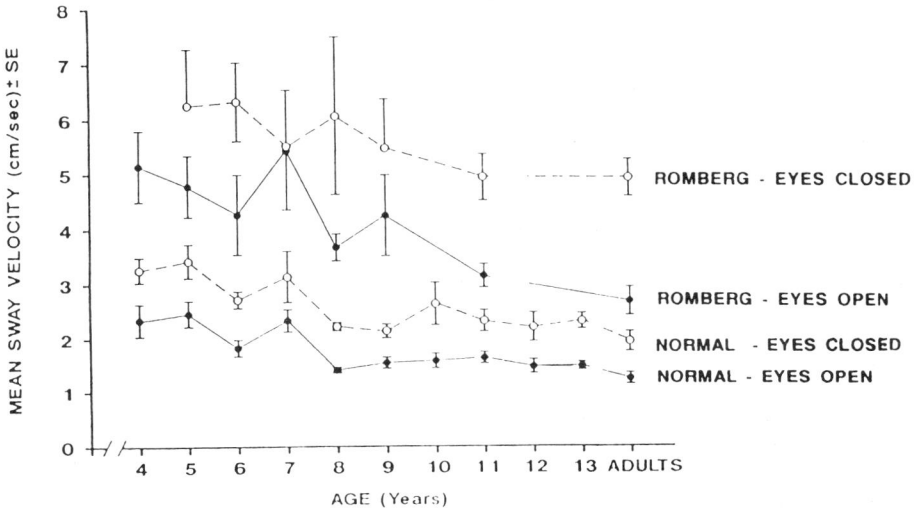

Figure 15.4. Mean sway velocity and standard error of the means for differing age groups. Normal and Romberg stance with open and eyes closed are each shown.

postural sway control. Children 7 years and younger may be more ballistic (open loop) in their postural adjustments while older children may rely more on sensory feedback monitoring in making postural corrections. Finally, sway velocity decreases with age and is sensitive to eye closure. Eye closure routinely increases velocity of sway. In the eyes closed condition there remains a decrease in sway velocity at 7 to 8 years. This suggests that the shift is to multi-sensory control, not just visual control. It is apparent, however, that vision is important in determining sway velocity and appears to be equally important throughout childhood.

In the second part of this chapter we have chosen to examine postural control in Parkinson patients. Parkinsonism inflicts both drastic and unique changes to the control of postural sway and movement. While some deficits may appear to be similar to performance of children, others are quite different. It is from these differences that we can learn both about the development of postural control as well as neurological deficit.

The Disease Process of Parkinsonism

Unfortunately space does not permit a full discussion of the nature, neuropathology and clinical symptoms of Parkinson's disease. Excellent sources are abundant on both the clinical syndrome of Parkinson's disease (Hefti & Weiner, 1988; Marsden, 1990) and the implications of the disease for motor control (Chan, 1986; Evarts, Teravainen, Beuchert, & Calne, 1979; Flowers, 1976; Rogers & Chan, 1988). As a brief summary, Parkinson's disease is a slow, progressive, neurodegenerative disorder that has four cardinal signs: tremor, rigidity, bradykinesia and gait and postural abnormalities. Any number of symptoms may be present in a patient, and there is high individual variability in clinical presentation. Parkinson's disease most often afflicts individuals between the ages of 58 and 61 years, but may present from 30 to 80 years (Factor & Weiner, 1988). The type of tremor seen is referred to as a "resting" tremor, since it is usually apparent at rest and disappears for the most part when the patient performs goal directed movements. While tremor will usually appear as a unilateral symptom, with time it tends to become bilateral and may affect the hands, legs, jaw, and tongue. While tremor is frequently embarrassing it is usually not extremely debilitating, although certainly it is worsened by stress. Bradykinesia is characterized by a slowness of movement with loss of automatic or stereotyped movements. Hesitation on movement initiation, loss of dexterity and early fatigue are also characteristic of movements (Flaherty, 1988).

Posture in a Parkinson's patient characteristically shows in flexion of the head, trunk and extremities. Patients appear to be stooped forward, and bent at the knees and elbows. Frequently the patient will appear to lean when sitting or standing and will appear to use a shuffling gait. Often there is a hesitation in starting to walk and because of the forward lean of the trunk the patient uses a propulsive gait, often appearing to move quickly to "catch up" with himself. Poor balance eventually results from loss of postural reflexes and in the advanced stages, the patient is not able to regain balance following even minor postural perturbations. The decreased or impaired equilibrium responses are aggravated by eye closure (Flaherty, 1988).

While drug interventions are for the most part effective in relieving many of the major motor symptoms (especially in the early stages of the disease) there are recognized adverse effects. These often include psychiatric side effects, dyskinesia, and fluctuations in motor control. These are often related to end of dosage failure, ineffectiveness of individual doses, and sudden on-off lack of control in movement.

There is ample evidence to indicate that Parkinson's disease patients have poorer quiet standing performance than normal subjects in terms of amplitude or area of sway, and this difference widens with eye closure. In a comparison of healthy young (< 40 years), healthy older individuals (< 85 years), and Parkinson's disease patients, Njiokiktjien and de Rijke (1972) found that with eyes open, over 30 seconds, these groups swayed an average of 0.78 cm, 1.51 cm and 1.92 cm respectively. With eye closure the median values increased to 1.2, 1.87 and 2.54 cm respectively. Thus Parkinson's disease patients exhibited more sway than healthy elderly, both with and without vision.

More recently, Kilbreath (1986) examined postural stability in 37 Parkinson's disease patients and healthy age-matched subjects. Subjects were tested for 20 trials, with feet spaced slightly apart. Her results indicated that Parkinson's disease patients swayed significantly more than healthy subjects and within Parkinson's disease patients the effect of eye closure was greater on antero-posterior than lateral sway. Interestingly multiple regression analyses showed that sway could not be predicted well by clinical features or cardinal signs present in the subject. This means that with eyes closed patients are very unstable in the forward-backward direction. Some patients are highly unstable even when other signs of disease are relatively mild.

Postural Control in a Parkinson Patient
In this section of the chapter we will illustrate stability limit changes in a stage 3 Parkinson's disease patient and examine how sway velocity varies over the course of a day. By definition, in stage 3 the patient exhibits bilateral movement problems and moderate gait impairment. In Figure 15.5 we see a typical response of a 6 year old, a normal 67 year old and a 67 year old Parkinson's disease patient (JN).[2] Each subject was asked to lean in four directions, with eyes open and with eyes closed. This test of size of functional base poses several unique problems when used in testing Parkinson's disease patients. First, if the subject is following instructions and leaning as far as possible there are isolated occasions when the patient will lean one direction and not be able to recover. Second, time of day, time of last dose, drug interactions with meals, exercise, stress, and confidence in abilities, all influence postural performance, independent of the ongoing change with disease process. As such there is tremendous heterogeneity between patients, likewise there is extremely high variability within subjects. For this reason we have attempted to consider patients as individual case studies and sought repeated measurements over the course of the day as well as between days.

---

[2]JN is a 67 year old Parkinson patient. Onset was at 60 years. He now suffers bilaterally. He exhibits resting tremor in the arms, legs, and jaw; dysarthrias; and certain dystonias even with medication. He occasionally exhibits on-off phenomena and end dose failure. At the time of testing his medication consisted of Sinemet, bromocryptine and cogentin. He was also assessed and began use of deprenyl.

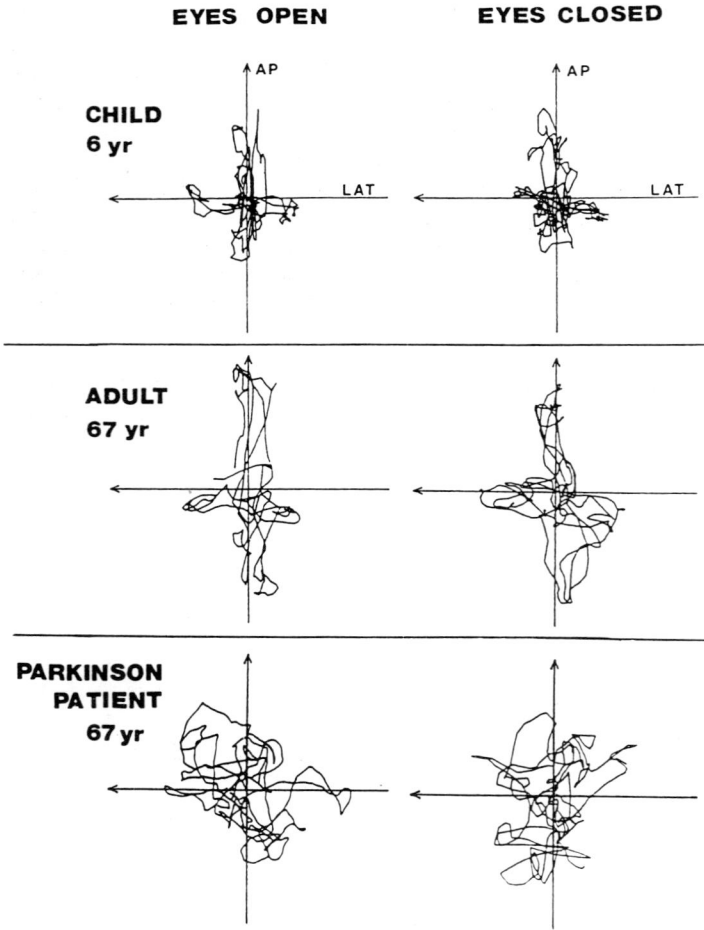

Figure 15.5. Sample stabilograms for a 6 year old child, a normal 67 year old, and a 67 year old Parkinson's patient. Stability limits are demonstrated with eyes open and eyes closed.

Having in mind the various limitations mentioned above, we can see from Figure 15.5 how variable JN is in comparison with a normal adult and a 6 year old child. At times the patient may appear to have relatively small stability limits, and be unable to recover once these are exceeded. At other times the limits appear to be exaggerated in comparison with normal subjects and while the maximum range is as high there is far more variability in the area of excursion over the same time periods. This is consistent with clinical observations that Parkinson's disease patients are far more reliant on visual feedback for both gait and arm movement control. This may in part be true because postural reactions involve the supraspinal equilibrium systems, whose activity is dependent upon an intact basal ganglia. Rogers and Chan (1988) and others have suggested that some of the deficits apparent in Parkinson's disease are in part related to task complexity, and problems in movement initiation and correction.

Sway velocity was also recorded for JN during 8 sessions from 9:30 a.m. to 4:30 p.m., on each of three days. Collapsing across days, mean velocities and standard errors are presented in Figure 15.6.

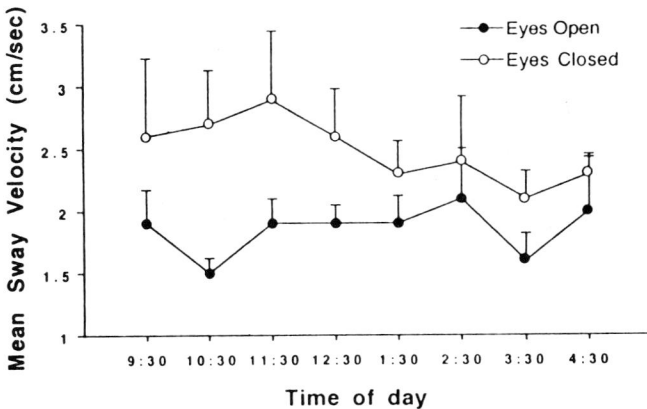

Figure 15.6. Mean sway velocity (cm/s) and standard errors for a stage 3 Parkinson's patient (JN) at eight times throughout the day. Means are for data gathered during three full days.

The data in this figure clearly illustrate the need to test patients repeatedly at varying times throughout the day and dosage schedule. Observations over several studies have shown that sway velocity is routinely higher and more variable than that seen developmentally or for age-matched controls. Velocity during eye closure is also routinely higher than with eyes open. While this is also true of adults, and particularly true of children, the absolute velocities seen in these other groups are far slower, and less variable. This is demonstrated by comparing the data for JN in Figure 15.6 with other ages in Figure 15.4. On both figures data are provided for normal stance eyes open and eyes closed.

The Dissociation of Development *vs.* Parkinson's Disease: Characteristics of Sway

At first glance postural sway in young children and Parkinson's disease patients appears similar. Both evidence increased antero-posterior and lateral sway and sway velocity. The simplest explanation of similarities is that children are experiencing neurological development and the accompanying changes in myelination, vestibular, proprioceptive and visual feedback. These in turn permit changes in postural control strategies. At the other extreme, Parkinson's disease patients have a deteriorating neurological system and are gradually losing vestibular, proprioceptive reflexes, etc. They too are forced to adjust postural control strategies to cope. This is an over-simplistic scenario, however, of one getting better, the other worse.

A closer look at changes in postural sway indicates several areas of disimilarity. First, in stability limits children show changes in both lateral and antero-posterior sway. Parkinson's disease patients also evidence deterioration in both directions but antero-posterior sway (forward - backward) is far more unstable. Even compared to age-matched controls Parkinson's disease patients are more unstable leaning forward - backward (Kilbreath, 1986).

Second, stability limits of children and adults are not necessarily affected by eye closure, presumably because they are able to rely on feedback from vestibular and proprioceptive sources in the absence of vision. Eye closure routinely affects the stability limits of Parkinson's disease patients, so much so that with eyes closed advanced patients often cannot recover from leaning in one direction.

Third, sway velocity reveals several differences between normal developmental changes and the neurological deficits associated with Parkinson's disease. In both children and patients sway velocity is increased and is also sensitive to eye closure. Figures 15.4 and 15.6 illustrate those similarities. If JN's data from Figure 15.6 are collapsed across the day and compared to Figure 15.4, JN's average sway velocity is comparable to data in the cluster of 4-7 year olds, both with eyes open and eyes closed. Sway velocity for JN is far more variable however. Recall that earlier in the summary of developmental changes it was suggested that in children eye closure increases sway velocity until around 7 years. The inference was that after this age there is a shift to multi-sensory control. In Parkinson's disease patients we might speculate that multi-sensory control is jeopardized first by the gradual loss of proprioceptive and vestibular information, then further diminished by the intermittent ability to respond to this information (related to the on-off phenomena of movement abilities with the disease).

Finally, the issue of reliability of sway performance must be addressed. Throughout this chapter we have referred to the high inter-subject variability of performance in young subjects and patients. A recent paper has demonstrated that although normal subjects generally show high inter subject variability of performance, intra subject measures of antero-posterior, lateral, total linear distance swayed, and sway velocity are highly reliable throughout the course of the day (Hattori, Starkes, & Takahashi, in press). Subjects aged 6 to 87 years were assessed three times throughout the day in normal, Romberg, and one foot stances, with eyes open and closed. As an example, test-retest reliabilities for sway velocity in normal and Romberg stances with eyes open and closed ranged from $r = 0.71$ to $r = 0.86$. In contrast, Parkinson's disease patients show very low test-retest reliability. Again, this is illustrated by JN's data in Figure 15.6. Intra-subject variability has many sources in these patients. There are day to day changes with the disease process and current stage of the disease. As well, blood dopamine level may effect performance and is in turn affected by types of medications taken, dosages and clearance rates, diet, activity level, excitation/agitation. Intra-subject variability may also stem from postural control strategies which are forced to change both throughout the day and as the patient deteriorates. These strategy changes may be both at the conscious and unconcscious level.

In children we have suggested that part of the variance in sway velocity may be accounted for by changes in strategies related to motor control. Recall that fast sway velocity may be related to a more ballistic form of control, whereas slow velocity may be more characteristic of sensory guided control processes. Certainly if one watches a Parkinson's disease patient lean, while they appear to concentrate on visual cues (and report doing so) they are inevitably forced to make constant, large ballistic corrections to return to vertical.

Gregoric and Lavric (1977) have also indicated abnormal postural regulation in both centre of pressure plots and the frequency spectra of sway with Parkinson's disease. They believe that these changes reflect two different pathophysiological mechanisms; the control of posture and control of equilibrium. At present there are several hypotheses as to why postural stability is affected in Parkinson's disease. Besides basal ganglia pathophysiology, neurotransmitters such as noradrenaline and others in the isodendritic core have been discussed. As yet however, none has been directly linked to stability problems in Parkinson's disease.

In future there are many behavioural questions to be addressed regarding Parkinsonism. We do not know how variable the patient control processes are throughout a dose, a day, or the course of the disease. Likewise we do not know how the stage of the disease interacts with postural control processes. Even those studies which have examined postural sway in Parkinson's disease patients rarely include extensive information on the stage of disease, clinical features and symptoms, age of onset, sensory assessment, extent of loss of postural reflexes, etc. In future the same kinds of behavioural questions need to be asked in the clinical setting, initially with a small number of highly delineated groups of patients, who have undergone extensive sensory and reflex assessment, and eventually with patients at different stages of the disease process.

By contrasting the performance of young subjects, subjects with neuropathology and age-matched controls, the dissociation of effects of maturation *vs.* deficit on the postural control system can be explained. We have attempted to illustrate the process with basic questions concerning eye closure and the relation of sway velocity to postural strategies. Certainly there is far more work to be done on each of these issues in future.

References

Bawa, P. (1981). Neural development in children - a neurophysiological study. *Electroencephalography and Clinical Neurophysiology, 52*, 249-256.

Butterworth, G., & Hicks, L. (1977). Visual proprioception and postural stability in infancy. A development study. *Perception, 6*, 255-262.

Chan, C.W.Y. (1986). Could Parkinsonian akinesia be attributable to a disturbance in the motor preparatory process? *Brain Research, 386*, 183-196.

Dichgans, J., Mauritz, K.H., Allum, J.J., & Brandt, T. (1976). Postural sway in normals and atactic patients: Analysis of the stabilizing and destabilizing effect of vision. *Agressologie, 17*, 15-24.

Evarts, E.V., Teravainen, H.T., Beuchert, D.E., & Calne, D.B. (1979). Pathophysiology of motor performance in Parkinson's disease. In K. Fuxe & D.B. Calne (Eds.), *Dopaminergic ergot derivations and motor function* (pp. 45-49). New York: Pergamon.

Factor, S.A., & Weiner, W.J. (1988). The current clinical picture of Parkinson's Disease. In F. Hefti & W.J. Weiner (Eds.), *Progress in Parkinson research* (pp. 1-10). New York: Plenum.

Flaherty, J.F. (1988). Parkinson's disease. In L.E. Young & M.A. Koda-Kimble (Eds.), *Applied therapeutics: Clinical use of drugs (4th edition)* (pp. 1349-1367). Vancouver: Applied Therapeutics.

Flowers, K. (1976). Visual closed-loop and open-loop characteristics of voluntary movement in patients with Parkinsonism and intention tremor. *Brain, 99*, 269-310.

Folkerts, N., & Njiokiktjien, C. (1972). L'influence de la L-Dopa sur la régulation des parkinsoniens [Influence of L-Dopa for controlling the symptoms of parkinsonism]. *Agressologie, 13C*, 19-24.

Forssberg, H., & Nashner, L.M. (1982). Ontogenetic development of postural control in man: Adaptation to altered support and visual conditions during stance. *Journal of Neuroscience, 2*, 545-552.

Gregoric, M., & Lavric, A. (1977). Statokinesimetric analysis of the postural control in Parkinsonism. *Agressologie, 18A*, 45-48.

Hattori, K., Starkes, J., & Takahashi, T. (in press). Age, diurnal variability, and the reliability of postural sway. *Japanese Journal of Human Posture*.

Hay, L. (1979). Spatial-temporal analysis of movements in children: Motor programs versus feedback in the development of reaching. *Journal of Motor Behavior, 11*, 189-200.

Hayes, K., Spencer, J., Riach, C., Lucy, S.D., & Kirshen, A. (1985). Age related changes in postural sway. In D. Winter, R. Norman, R. Wells, K. Hayes, & A. Patla (Eds.), *Biomechanics IX-A* (pp. 383-387). Champaign, IL: Human Kinetics.

Hefti, F., & Weiner, W.J. (1988). *Progress in Parkinson research*. New York: Plenum.

Kilbreath, S.L. (1986). *Postural instability in patients with Parkinson's disease*. Unpublished master's thesis, University of Western Ontario, London, Ontario.

Koozekanani, S., Stockwell, C., McGhee, R., & Firoozmand, F. (1980). On the role of dynamic models in quantitative posturography. *IEEE Transactions on Biomedical Engineering, BME27*, 605-609.

Lee, D.N., & Aronson, E. (1974). Visual proprioceptive control of standing in human infants. *Perception & Psychophysics, 15*, 529-532.

Lee, D.N., & Deming, D. (1987, June). *Correlation between age and the size of normalized static support base while standing*. Paper presented at the meeting of the North American Society for the Psychology of Sport and Physical Activity, Vancouver, Canada.

Lucy, S.D., & Hayes, K. (1985). Postural sway profiles: Normal subjects and subjects with cerebellar ataxia. *Physiotherapy Canada, 37*, 140-148.

Marsden, C.D. (1990). Parkinson's disease. *Lancet, 335*, 948-952.

McCollum, G., & Leen, T. (1989). Form and exploration of mechanical stability limits in erect stance. *Journal of Motor Behavior, 21*, 225-244.

Murray, M., Seireg, A., & Sepic, A. (1975). Normal postural stability and steadiness: Quantitative assessment. *Journal of Bone and Joint Surgery, 57A*, 510-516.

Nashner, L.M., Black, F.O., & Wall, C. (1982). Adaptation to altered support surfaces and visual conditions of stance in patients with vestibular deficits. *Journal of Neuroscience, 2*, 536-544.

Nashner, L.M., & Grimm, R.J. (1977). Analysis of multiloop dyscontrols in standing cerebellar patients. *Progress in Clinical Neurophysiology, 4*, 300-319.

Njiokiktjien, C., & de Rijke, W. (1972). The recording of Romberg's test and its application to neurology. *Agressologie, 13C*, 1-7.

Njiokiktjien, C., de Rijke, W., Kieker-Van Ophen, A., & Voohoeve-Coebergh, O. (1978). A possible contribution of stabilography to the differential diagnosis of cerebellar processes. *Agressologie, 19B*, 87-88.

Njiokiktjien, C., & Van Parys, J. (1976). Romberg's sign expressed in a quotient II - pathology. *Agressologie, 17D*, 19-24.

Odenrick, P., & Sandstedt, P. (1984). Development of postural sway in the normal child. *Human Neurobiology, 3*, 241-244.

Ornitz, E. (1983). Normal and pathological maturation of vestibular function in the human child. In R. Romand (Ed.), *Development of auditory and vestibular systems* (pp. 479-528). Santa Clara, CA: Academic Press.

Riach, C., & Hayes, K. (1987). Maturation of postural sway in young children. *Developmental Medicine and Child Neurology, 29*, 650-658.

Riach, C., & Starkes, J. (1989). Visual fixation and postural sway in children. *Journal of Motor Behavior, 31*, 265-276.

Riach, C., & Starkes, J. (1991a). *Sway velocity as an indicator of postural control strategy in children.* Manuscript submitted for publication.

Riach, C., & Starkes, J. (1991b). *Stability limits of quiet standing postural control in children and adults.* Manuscript submitted for publication.

Rogers, M.W., & Chan, C.W.Y. (1988). Motor planning is impaired in Parkinson's disease. *Brain Research, 438*, 271-276.

Shambes, G. (1976). Static postural control in children. *American Journal of Physical Medicine, 55*, 221-252.

Shumway-Cook, A., & Woollacott, M. (1985). The growth of stability: Postural control from a developmental perspective. *Journal of Motor Behavior, 17*, 131-147.

Starkes, J., & Riach, C. (1990). The role of vision in the postural control of children. *Clinical Kinesiology, 44*, 72-77.

Whitney, R.J. (1962). The stability provided by the feet during manoeuvres whilst standing. *Journal of Anatomy, 96*, 103-111.

Williams, H., Fisher, J., & Tritschler, K. (1983). Descriptive analysis of static postural control in 4,6, and 8 year old normal and motorically awkward children. *American Journal of Physical Medicine, 62*, 12-26.

Acknowledgment

Preparation of this article was supported in part by grant #A3700 from NSERC to the first two authors. The authors wish to thank Dr. A.R.M. Upton for his referral of Parkinson patients, and Ms. R. Sears for her assistance in data collection. We would also like to thank those patients and subjects who gave willingly of their time to participate in the studies.

VISION AND MOTOR CONTROL
L. Proteau and D. Elliott (Editors)
© 1992 Elsevier Science Publishers B.V. All rights reserved.

CHAPTER 16

THE ROLE OF VISION IN THE PLANNING AND GUIDANCE OF LOCOMOTION THROUGH THE ENVIRONMENT

JOHN CORLETT

Faculty of Human Kinetics, University of Windsor
Windsor, Ontario, N9B 3P4

Organisms that remain stationary, such as the Sea Squirt (Ciona intestinalis) in the adult phase of its life cycle, or the 350 species of carnivorous plants, have relatively few sensory and perceptual needs. Because they cannot flee from predators or chase prey themselves, it is not important for them to know that such adversaries are in the general vicinity. Because they cannot explore their surroundings by moving through them, they require no exproprioceptive mechanisms to help them plan where to go next and how best to do it. Having information about distant conditions would be of no advantage since their techniques for either feeding or self-preservation are rapid, reflexive, and do not require any preparation prior to being initiated. Such organisms are not totally passive. For example, they do feed, when currents fortuitously bring food particles close enough to be filtered or when insects make fatal blunders in choosing a landing site (of course, they have been shaped by evolution to take advantage of water flow patterns in the case of the former and to encourage such mistakes in the case of the latter) Still, their lives are not beset by the difficulties of having to plan and perform movements that will take them from one place to another.

Humans, on the other hand, are regularly confronted with decisions involving perceptions of environmental statics and dynamics. How these conditions threaten or protect us, and afford or deny us opportunities to navigate natural and artificial landscapes is of considerable importance, even to the point of influencing an individual's chances of survival. A simple act such as stepping off a curb to cross the street involves estimations of how far away the other side of the road is, how quickly one can or wishes to walk or run while avoiding traffic, how far away oncoming cars are in both directions, and how fast they are travelling. The number of pedestrian fatalities that occur each year suggests that at least some of these calculations are not being performed accurately, with tragic results. Not all such decisions are as life-threatening, of course. A soccer player deciding on the best route to a goal scoring position will not pay as heavy a price as the errant pedestrian for a miscalculation of distance and optimum locomotor output. Still, we do move through our environment every day for many purposes, be they utilitarian or recreational, and, in context, it is always important to do it accurately and consistently.

This chapter addresses two fundamental problems regarding the role of vision in locomotor navigation. First, visual information is usually available to us not only before embarking on a journey, even one as brief as a meter or two, but also while we walk or run, whether to a specific destination or not. We sample relevant (and perhaps irrelevant) cues about what exists underfoot, to the sides, and ahead along our intended path so that an original plan for movement through the environment can either continue or be modified appropriately. These visual processes that guide changes of path or gait during locomotion are demonstrably important. The effects of failing to adequately perform such updating is a common experience: most people have, through inattention, tripped over an imperfection in the walking surface that could have been avoided easily by walking around. We do not dwell upon the motor control mechanisms used to actually make such changes although these have been the subject of recent interesting study (Patla, Robinson, Samways, & Armstrong, 1989; Warren, Young, & Lee, 1986). We do , however, evaluate to what extent goal directed locomotion is reliant on relatively continuous or intermittent visual monitoring en route from a starting location to the goal itself.

Second, vision is used to *plan* a route from an egocentric viewing position to a desired travel endpoint. Environmental cues are inevitably available in greater complexity than can be completely perceived, especially given the short scanning times usually allowed by normal navigation circumstances. Walking successfully from one's front door to the car in the driveway implies that a considerable amount of important information has been successfully processed, much of it in ways that do not necessarily enter consciousness. For example, it is helpful to know if the terrain is slippery or not (perhaps due to a fall of rain or snow) , if a child has left a toy lying on a step, and if the car is in fact in the driveway at all or whether it was left on the street overnight. Knowing all this (and more), it is necessary to decide how to respond to these cues. How can gait be best programmed (perhaps by shortening or lengthening step length) to account for a rain-soaked sidewalk?

It is the purpose of this chapter to summarize what is known about how vision guides locomotion by updating of one's position from a starting point to a selected goal and to discuss the role of vision in the planning of locomotion prior to its onset.[1] The relevant literature is organized so as to discuss two main questions. First, to what degree is visual information, beyond that originally available during the initial planning of goal approach, required during target directed locomotion? Specifically, is vision used relatively continuously or intermittently and how is the sampling of available information accomplished? Second, which features of the environment are important in preparation for movement toward a goal? In particular, which cues underlie the mental representations that allow goal locations to be accurately determined and how is this information remembered while en route to the goal?

In a chapter such as this, it is often the case that the literature to be surveyed is voluminous and the specifics of particular studies kept necessarily brief. We take a different tack by concentrating in greater detail on fewer papers, most of which have used a "blind walking task" in one form or another as their method of investigating the planning a visual guidance of locomotion through the environment.

Introduction

Spatial ability has long been a topic of scholarly interest to researchers in numerous disciplines ranging widely in levels of explanation of human behaviour. In each case, the processes whereby the space in which we live is mentally represented provides useful insight into the organization of the human mind. For example, in the neurosciences, the single cell electrophysiology studies of Mountcastle (1978) and Hubel and Wiesel (1979) demonstrated that the structure of the neocortex is, in its own way, a map of the space around us. In psychology, Shepard and Metzler (1971) showed that mental imagery is a powerful explanatory construct for the interpretation of mental object manipulation. In anthropology, Gladwin (1970) in studies of the remarkable navigational skills of the Puluwat Islanders showed that spatial ability has sociocultural meaning and value. The understanding of spatial ability is also an area of important practical concern in child development and education. Piaget, even though viewing psychological development as an orderly process following general underlying principles, recognized the special nature of spatial development and devoted considerable research to the topic (Piaget & Inhelder, 1956).

---

[1]If the posing of these questions seems to be in reverse order from what makes better intuitive sense, it is because the recent literature has developed in this fashion and the thread of the various arguments is more easily followed if we approach matters in this way.

For those interested specifically in movement skills, the role of vision has been a particularly fertile area. Most vision-related research in the movement sciences, from the very early to the very recent, has been linked to aiming movements of the upper limbs (e.g., Jeannerod & Prablanc, 1983; MacKenzie, Sivak, & Elliott, 1989; Woodworth, 1899). From the numerous studies done in this area we have learned much about the way in which the eye and the hand coordinate their efforts to produce accurate goal directed movements of the arms, hands, and fingers. However, our understanding of the contributions of vision to movement through the environment has been given less attention. Indeed, the problem of goal directed locomotion of the kind in which we engage on a day to day basis has only recently (that is, within the last decade) received more than passing interest from those in the movement sciences. Despite the relative paucity of studies available when compared with aiming movements of the upper limbs, those that have appeared in the literature are potentially valuable to those trying to fathom both the theoretical and practical aspects of movement control, memory, and mental representation.

For purposes of focussing the discussion, two central assumptions are made about the planning and guidance of locomotion. First,it is assumed that we are concerned with an individual moving from his or her own starting position, Point A, to Point B, implying an egocentric perspective for both the visual appraisal of the environment and for the onset of movement through it. We all see the environment from our own point of view and are presumably best equipped to estimate distances and locations and to plan locomotor plans on this basis. Certainly, there are occassions when we are confronted with the need to make a non-egocentric judgement. For example, at times it might be useful to know whether we are closer to something than someone else and, therefore, to make a sensible appraisal of who would arrive first. ("Get me the shovel, will you?" "Get it yourself, you're closer.") Relevant as it might be to everyday situations, however, we do not deal with problem of non-egocentric spatial ability in this chapter.

Second, we are not concerned with the large body of literature dealing with the spatial ability associated with cognitive mapping (Tolman, 1948) of very large scale environments (e.g., Hazen, Lockman, & Pick, 1978). Lynch (1960) suggested that an overwhelming reason for our development of mental maps is the fear of getting lost. In this chapter, we do not find our subjects moving in such threatening environments but centre the discussion on movement through relatively short distances over which it would be very difficult to lose one's way. For example, although a toe might be stubbed while walking between the bedroom and the bathroom in the middle of the night, one is unlikely never to return from the excursion. It is to this size of environment that we confine ourselves, one bigger than the immediate action space in which we reach for things (Kolb & Whishaw, 1985) but smaller than the cities, fields, and forests of geographic space. These two constraints being put in place, let us now consider a paper by James Thomson (1980), from which a small growth industry of research, and some controversy, developed in the subsequent decade.

Visual Guidance of Locomotion

Thomson (1980) began with a simple question that was, in fact, the title of the paper. How do we use visual information to control locomotion? Thomson lamented the fact that psychologists had largely ignored the linkages between action and the visual perceptions from which they arise. This, in his opinion, gave rise to the situation that:

> "while we have no shortage of theories purporting to explain our ability to distinguish crosses from circles or red objects from green objects ... when it comes to explaining how we walk through a doorway without bumping into the wall, the research effort is markedly less impressive. Students of motor control, on the other hand, are prone to produce theories which fail to take adequate account of the information available to the system in the first place and, indeed, tend to ignore vision altogether." (p. 247).

Presumably, Thomson was referring to whole body movements such as locomotion since, as mentioned earlier, there was already at that time a considerable body of literature describing the role of vision in upper limb movements such as catching, throwing, and reaching. In any event, Thomson (1980) presented data that suggested that the successful performance of locomotion toward a goal was not necessarily dependent on the continuous devotion of attention to visual monitoring. The premise upon which the study was conducted was, as stated in the paper's abstract, that "we are accustomed to thinking that we need to use vision all the time to guide us as we walk about in the world" (p. 247). This premise is not strictly true: we are aware at an anecdotal level if not a scientific one that it is possible while riding a bicycle to take one's eyes off the road for a moment without precipitating an accident and that a runner in a race can look back over his or her shoulder for the competition without falling down or becoming disoriented. Still, the question was a good one in that it considered the *degree* to which we are dependent upon vision during movement through the environment.

Thomson's experimental paradigm and data collection method was refreshingly simple, requiring only a stopwatch and a tape measure. Subjects viewed targets, one at a time, ranging from 3 m to 21 m away for a period of 5 s and then walked to them with eyes closed. Upon stopping at where they thought the target was, their distance from the actual target was measured.

Since no further visual updating was allowed during walking, the accuracy (the definition of which is discussed later) with which subjects approached the targets must have been due to the memory of the original perception of where the target was. The basic result was that for targets closer than 9 m to the starting position, errors were very low. This finding held true for straight line paths and for routes involving avoidance of obstacles along the way and was also seen when subjects were asked in some trials to stop unexpectedly and throw an object the remainder of the distance to the target. For targets farther than 9 m away, however, errors increased dramatically. There were, in Thomson's view, two possible explanations for this decrease in accuracy for the more distant targets.

One possibility was that the actual perception of distances farther away than 9 m was poor and that subjects never knew where the target was in the first place. This was not an appealing interpretation and it was not pursued. Such an explanation is contradicted by a wide variety of experiences such as drivers applying the brakes in response to events even hundreds of meters distant and competitive archers using appropriate trajectories to hit targets considerably farther away than 9 m. Alternatively, it was suggested that subjects did know where the target was at the start of their locomotion toward it but that this information decayed as time passed. This hypothesis was tested by varying the time taken to reach targets by having subjects either change their locomotion speeds (by running) or by waiting for several seconds after closing their eyes before starting to walk to the target. In this way, targets of different distances away could be made similar in terms of the time taken to reach them. When these manipulations were introduced, a cutoff point at which there was a sudden increase in error was seen again; however, not necessarily at 9 m. Instead, when all conditions were considered, the critical point appeared to be time-based and not distance-based. Specifically, targets that took more than 8 s to reach were subject to much larger errors than those that could be reached in less than 8 s.

This could be taken as evidence that vision is required only intermittently as we move from place to place. After all, if it is possible to get into the ballpark fairly well without having had vision available for as long as 8 s, then it seems unnecessary to postulate a continuous demand for visual updating of one's position during locomotion. There is an analogy to upper limb positioning in which the evidence suggests that moving an arm to a target occurs in two phases. The first of these is simply to move most of the distance in the right general direction while the second is the fine tuning that ultimately produces accurate placement of the limb. Only the second phase is dependent on vision (e.g., Jeannerod & Prablanc, 1983) for successful task completion, meaning that most of the movement's extent can be made without visual feedback. This line of reasoning has featured prominently in various theories of motor programming in which explanations for the effects of movement time, amplitude, and force on movement variability of positioning of the upper limbs have been advanced (e.g., Newell, Carlton, & Carlton, 1982).

Evidence for a similar mechanism was provided in the locomotor domain by Lee, Lishman, and Thomson (1982) in their study of the sport of long jumping. During the long, 20 stride run to the take off board (which must be hit with a high degree of accuracy), strides taken early in the approach were uniform in length and varied little from trial to trial. The last five strides before the take off board, however, showed much greater variability from jump to jump, demonstrating that these were being used as compensatory strides for the accumulation of small positional errors that occurred during the early strides. The final stride showed the greatest variability and this final footfall was clearly under visual control. An analysis of the task suggests strongly that vision is more important during some parts of the performance than others. From the point of view of the performer, the need to reduce the number of visual updates is critical. Because the speed at which one is running at the take-off point largely determines the distance jumped, any demands that might reduce speed, such as the need for visual attention to be paid to the position along the runway, are undesirable. Leaving all the visual updating to the end is no doubt an optimal strategy. This study also showed that the visual guidance used during the "run-throughs" done by jumpers had little in common with the actual visual control strategies used in actual competitive jumps. Nevertheless, attendance at any long jump competition will still find jumpers performing their run-throughs under the approving eyes of their coaches.

Returning to Thomson (1980) and his suggestion of visual intermittency during locomotion, there is a fundamental problem in trying to interpret these results. The specification of what was meant by error, and therefore accuracy, was not given in the text and both constant error (CE, the signed difference between the target distance and the actual distance walked) and its standard deviation were shown graphically. Targets farther away than 9 m produced a slight tendency to increased overshooting of the target distance when compared with shorter distance targets. Greater changes were observed for the standard deviation and it is presumably upon these that interpretations were made.

Fortunately, a later, more comprehensive paper describing in greater detail these same experiments appeared (Thomson, 1983). It is to this paper that several responses have appeared in the literature, many of which render the original findings controversial. There are a number of reasons why this has taken place, and the analysis and interpretation of error (and hence of accuracy) has been foremost among these. In particular, Thomson (1983) explained that the dependent measure of interest in interpreting his findings was the standard deviation of CE taken over all trials (thus combining within-subject variability and between-subject differences in bias). This creates one difficulty of a purely semantic nature: it is common in the literature reporting studies of arm movements to use CE as a measure of *accuracy* and variable error (VE, the variability around the mean of a subject's repeat trials at a given distance) as a measure of intertrial *consistency*. Thomson strikes out on his own in this regard in that he refers to accuracy when talking about an intertrial variability measure. In purely pragmatic terms, this might not be advisable: for example, when considering many day to day tasks (such as jumping over a puddle on the street) it is more important to be accurate than it is to be consistent. The former keeps one's shoes dry, the latter could conceivably result in three consecutive soakings in the same place in the puddle (if, indeed, one were to try to jump over the same puddle three times). The point is that it is not readily apparent that Thomson was really talking about accuracy at all but was considering consistency instead.

A further difficulty associated with Thomson's (1983) statistical analysis was raised by Elliott (1986) who pointed out that the way in which Thomson had combined between- and within-subjects variances in his calculation of standard deviations of CE was flawed. Specifically, by combining the variances as he did, thus greatly increasing the degrees of freedom, Thomson had made it impossible to determine whether the differences being reported as significant were due to differences between experimental conditions or due to differences between subjects. In Elliott's words, the "between-subjects variability has a profound influence on error estimates" and "Thomson's choice of method throws his findings into question, because there may be unknowns in his work that affect between-, but not within-subjects variability" (p. 391).

Statistics aside for the moment, Elliott (1986) challenged Thomson's (1983) conclusions that time was the critical factor in determining accuracy of performance in the blind walking task. Specifically, Elliott argued that it was distance and not time that caused impairments in estimates of target positions. Indeed, his results showed that for the same distance (whether it be 3, 6, 9, 12,or 15 m), neither accuracy nor consistency were systematically worse when the onset of walking is delayed by as much as 4 s. Furthermore, Elliott (1986) maintained that, in his Experiment 2, the absence of vision during walking produced statistically significant increases in both CE and VE when compared with trials during which vision was allowed. This, he suggested, was contrary to Thomson's position that accurate approaches to goals could be achieved in the absence of vision which in turn supported intermittency of visual guidance of locomotion. Elliott extended this to conclude that his results supported a continuous need for visual updating of position while proceeding to a target.This seems, in retrospect, to be partially unwarranted. First, as we will see from Thomson's response, a common definition of what is meant by accuracy and how it relates to vision's role is lacking in this point-counterpoint debate. Second, from a research design point of view, to fail to support the intermittency thesis is not necessarily the same as supporting continuity.

Thomson (1986) was quick to respond. First, he suggested that Elliott's experiments were flawed because his subjects, in almost all conditions, had walked too slowly toward the targets to make any meaningful judgments about the existence of an 8 s critical period. This is indeed the case: in Elliott (1986), there are only three pairwise comparisons that can be made between a given distance being walked in less than and in more than 8 s and two of these comparisons are at the very short 3 m distance. Overall, Thomson (1986) argued, the results merely showed that, for a set of walking times greater than 8 s, the longer walking times produced greater impairments than the shorter times. Elliott (1986) did not test the existence of a critical 8 s barrier.

Second, he asserted that his 1983 paper did not show, nor was it intended to show, that performance in the absence of vision was *equal* to that with vision but rather that a reasonable *approximation* of sighted performance could be accomplished, for a period of about 8 s, when visual input was unavailable. His response, then, to Elliott's (1986) Experiment 2 comparing vision and no-vision conditions, was that "the finding of a significant difference between vision and no-vision conditions is really irrelevant to this conclusion" (p.393). It is true that Elliott's subjects in the no-vision condition in Experiment 2 were missing targets as far away as 8 m by only about 20 cm, thus suporting the relatively high degree of accuracy that Thomson says can be achieved on the basis of a mental representation. Further to this argument, however, Steenhuis and Goodale (1988) said that, since Thomson's (1983) article "there have been no other demonstrations of a highly accurate short-term memory for target distance during locomotion" (p. 400). It might be argued that this begs the question of what is meant by "highly accurate". Certainly, when Steenhuis and Goodale (1988) were writing, there were numerous examples in the literature that subjects could perform the blind walking task with average errors in distance estimation of as low as 2 - 3% (e.g., Corlett, Patla, & Williams, 1985). This surely must qualify as accuracy of some kind.

Third, the way in which Elliott's subjects were prepared for experimental trials was deemed to be unacceptable. For practice at the blind walking task to be effective, said Thomson, it should be conducted with specific goal distances being used and with feedback provided as to how accurately the practice target distances were estimated. In Thomson's opinion, "if subjects are to be persuaded to walk sufficiently quickly and normally for this kind of experiment to be meaningfully conducted, a proper practice routine is necessary" (p. 393). In this regard, he concluded that he "really cannot see what is the point of a practice session in which the subjects receive no feedback" (p. 393).

Thus armed with possible reasons why he had failed to replicate Thomson's results, Elliott (1987) once again set out to test the hypothesis that time was the critical feature in determining blind walking performance, while simultaneously investigating the methodological differences pointed out by Thomson (1986). To this end, in addition to using distance as an independent variable to test the prediction that time to reach a target is the key feature in accuracy, Elliott (1987) manipulated both walking speed and practice conditions . The results were quite clear. First, although receiving feedback during practice did not play a statistically significant role in walking performance, it did result in a nonsignificant trend toward undershooting near targets.[2]

More important to the discussion, Elliott (1987) again found no evidence to suggest that time was important in determining either CE or VE scores. Predicting from Thomson's point of view, one would expect from Elliott's (1987) design, significant main effects due to the time-based independent variables (delay, distance, and walking speed) while from Elliott's (1986) position, one would expect only a significant main effect due to distance. Distance was the only independent variable to produce a significant main effect for CE (accuracy) or for VE (consistency). Furthermore, Steenhuis and Goodale (1988) have shown that times to reach targets up to 30 s still afford reasonable accuracy. Taken with Elliott's studies, it would appear to be reasonable to conclude that the 8 s barrier was an artifact of some aspect of Thomson's original methods and that memory for environmental information has the same characteristics as short term memory for other kinds of verbal and motor tasks (Steenhuis & Goodale, 1988).

Where does this leave us with respect to the need for continous or intermittent visual information during target-directed locomotion? In the flurry of research activity surrounding the presence or absence of a critical cutoff time and disputes over data collection and analysis methods, the original question asked by Thomson (1980) seems to have been lost. It would be fair to say that, on the basis of all the data in the papers summarized thus far, there is

---

[2]As an aside, in any studies my colleagues and I have done, feedback has never been used during pre-experimental practice for fear of biasing subjects via proactive interference at certain distances. Practice has been given solely for the purpose of allowing subjects to appreciate that they are in no danger while walking with eyes closed and that they can do so at normal speeds. Elliott (1987) vindicated this position and stands in contrast to the position taken by Thomson (1986) regarding the role of practice prior to experimental trials. Steenhuis and Goodale (1988) have also addressed this point and reported that the practice effect was not the reason for failing to duplicate Thomson's (1983) findings.

good reason to believe that vision is not necessary on a continuous basis for relatively accurate target approaches to be made. While the length of time that we can walk before losing mental sight of where we are going is probably not limited to 8 s, it is nevertheless apparent that intermittent visual updating of our position is sufficient for accurate goal directed locomotion. Laurent and Thomson (1988) have, in fact, provided further evidence for this, even to the extent of showing that the mechanics of locomotion itself are tied into the processes by which visual information is perceived. At the risk of coming down firmly on the fence of what has been a contentious issue, it would seem prudent to conclude that Thomson (1980, 1983, 1986) was correct in insisting that he had demonstrated the intermittent nature of the role of vision in guiding locomotion. Subjects, after all, could do quite well while walking blind and even subjects in experiments performed by Thomson's critics have demonstrated this. However, Elliott (1986, 1987) and Steenhuis and Goodale (1988) were also correct, not necessarily for disproving the possible intermittency of vision during locomotion, but for successfully challenging the time constraints placed on that intermittency by Thomson. This, in turn, has the important effect of forcing a re-evaluation of the memory mechanisms associated with task performance and we will return to this later.

Before leaving this debate, it should be noted that a further problem exists in interpreting how people perform the blind walking task. This is the alleged difference between memory for "near space" distances (up to about 5 m) and "far space" distances (greater than 5 m) suggested by Thomson (1983). He found that for near space distances, the consistency with which targets could be accurately approached was less affected by the total time taken to walk to them than were far space targets. This suggested that "for distances up to 5 m, information about the location of the target is internalized in a relatively stable form in that it is not unduly affected by the passage of time" (p. 436). For targets in the far space, however, "accuracy is high as long as no more than approximately 8 s elapse between excluding vision and reaching the target"... (p. 436). To interpret these results, Thomson (1983) postulated that, when targets are in the near space, a motor program of instructions is used to represent the actions appropriate to approaching the goal. On the other hand, when targets are located in the far space, he theorized that a more general, and less stable, representation of distance information that is independent of specific motor instructions is used to guide locomotion to the target.

Several questions arise from these proposals. First, what kind of motor program is being suggested for near space performance? Thomson (1983) acknowledged that the term "motor program" is usually reserved for short duration, open loop movement plans and that what is proposed for blind walking is not of the same ilk. The reference, in fact, appears to be more of an analogy than a strict application of motor program theory. Second, what kind of memory strategy is used to guide blind walking? Relying on the introspections of his subjects, Thomson (1983) suggested that some form of visual imagery was used with subjects seeing themselves moving toward the target. The details of this imagery were not made clear, however. Third, why would two distinct modes exist for the performance of the same task? Thomson (1983) acknowledged the need for an explanation when he stated that "the ways in which they interact will require a good deal of future investigation" (p. 442). This is true for the first two questions, also.

Overall, it could be argued that the most interesting questions raised by the Thomson (1983) paper and by those who have responded to his initial findings, are those related to how we remember where the target is when we can no longer see it. The topic of mental representation of possible motor programs and mental images is, therefore, essential for our understanding of how we remember the salient features of the environment as we prepare to move through it. It is to this topic that we now turn.

Mental Representation of Environmental Information

Strelow (1985) used the term mobility to refer to the "skill of traveling through the spatial environment, avoiding obstacles, and traveling directly or indirectly towards goals" (p. 226). Of particular interest to our understanding of how environmental information is represented in mind was Strelow's point that while we most often think of spatial abilities as being highly visual in origin, mobility "can be performed in the total absence of vision" (p. 227) as the blind demonstrate everyday. Strelow argued that a single process is unlikely to account for all of the problems identified by Gibson (1958) as being associated with mobility. However, many of these problems, such as backing up and avoiding obstacles, are not relevant to the blind walking task as used most often in the literature. In fact, all that subjects are required to, do from a large list of possibilities, is to start and stop. The task, then, is relatively impoverished when compared with normal mobility but still must require some form of cognition for performance to be as accurate as most experiments have shown it to be. What, then, are the theortetical options available to us that might serve as a means of understanding how mental representations serve the performance of subjects in the experiments of Thomson, Elliott, and others?

For Gibson (1979), the structure of the space through which we move, even including that which is not currently visible, is perceptually independent of any unique point of view. Layout, as he called it, is an invariant and does not change simply because our view is temporarily blocked. This is consistent with the comments of Thomson's (1983) subjects who reported using a mental picture of what was no longer visible to guide them to the estimated target position. In Gibsonian terms, the knowledge of the layout (simple as it is in such a task with only a single location and no obstacles) was available from the initial scanning of the visual array and as subjects moved with eyes closed, they presumably carried with them an invariant knowledge of how close they were to a target unseen but still where it was when last glimpsed. However, is it necessary for this knowledge to be visual in nature? If so, the congenitally blind would never be able to exhibit mobility skills and Strelow (1985) has argued that this clearly is not the case.

Kosslyn (1983) has stressed that mental depiction need not be thought of as synonymous with visual imagery. While mental representations of layout (or anything else) could be pictorial or visual in nature, they might also be propositional, or verbal (e.g., Acredolo, 1979; Hintzman, O'Dell, & Arndt, 1981). It is worth noting that my personal bias is toward a propositional explanation of how subjects accurately reach target locations. The reason for this is personally anecdotal in that, when I have tried to perform the blind walking task myself, I look at the target, then close my eyes, and wait for the image to appear that will guide me forward. It never comes. I am not advocating Wundtian introspection as the way to proceed in investigating the problem. I am merely suggesting that I am predisposed toward an explanation couched in non-visual terms since I cannot seem to conjure up the pictures that are the standard currency of discussions of mental imagery. This being said, we will consider a series of experiments done in our laboratory (actually in the fields, walkways, and parking lots outside of the building housing our laboratory) that have attempted to ascertain what subjects perceive and subsequently use to guide their blind locomotion.

Let us begin by considering imagery and its alleged role in the performance of the blind walking task. First, Corlett, Anton, Kozub, and Tardif (1989) demonstrated that performance on the blind walking task was not significantly correlated with subjects' classifications as low, medium, or high imagers as determined by the Movement Imagery Questionnaire (Hall, Pongrac, & Buckolz, 1985). Even low imagers were able to produce fairly accurate estimates of a target position 9 m away. This study was not conclusive, however, for several reasons. First, it is not necessarily the case that there is one kind of imagery (Harshman & Paivio, 1987) and that the kind of imagery measured by the MIQ would be used in the performance of mobility tasks. Second, the task is visually undemanding with relatively little information to be extracted in the few seconds allowed. It might simply be the case that, if the task is imagery-based, even those with low imaging ability can handle it with little difficulty.

In a second study of the role of visual imagery, Corlett, Kozub, and Quick (in press) confronted the problem directly by using an interference approach.

They manipulated the activities occurring in the time interval between the end of viewing and the start of walking. In their experiment, they used a 10 s unfilled retention interval as a control and created three experimental conditions with which to compare it. In two of the conditions, the interpolated task was imagery based. The first of these involved subjects reporting the scene they would see when scanning from left to right from a well-known point of view on the university campus. In the second imagery condition, the task was more dynamic, requiring subjects to report in real time their route from the front door of the university library to a particular location inside. The third experimental condition was a computational one in which subjects merely counted backwards by sevens from 200. The three interference tasks had all been designed to be similar in terms of their capacity interference characteristics.

It was hypothesized that, if the blind walking task was reliant upon visual imagery, the two imagery interfering activities would produce decremental effects on CE and VE whereas the computational task would not. In fact, neither of the imagery tasks produced a significant effect but the counting task did cause significant increases in the amount of undershooting the target, as exemplified by increasingly negative CE scores (no VE effects were found) The authors concluded that these results cast serious doubt on whether the blind walking task was really performed with reliance on visual imagery. Their findings were much more consistent with a propositional strategy that involved the counting of steps or time needed to reach the target. As mentioned, the original task as designed by Thomson (1983) makes relatively few perceptual demands, with only a single target available to view and a fairly long time to have a look. In fact, Corlett, Patla, and Williams (1985) showed that adult subjects could make accurate estimates of target distance with viewing times as short as about 1 s.[3]

Corlett (1986) conducted an experiment with adults and 10 yr old children in which the amount and arrangement of visual information seen by subjects in the viewing phase of the task was manipulated. Various landmarks were placed at different locations between the subjects and the target and, in some trials,

---

[3]To digress briefly, it might be the case that even shorter exposures are sufficient but the technology needed to control accurately such exposures was not available at the University of Botswana where our original data collection was done. It might be worthwhile to examine the lengths of exposure times and the visual fields to which the target array is presented. The lower limits of scanning time that afford successful walking performances and the differences, if any, between short duration exposures to right and left hemispheres might clarify what kinds of mental processes can and cannot produce accurate performance under severely time-constrained conditions. One might even speculate that conducting these experiments with split brain subjects would tell us much of interest about the links between visual perception, spatial representation, and action planning.

past the target position. The target distance was always 12 m and single or multiple landmarks were placed in various combinations of 4 m, 6 m, 8 m, 10 m, and 14 m from the subjects' viewing position. It was found that target distance was not reproduced with the same accuracy in all conditions of visual presentation. Considering first the adults, a single landmark 4 m from the subject was found to have no effect on subsequent walking accuracy (CE) or intertrial consistency (VE). These data are consistent with Thomson's (1983) near space idea. If the distance of close targets (and presumably landmarks) are represented in a stable fashion, then knowing where the 4 m target is located would be of little help because the landmark would be contributing little that could not be represented without it. On the other hand, a single landmark at 8 m did improve accuracy and consistency of walking by adults to 12 m targets. This distance is beyond the proposed 5 m near space boundary and would be expected to help subjects gauge where they were. Subjects, in fact, reported trying to reach the landmark accurately and then walk a couple of steps further. No difference was found between the single 8 m landmark and a condition in which a 4 m and an 8 m landmark were used together, providing further evidence that the 4 m landmark was not useful. A single landmark at 14 m, 2 m past the target, did not, on its own, produce a beneficial effect on CE or VE but multiple landmark conditions using 8 m and 10 m distances resulted in reductions of both kinds of error. The landmark condition that produced the most beneficial effects of all on performance was the framing effect created by placing an 8 m landmark before the target and a 14 m one past the target position. The results taken together clearly demonstrated that distance was not merely perceived on its own, but that locomotion was most accurately served by an estimate of target position that was contextually dependent.

A separate experiment was reported by Corlett (1986) in which several of the conditions shown to have influenced adult performance were used to investigate children's responses. This was considered important because Corlett, Patla, and Williams (1985) showed that children were affected differently than adults by manipulations of the time allowed to view the target and the length of the delay enforced between the end of the viewing phase and the start of the walking phase. Thomson (1986) suggested that this was indicative of differences between adults and children in their assessments of what is near space and far space. Again, Corlett (1986) found that children did not respond as adults did. Specifically, the conditions that best aided accuracy and consistency were those displaying only a single landmark with the 8 m one being slightly better than the 4 m one. The multiple landmark conditions were of no help in reducing either CE or VE when compared with a no-landmark condition and the framing condition that produced the least error with adult subjects was, in fact, the visual environment that produced the greatest errors by children.

A possible reason for this was thought to be that children were simply not able to process the information available in the multiple landmark conditions. Evidence that this was true was shown by a small follow-up experiment in which a group of children was given a more extensive task orientation in which the benefits of the framing features of the 8 m -14 m condition were explained. With this task orientation, CE and VE were both reduced but not to the levels of the single 8 m landmark. Fodor and Pylyshyn (1981) contended that our knowledge of how selective attention is controlled was sparse. Picking up on this, Corlett (1986) suggested that when confronted with a goal-directed walking problem, our first response is to pick up information near to the goal itself since this appears to be of the most use, at least for adults. Children seemed more inclined than adults to stick to the basic information and less likely to benefit from more complex visual scenes, a conclusion in keeping with the sentiments of Gibson (1979) that perceptual learning is the education of attention. What is clear is that the question posed in the final sentence of Corlett (1986) has not yet been answered: "what do we actually look at when planning locomotion?". For those blessed with sophisticated eye scanning technologies in their laboratories, this would be a potentially fruitful avenue of future research.

The studies conducted between 1980 and 1986 by Thomson, Elliott, and Corlett and their co-workers all involved walking on a flat surface that, of itself, exerted no effect on the results. Also, the target distance and its location were not at odds with each other since walking was always to the target location itself. The differentation between distance and location featured prominently in discussions of what kind of kinesthetic information was used in arm positioning tasks (Laabs, 1973). This led Corlett and Patla (1987) to ask whether target distance was in some way perceptually distinct from target location and also whether walking in a plane other than the horizontal one would affect the accuracy of goal approach. Some of their experimentation was largely methodological in focus since their major study would rely on subjects turning between viewing and walking phases of the task, with walking taking place on non-level surfaces in some trials. They showed that, after viewing a target 6 m away and then closing one's eyes, a turn of 130°, 180°, and 360°, had no effect on subsequent accuracy of walking the distance to the target originally seen. In the second experiment, seeing the target on a level plane and then walking to it on the level produced a slight undershooting of the target. Conversely, seeing and walking in either a downhill or uphill fashion produced statistically significant (relative to the level-level condition) overshooting of the target distance of 6 m. From these results, Corlett and

Patla (1987) concluded that "the distance estimate is not established in any particular context of the direction to be walked subsequent to scanning" (p. 89) and that distances are not perceived merely as functions of recession along the ground (Turvey & Solomon, 1984). Rather, because "distances from subject to target along the ground are identical for all conditions but locomotor estimation errors are not" (p. 93), they concluded that factors other than distance and location might be included in the creation of a plan of action to reach the perceived target position. Their Experiment 3 addressed this possibility more fully.

Using a ramp with an incline of 24°, three orientations of target viewing - level, up, and down - were combined with the same three orientations of walking to the target to create 9 experimental conditions in a factorial design. Although some combinations required subjects to make a 180° turn between viewing and the start of walking, this had already been shown to have no effect on CE. The results were complicated and not fully explicable in terms of possible perceptual and/or motor mechanisms. What was shown was that accuracy was greatest when viewing and walking occurred on the same plane and that dissimilar orientations produced a distinct pattern of inaccuracy. Specifically, when the viewing phase occurred in a plane that would require more effort to reach the target than the plane in which walking actually occurred (for example, upward viewing and level walking), subjects showed a tendency toward overshooting the target. Conversely, when the opposite was the case (for example, downward viewing and level walking), subjects showed a tendency toward undershooting of the target.

In explanation, it was suggested that "features such as perception of effort to accomplish the upcoming locomotion may be a part of the coding process, and [these] are not properly coded during level scanning for upcoming non-level walking trials" (p. 93). There were, however, some anomalous results left unexplained and the effects on gait of walking in non-level planes were not reported. In general, though, the study indicated that when assessing the environment for the purpose of moving through it, factors other than just spatial relationships between objects might be involved.[4]

Having left the possible role of distance and location estimation in the mechanics of locomotion only partially explained (Corlett & Patla, 1987), we (Corlett, Byblow, & Taylor, 1990) conducted a similar experiment in which

---

[4]It is worth noting that this same experiment was conducted separately using children as subjects. Although presented at a conference (Corlett & Patla, 1986) the results have never been published because they were impossible to interpret, showing no apparent pattern that could be explained in valid perceptual-motor terms. This is an experiment that should be repeated, perhaps with better success, since it would provide insight into the process of perceptual learning and how it is linked to the strategies by which children plan action.

the mechanics of locomotion were constrained not by the exigencies of walking uphill or downhill but by resistances applied to the body or legs while walking on the level. This was done by having subjects walk forward while being held back by thick rubber bands (used by athletes to develop leg power during sprinting). This eliminated some of the confounds present in the previous experiment. The results of this study demonstrated that estimation of target position did depend on the way in which subjects perceived that they would be allowed to proceed to the target. However, this was not mediated via a perception of the effort involved as had been suggested by the upward and downward inclines.

Instead, the key factor was the certainty with which subjects knew the constraints that would be placed on walking after the target had been seen. Prior to viewing the target, subjects walked to the viewing position itself: this locomotion was either resisted or not. Subjects viewed the target, then closed their eyes and walked as the paradigm usually demands. In some trials, subjects proceeded under the constraints (or lack thereof) that they had been given prior to viewing. In other trials, the amount of resistance was changed after a few steps had been taken. When the resistance was applied to the body, no changes in CE or VE were found, regardless of previewing condition This method of applying resistance did not produce significant changes in subjects' gait, or affect accuracy or consistency. It did however make it more difficult for the subject to maintain their normal gait.

When the same experimental design was used but with resistance applied to the legs themselves, thus causing changes in gait, decrements in accuracy were found. More specifically, when gait was changed from what was expected from the pre-viewing phase resistance application, subjects became less accurate. This was true whether gait went from being resisted to being unresisted or vice versa. From these results, it was concluded that the perception of effort thesis advanced by Corlett and Patla (1987) was an incomplete interpretation of the data resulting from the gait differences that might have resulted from level, upward and downward walking. More importantly, the results were taken as further support for the position that preparing a plan of action to reach a target based on visual information involved more than just target distance and/or location. In this case, the degree of certainty surrounding the kind of locomotion that would be allowed played a role. In its simplest terms, this means that a given target position will be estimated with some thought given to how one will be able to move to it. A target 10 m away might look closer if it is seen at the other end of a patch of ice over one which might slide on a winter's day and might look farther away if walking to it involves slogging through mud or heavy sand.

The literature seems to have established that information other than distance or location is important in planning how to move from one place to another. One factor that is common in all the experiments reported here is that they have had subjects walking over a surface to reach where they wanted to go. Indeed, the alternatives are few, jumping being the only one that comes immediately to mind if we accept that flying is unlikely for subjects of our own species. This being so, we know virtually nothing about the way in which we perceive surface characteristics and how these might affect our assessments of how we can move toward goal objects. Thus, the whole area of surface conditions would be fertile ground for further study. Situations in which accuracy is compared for unstable and stable surfaces, slippery and dry surfaces, and safe and less safe surfaces (without endangering life and limb of subjects, of course) are good possiblities. Of special interest would be such studies conducted with children, preferably very young ones, whose experience of mobility is not yet well-established.

Epilogue

To where have we come in our appraisal of the planning and guidance of locomotion? Let us ask again the questions that this chapter was supposed to address. First, to what degree is visual information, beyond that originally available during the initial planning of goal approach, required during target directed locomotion? Specifically, is it possible to engage in accurate target-directed locomotion when there are discontinuities in the presentation of visual information? The comments of Steenhuis and Goodale (1988, p. 413) are interesting here:

> "... we would maintain that precision in locomotor control is necessary only in the immediate vicinity of a target. This concept is supported by evidence suggesting that updated visual information is most likely to be obtained at or near an intended target. It seems paradoxical that most other demonstrations of discontinuous visual feedback (aside from Thomson's) suggest that there are two stages to a movement, an initial inaccurate stage followed by a corrective stage."

This sounds very much like the premise upon which Thomson (1980, 1983) conducted his original experiments, does it not? What, then, is the argument? It is surely not to do with continuity or intermittency. Indeed, it is worth asking whether or not continuity and intermittency are usefully phrased as opposites and at what level of explanation the idea of continuity becomes unhelpful. Given that light itself can be thought of as existing in quantum packages and that the retina consists of discrete receptor cells, each sending electrical signals ultimately to the brain in the form of individual action potentials, any visually-based phenomenon is inevitably intermittent when taken to its neurological level. Is anyone really arguing for a continuous need for visual information during goal-directed walking? I do not think so. Everyone's data, whether intended to be so or not, seems to agree that subjects are able to approach a target within reasonable error tolerance even when vision is not available for many seconds at a time. The dispute, then, centres on the length of time represented by "many seconds" and the nature of the memory system that keeps the relevant information available. This was the second question being asked in this chapter.

It now seems clear that Thomson's assertions that time was the critical factor in determining the accuracy of a non-visually guided approach to a target, and that 8 s was a cliff-like barrier beyond which subjects fell into a chasm of inaccuracy, have not been borne out. Distance rather than time seems to be the experimental variable contributing most greatly to inaccuracy and memory for target information does not degrade abruptly but rather decays gradually. Furthermore, there is evidence that the kind of visual representation underlying task performance is not necessarily grounded in visual imagery as Thomson suggested. It would, however, be a mistake to conclude that Thomson had it all wrong. For example, titles of several papers notwithstanding, disagreements have really been about the characteristics of vision's intermittent role and not about whether vision's role is really intermittent or not. Thomson's 1980 and 1983 studies prompted a surge in related research activity, suggesting that there is a lively interest in the area of visual control of locomotion, not just from a physiological control approach but from a behavioral viewpoint as well. The arguments have, in some cases, been at crossed purposes and the disparate use of critical terms such as accuracy and intermittency have clouded the commonalities present in the data of various studies. Still, it would be unfortunate to assume the questions have all been answered and that such a simple paradigm has already outlived its usefulness. Combined with analysis of eye movements and gait kinematics, studies of blind walking accuracy still have much to tell us about mobility, spatial ability, and our interactions with the locomotor environment.

References

Acredolo, L.P. (1977). Developmental changes in the ability to coordinate perspectives of a large-scale space. *Developmental Psychology, 13,* 1-8.

Corlett, J.T. (1986). The effect of environmental cues on locomotor distance estimation by children and adults. *Human Movement Science, 5,* 235-248.

Corlett, J.T., Anton, J., Kozub, S., & Tardif, M. (1989). Is locomotor distance estimation based on visual imagery? *Perceptual and Motor Skills, 69,* 1267-1272.

Corlett, J.T., Byblow, W., & Taylor, B. (1990). The effect of perceived locomotor constraints on distance estimation accuracy. *Journal of Motor Behavior, 22*, 347-360.

Corlett, J.T., Kozub, S., & Quick, H. (in press). The effect of interfering activities on memory for distance to be walked. *Journal of Human Movement Studies..*

Corlett, J.T., & Patla, A.E. (1986, October). *Developmental aspects of visual estimation of distance in level and non-level planes.* Paper presented at The Canadian Society for Psychomotor Learning and Sport Psychology, Ottawa, Ontario.

Corlett, J.T., & Patla, A.E. (1987). Some effects of upward, downward, and level visual scanning and locomotion on distance estimation accuracy. *Journal of Human Movement Studies, 13*, 85-95.

Corlett, J.T., Patla, A.E., & Williams, J.G. (1985). Locomotor distance estimation of distance following visual scanning by children and adults. *Perception, 14*, 257-263.

Elliott, D. (1986). Continuous visual information may be important after all: A failure to replicate Thomson (1983). *Journal of Experimental Psychology: Human Perception and Performance, 12*, 388-391.

Elliott, D. (1987). The influence of walking speed and prior practice on locomotor distance estimation. *Journal of Motor Behavior, 19*, 476-485.

Fodor, J., & Pylyshyn, Z.W. (1981). How direct is direct visual perception? Some reflections on Gibson's "Ecological Approach". *Cognition, 9*, 604-611.

Gibson, J.J. (1958). Visually controlled locomotion and visual orientation in animals. *British Journal of Psychology, 49*, 182-194.

Gibson, J.J. (1979). *The ecological approach to visual perception.* Boston: Houghton-Mifflin.

Gladwin, T. (1970). *East is a big bird: Navigation and logic on Pulawat atoll.* Cambridge: Harvard University Press.

Hall, C., Pongrac, J., & Buckolz, E. (1985). The measurement of imagery ability. *Human Movement Science, 4*, 107-118.

Harshman, R.A., & Paivio, A. (1987). Paradoxical sex differences in self-reported imagery. *Canadian Journal of Psychology, 41*, 287-302.

Hazen, N.L., Lockman, J.J., & Pick, H.L. (1978). The development of children's representations of large-scale environments. *Child Development, 48*, 623-636.

Hintzman, D.L., O'Dell, C.S., & Arndt, D.R. (1981). Orientation in cognitive maps. *Cognitive Psychology, 13*, 149-206.

Hubel, D.H., & Wiesel, T.N. (1979). Brain mechanisms of vision. *Scientific American, 241*, 44-53.

Jeannerod, M., & Prablanc, C. (1983). The visual control of reaching movements. In J.E. Desmedt (Ed.), *Motor control mechanisms in man* (pp. 13-29). Paris: Masson.

Kolb, B., & Whishaw, I.Q. (1985). *Fundamentals of human neuropsychology.* New York: WH Freeman and Company.

Kosslyn, S.M. (1983). *Ghost in the mind's machine.* New York: Norton & Company.

Laabs, G.J. (1973). Retention characteristics of motor short term memory cues. *Journal of Motor Behavior, 5,* 249-259.

Laurent, M., & Thomson, J.A. (1988). The role of visual information in control of a constrained locomotor task. *Journal of Motor Behavior, 20,* 17-37.

Lee, D.N., Lishman, J.R., & Thomson, J.A. (1982). Regulation of gait in long jumping. *Journal of Experimental Psychology: Human Perception and Performance, 8,* 448-459.

Lynch, K. (1960). *The image of the city.* Cambridge: MIT Press.

MacKenzie, C.L., Sivak, B., & Elliott, D. (1988). Manual localization of lateralized visual targets. *Journal of Motor Behavior, 20,* 443-457.

Mountcastle, V.B. (1978). An organizing principle for cerebral function: The unit module and the distributed system. In G.M. Edelman & V.B. Mountcastle (Eds.), *The mindful brain* (pp. 7-50). Cambridge: MIT Press.

Newell, K.M., Carlton, L.G., & Carlton, M.J. (1982). The relationship of impulse to response timing error. *Journal of Motor Behavior, 14,* 24-45.

Patla, A.E., Robinson, C., Samways, M., & Armstrong, C.J. (1989). Visual control of step length during overground locomotion: Task specific modulation of the locomotor synergy. *Journal of Experimental Psychology: Human Perception and Performance, 15,* 603-617.

Piaget, J., & Inhelder, B. (1956). *The child's conception of space.* London: Routledge & Kegan Paul.

Shepard, R.N., & Metzler, J. (1971). Mental rotation of three dimensional objects. *Science, 171,* 701-703.

Steenhuis, R.E., & Goodale, M.A. (1988). The effects of time and distance on accuracy of target-directed locomotion: Does an accurate short term memory for spatial location exist? *Journal of Motor Behavior, 20,* 399-415.

Strelow, E.R. (1985). What is needed for a theory of mobility: Direct Perception and cognitive maps-lessons from the blind. *Psychological Review, 92,* 226-248.

Thomson, J.A. (1980). How do we use visual information to control locomotion?. *Trends in Neurosciences, 3*, 247-249.

Thomson, J.A. (1983). Is continuous visual monitoring necessary in visually guided locomotion? *Journal of Experimental Psychology: Human Perception and Performance*, *9*, 427-443.

Thomson, J.A. (1986). Intermittent versus continuous control: A reply to Elliott. *Journal of Experimental Psychology: Human Perception and Performance*, *12*, 392-393.

Tolman, E.C. (1948). Cognitive maps in rats and man. *Psychological Review, 55*, 189-208.

Turvey, M.T., & Solomon, J. (1984). Visually perceiving distance: A comment on Shebilske, Karmiohl, and Proffitt, (1983). *Journal of Experimental Psychology: Human Perception and Performance*, *10*, 449-454.

Warren, W.H., Young, D.S., & Lee, D.N. (1986). Visual control of step length during running over irregular terrains. *Journal of Experimental Psychology: Human Perception and Performance*, *12*, 259-266.

Woodworth, R.S. (1899). The accuracy of voluntary movement. *Psychological Review, 3*, (Monograph Supplement), 1-119.

VISION AND MOTOR CONTROL
L. Proteau and D. Elliott (Editors)

CHAPTER 17

LOCOMOTOR AUTOMATISM
AND VISUAL FEEDBACK

JEAN PAILHOUS and MIREILLE BONNARD

Université d'Aix-Marseille II, Faculté de médecine
Cognition & Mouvement, URA CNRS 1166
IBHOP, Traverse Charles Susini, 13388 Marseille, France

Locomotion and Displacement

Locomotion is both the act of moving from place to place and the means by which that displacement is accomplished. The former aspect has given rise to numerous studies in the field of cognitive psychology, including research on spatial representation, pointing tasks, and localization (see Corlett, this volume), as well as in psychology and the psychophysics of visual perception, including research on motion sensitivity and the perception of the velocity and direction of optical flow (see Mestre, this volume). The latter aspect of locomotion, which deals with sensorimotor integration, has been the subject of many studies in neurophysiology and neurobiology, and has provided evidence of the role of neuronal circuits in producing locomotor rhythms and patterns in various species (Grillner, 1981; Katz & Harris-Warrick, 1990; Rossignol, Lund, & Drew, 1988; Stein, 1978). These studies have emphasized neuronal specialization, neuromodulation, and reflex adaptation in locomotor systems. In vertebrates, two main findings have emerged. First, if the spinal cord is isolated from the rest of the central nervous system (and thus from the visual centers), well-organized stepping movements are observed in the hindlimbs when placed on a moving treadmill, provided the animal is supported to prevent falling due to severe loss of balance. In this case, both the synergic organization of leg movement and across-leg coordination are preserved. Secondly, if the animals is deafferented (and even in fictive locomotion if it is curarized), the pattern of muscle activity still remains. These results show that the locomotor command (rhythm and pattern) is powerfully and centrally organized - it is not a reflex chain - and that this organization is spinal. From this evidence, it could be concluded that there is no tight relationship between locomotion and sensory information (in the light of studies on deafferentation) and even less connection between locomotion and vision (since the central organization is spinal). However, three main experimental findings further qualify this point of view:

(a) Spinal locomotion is generally only observed following the administration of chemical, electrical, or mechanical aids, which can be thought of as partial substitutes for the supraspinal activity occurring during spontaneous locomotion in intact animals.

(b) The locomotor patterns observed under these conditions, although clearly identifiable, are less well organized than those found in spontaneous locomotion. More specifically, they are somewhat less stable when the animal is spinalized, and much less stable when it is also deafferented.

(c) Finally and above all, spinal locomotion is not very adaptive. Accordingly, when speaking of locomotion, mastication, and respiration, Rossignol et al. (1988, p. 262) stated:

> "Open-loop operation might be sufficient in some undemanding conditions, such as walking on an infinitely smooth surface, chewing gum or breathing quietly. However, closed-loop operation is essential when there are demands to change the frequency of the patterns, or the timing of elements within the patterns, or when the amplitude of the movement (depth of breath, step length, mouth opening) must be altered to accomodate particular demands".

It should be noted that these results, obtained in spinal animals, do not inform us directly about the role of these elementary circuits in intact animals "when higher centers may contribute to a degree presently unknown" (Grillner, 1981, p. 1199). In human beings, although the existence of specialized locomotor centers has not been clearly demonstrated at the spinal level, a great deal of behavioral evidence (in particular, research on locomotor development and its pathologies) has shown that locomotion is an automatism that is organized at sub-cortical levels in the central nervous system.

However, saying that locomotion has an elementary structure at the neurobiological level and strong stereotypical properties at the behavioral level cannot be taken to contradict the fact that locomotion is an adaptive behavior, one that is markedly under the control of higher centers. For example, we can intentionally move faster, change direction, avoid obstacles, and put our feet wherever necessary. What are the links between this extraordinary diversity of natural walking and its synergic, stereotypical organization? Although these links are not very well understood at either the neurobiological or behavioral level, one can nevertheless refer to a simple description: there are variations in locomotion that do not modify the structural properties of basic synergy, that is, its invariant properties are maintained. We shall call such variations in basic synergy "modulations". They may be contrasted to variations which transform that synergy, such as hopping or paraplegic walking. The real adaptive capabilities of locomotion lie in such modulations, since the synergy relieves on higher centers for a large part of the movement organization task (Bernstein, 1967). Such modulations have been widely studied, either through peripheral elicitation (treadmill velocity changes, cutaneous stimulation) or central activation (orders to walk faster, mesencephalic stimulation). However, the links between locomotor organization and vision, which are so important to the control of direction and velocity, are not well understood. Some studies have been conducted to examine changes in velocity and even in direction (Clarac, 1984; Shik & Orlovsky, 1976; Thelen, Ulrich, & Niles, 1987) but they have never focused on the visual aspects of locomotor control.

One of the most noteworhty effects of natural locomotion (i.e., motion involving a real displacement) is that it produces an apparent movement of the entire visual field in the opposite direction of the displacement. (Note that the treadmill situation, so widely used in this field, does not satisfy this condition). Gibson showed that this apparent movement (called "optical flow", Gibson, 1958) has all the kinematic properties of the displacement of the whole body. Vision should thus have both a proprioceptive function (knowledge of one's direction and displacement speed) and an exteroceptive function (knowledge of one's surroundings). Its proprioceptive function was described by Lishman and Lee (1973), who called it exproprioceptive (Lee & Lishman, 1977). Without any doubt, visual information is used to intentionally modulate locomotor output in compliance with environmental constraints. The locomotor pattern can thus be modulated in advance so that the foot can be placed on a visual target (Lee, Lishman, & Thomson, 1982; Patla, Armstrong, & Silveira, 1989; Warren, Young, & Lee, 1986) or an obstacle can be jumped (Hay & Schoebel, 1990). Warren et al. (1986) showed that the force pattern is specified by the optical flow in subjects running over irregular terrain. This model has been generalized and applied to the visual control of flight in insects (Warren, 1988). In a comparable but more ecological situation (in which walking speed was not set by a treadmill), Patla, Robinson, Samways, and Armstrong (1989) demonstrated the importance of central pattern generator and peripheral (effector system) factors in making synergy modulations to adapt stride amplitude to the environment. The integration of visual information in high level loops (cortical) seems to be necessary to perform these intentional regulations of locomotion. Indeed, cats with lesions of the bulbar pyramids are known to be able to walk on a flat surface, but unable to walk along a horizontal ladder or along a curved surface (Liddell & Phillips, 1944). This suggests that the motor cortex plays a major role when precise visual control over locomotion is required. Moreover, this hypothesis is supported by recent neurophysiological studies comparing walking on a flat surface to walking on a horizontal ladder (Armstrong, 1988), which found an increase in the peak discharge of the motor cortical cell (including the pyramidal tract neurons) in the latter case. The same results were reported by Drew (1988), whose animals had to modify their gait to step over different kinds of obstacles fastened to the moving treadmill, and thus moving toward the animal. In Drew's studies, the increased discharge of pyramidal tract neurons occurred 50-100 ms before the major changes in the timing and amplitude of EMG activity in the cleidobrachialis (elbow flexor), changes which are required to adapt leg movement and perform the task; the author suggests that in this case, the cortical influences occur directly at the spinal level rather than being transmitted through sub-cortical structures.

In the following sections, we shall attempt to provide evidence of a more automatic kind of integration of visual information into the sensorimotor loops that control locomotion, showing how vision (especially optical flow, for our purposes) is able to trigger and modulate the locomotor command or motor commands associated with locomotion (including postural commands). Then we shall deal with the head movements observed during locomotion, since they largely determine the morphology of optical flow; we shall focus on their link to leg movements and their modulations.

Locomotion as an Optomotor Response

Motor responses of the eye (OKN), the eye-head system (VOR), and the whole body, which are triggered by a visual stimulus are called optomotor responses. From a neurophysiological point of view, the structures that integreate these visual stimuli have been found in the lower levels of the central nervous system; they also receive vestibular inputs which are normally induced by accelerated head movements, and often somato-sensory inputs. Among others, these are the vestibular cerebellum (floculus: Waespe & Henn, 1981, and nodulus: Precht, Simpson, & Llinas, 1976), the Deiter's nucleus (Henn, Young, & Finley, 1974) and the colliculus (Bisti, Maffei, & Piccolino, 1974). To our knowledge, little direct information is available about the activity of these structures during locomotion. The only exception is the Deiters' nucleus, almost all of whose neurons are known to increase their discharge during locomotion, some showing a phasic pattern of discharge linked to the locomotor cycle (Mori, Matsuyama, Takakusaky, & Kanaya, 1988; Orlovsky, 1972). Moreover Kanaya, Unno, Kawahara, and Mori (1985) showed that the percentage of neurons showing this phasic discharge increases with walking speed (see Figure 17.1).

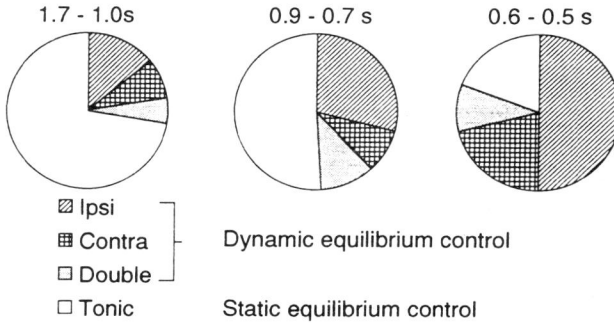

Figure 17.1. Relative proportions of Deiters' neurons in relation to the periods of cycle time. In each circle, open areas represent the proportion of Deiters' neurons exhibiting tonic discharges. The hatched areas and cross-hatched areas represent the proportions of Deiters' neurons exhibiting bursting discharges phase-locked with those of the ipsilateral gastrocnemius and the contralateral gastrocnemius muscle, respectively. The dotted areas represent the proportion of Deiters' neurons exhibiting double bursting discharges in a single step cycle. All of the Deiters' neurons in each of the circles were obtained from the same neuron population (from Kanaya et al., 1985; reproduced with permission).

The discharge of these cells is known to be phase-locked with the EMG activity of the leg extensor muscles. It is also known that during locomotion, the amplitude of the vestibular reaction of Deiters' neurons is substantially lower in walking animals than in resting ones (Orlovsky & Pavlova, 1972).

From a behavioral point of view, automatic servo-loops linking visual flow and locomotion have been brought to the fore in animals. Davis and Ayers (1972) showed in several invertebrate species that locomotion can be triggered by large visual patterns moving backward under a Plexiglas surface supporting the animal; the animal starts walking in the opposite direction, and its walking speed is modulated by the velocity of the moving visual patterns. In some drosophilas, locomotor output is also modulated by optical flow (Göetz & Wenking, 1973). In other species, locomotion can serve the purpose of stabilizing the visual scene; for example Rock and Smith (1986) showed that, unlike the locomotion described by Davis and Ayers, the triggering and control of locomotion speed in tropical fish goes in the direction of the flow, and therefore stabilizes the visual scene (see Table 17.1).

Table 17.1
Optomotor response and induced self-motion in tropical fish (from Rock & Smith, 1986; reproduced with permission).

Experiment	Condition	Response	Mean latency (s)
1	Tank rotated; drum not visible	10/10 fish remain passive	-
2	Tank rotated; stationary striped drum	9/10 fish swim against current	16.0
3	Optomotor condition: tank stationary; striped drum rotated	9/10 fish swim with drum (OMR)	18.0
4	Tank stationary; drum with circles rotated	10/10 fish remain passive (no OMR)	-
5	Double drum: inner drum rotated	2/13 fish swim with drum	-
	Double drum: outer drum rotated	0/13 fish swim with drum (OMR)	52.0

Rock and Smith emphasized that tropical fish do not have autonomous eye or head movements, and therefore must locomote to stabilize the visual scene. We might add that fish may be driven by external forces (currents) to achieve locomotor activity aimed at stabilizing the visual field. Curiously, the treadmill paradigm leads human subjects (for whom this situation is rarely an ecological one) to produce an optomotor response in order to stabilize the visual scene. This creates a good situation for investigating cycle properties, although it does not allow for studying the efficiency modulations related to the displacement function ordinarily performed by locomotion. Over-ground experiments were conducted in our laboratory in order to determine if optical flow can be processed automatically in human beings, that is, independently of the subject's will. Our goal was to find out whether changes in optical flow are able to induce unintentional modulations of human gait. The optical flow was changed in two ways. Firstly, the visual field was reduced to 12-degree perifoveal field (Bonnard & Pailhous, 1989); indeed, from a visual point of view, locomotion results in the slippage of the whole visual scene on the retina. In his studies on the accessory optic system (which drives visual self-induced motion in the direction of the vestibular nuclei), Frost (1985) showed how the cells of this sytem only discharge if the size of the stimulus is large (more than 30 degrees). Secondly, artificial optical flows moving from front to back (backward condition) or from back to front (forward condition) were projected on the floor, adding to the natural optical flow induced by locomotion. This caused the subjects to have the visual illusion that they were moving faster than they actually were (backward condition), or that they were moving backward instead of forward (forward condition) (Pailhous, Ferrandez, Flückiger, & Baumberger, 1990). In the latter experimental conditions, the artificial optical flow shared a great number of the kinematic properties with the natural visual flow produced in straight line walking. Such flow radiates from the focus of expansion, and its velocity gradients are consistent with the structure of the surrounding space. Thus, the visual resultants of this artificial flow and natural flow did not contain contradictory information on the displacement path (subjects walked in a straight line) or on the structure of the space in which they moved (the horizontal ground). In both experiments, the subjects were asked to maintain the same walking speed regardless of the visual conditions. This is a simple task insofar as it can be accomplished by maintaining the same motor commands. The results of both experiments converge: unintentional modulations of locomotor parameters (frequency and amplitude) were always observed, as shown in Figure 17.2.

Stride length always decreased, whether the condition involved backward optical flow, forward optical flow, or restricted vision. The effect on cadence varied, causing decreases in the backward and restricted vision conditions, and increases in the forward condition. Experimental variations of visual flow were shown here to cause unintentional modulations in locomotor parameters (stride length and cadence). Artificial visual flow cannot automatically trigger and control human locomotion, as it does in lobsters (Davis & Ayers, 1972). On the contrary, the cognitive capabilities of human beings enable them to trigger locomotion following verbal instructions, and to intentionally repeat it, as they do with arm movements for example. However, like perturbations in articular and muscular proprioceptive sensations (Taub, 1976), artificial visual flow modulates the spatial and temporal parameters of locomotion despite the subject's intentions. Maintaining the same walking speed looks very simple in this experiment since the subjects' task was only to intentionally maintain the same motor commands on the basis of somato-sensory information, which was not experimentally manipulated. However, the subjects were not able to perform this task when the optical flow was modified. Motor command alteration was slight, but its functional role must be stressed: thanks to the repetitiveness of the locomotor program, people may intentionally modulate their locomotion by acting often on slight and systematically repeated modifications.

The observed modifications confirm that the visual perception of motion plays a proprioceptive role. The functional duality of visual flow thus clearly appears: (a) it is known to act as an afference allowing for stabilization of the visual scene on the retina during rotation (vestibulo-ocular reflex, optokinetic reflex), and (b) it also acts as an afference of locomotion leading to the translation of the head. From a more fundamental standpoint, these studies show that the proprioceptive function is plurimodal (e.g., the combination of the afferences of the movement). The control of stable locomotion involves afferences that come from the lower limbs as well as from vision (and the labyrinths). However, when the visual flow is varied experimentally, stride length, for example, decreases significantly in all conditions. Subjects therefore ought to sense this decrease in the length of their strides. Yet, in reality, they think they are always walking the same way, and are unaware of these modifications. It looks as though visual perturbations contaminate the other afferences, such as the muscular and articular afferences coming from the lower limbs. At the behavioral level, such contamination reveals that the proprioceptive control of movement is of a plurimodal nature: when one afference is disrupted, the disruption is carried over to the others. Such proprioceptive contamination (which we can assume to occur if we look at the multimodal nature of the various nervous structures, as mentioned above) is not well understood. We might suggest two possible mechanisms as responsible for the contamination of locomotor parameters by visual flow.

First, the motor order sent to the lower limbs should be modified and, by the collaterals, somesthesic sensitivity would be changed (McCloskey, 1981); this modification may be due, for example, to cognitive factors. The sensation of going fast in the backward condition (due to a gain in the visual afference) would be compensated for by low reference values on muscular and articular afferences. The absence of a modulation in velocity in the forward condition shows that this factor does not act alone. Secondly, the postural conditions through which locomotion is produced may be changed; if so, the motor order is unchanged, but since the subject's posture changes (he/she is bending forward, for example) and his/her leg movement is ballistic, the amplitude of the movement (stride length) is shorter.

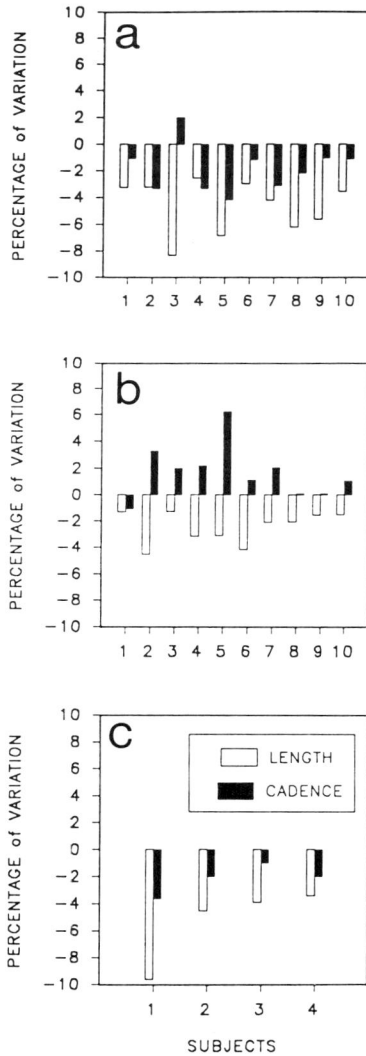

Figure 17.2. Differences (expressed as percentages) in stride length and cadence between the control condition and the experimental conditions, where the optical flow was manipulated. (a): backward versus control condition; (b): forward versus control condition (from Pailhous et al., 1990); (c): restricted versus control vision.

Studies conducted by Young (1988) argue in favor of the latter hypothesis by providing evidence of the links between trunk balance in running, and visual flow processing. In this research, subjects ran on a treadmill placed in a movable room. By tilting the room by about 3 degrees from right to the left or left to right (roll), a lateral inclination of the runner's trunk identical to the inclination of the moving room was observed. When the room stopped moving, the initial posture of the trunk was restored. By tilting the moving room from front to back or vice versa (pitch), the runner's trunk was tilted backward for a short period, and then the initial posture was restored (see Figure 17.3).

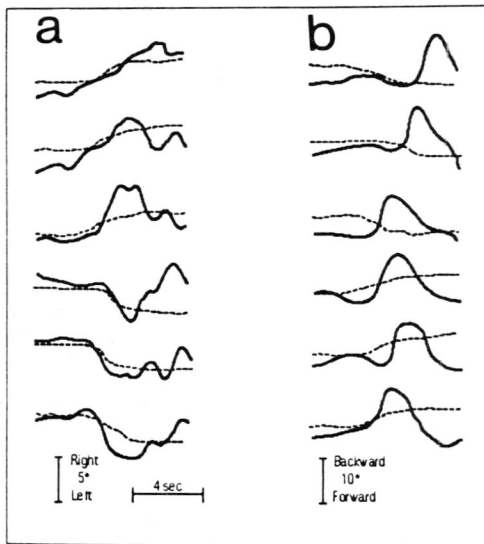

Figure 17.3. Results for the roll experiments in the "tilting room". (a) Results in the roll condition. Room roll (dashed line) and perturbation to subject's trunk roll (solid line) are presented as a function of time. (b) Same for the pitch condition. (Reproduced with permission from Young, 1988).

Balance control during locomotion thus appears to be highly dependent upon visual flow processing. This finding is consistent with studies by Stoffregen, Schmückler, and Gibson (1987), who reported that young children (below the age of 2 years) walking in a moving room fall down on 64% of the

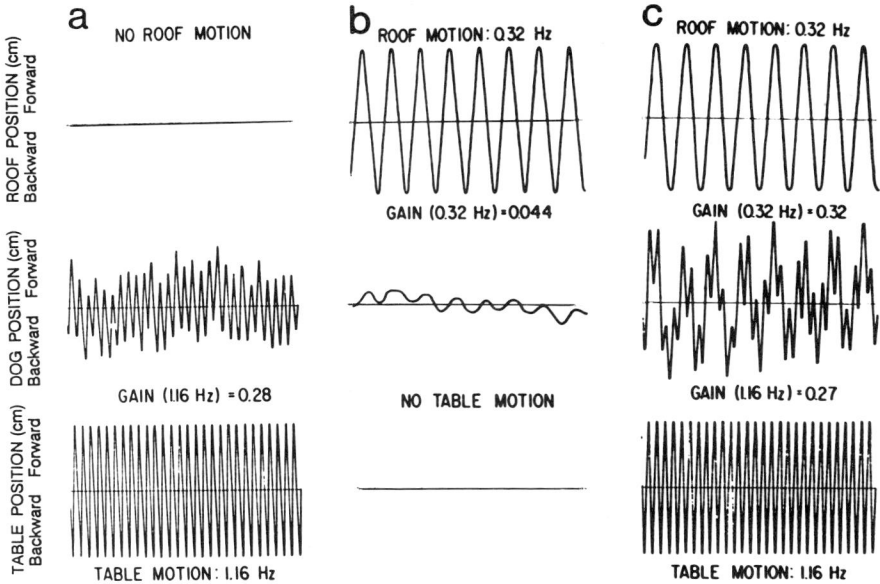

Figure 17.4. (a) Postural response to table motion (1.16 Hz, 4 cm) alone. (b) Postural response to roof motion (0.32 Hz, 4 cm) alone. (c) Postural response when both roof (.32 Hz, 4 cm) and table (1.16 Hz, 4 cm) were moved simultaneously. (Reproduced with permission from Talbott & Brookhart, 1980).

trials. Links between visual flow and static balance are well known (Lestienne, Soechting, & Berthoz, 1977; Paulus, Straube, & Brandt, 1984), and it appears that an equivalent mechanism also acts during locomotion. A very nice experiment by Talbott and Brookhart (1980) illustrates visuo-motor integration allowing for postural control in conditions of varying difficulty. When the quiet stance of a dog is perturbed by rhythmic and predictable oscillations of the table supporting the animals (1.16 Hz, a rhythm close to the locomotor rhythm) without visual perturbation (Figure 17.4a), the animal reacts with a postural response whose frequency is equal to that of the perturbation, but whose amplitude is lower (gain: 0.28 Hz). If the animal is subject to a rhythmic and predictable perturbation of the optical flow (0.32 Hz) without any table motion, a postural response is also observed, with a frequency equal to the visual perturbation frequency, but at a much lower amplitude (gain: 0.44 Hz, see Figure 17.4b). If these two perturbations (dynamic and visual) are combined, the postural response of the dog contains both frequencies (1.16 Hz and 0.32 Hz). Note, however, that while the postural response to the table movement has the same amplitude, the postural response to the visual perturbation is highly modified, its gain increasing to 0.32 Hz, and the amplitude of the response being seven times higher than the amplitude observed when the table is not moved (see Figure 17.4c).

In this sensory-conflict experiment, the postural response to the combined perturbation (table and vision) was similar to that observed in a control condition where the table was moved at both frequencies and amplitudes. This potentialization of the reponse to visual flow achieved by moving the supporting table shows that the importance of visual information depends on the sensorimotor activity in which the subject is involved. If that activity is simple (maintenance of posture on a stable support), the role of visual information is weak, but for more complex activities (maintenance of posture on a moving support) it becomes much greater.

Head Movement During Locomotion
It could be thought that head movement during locomotion is quite simple. For example, the head moves, at a constant velocity, along a rectilinear trajectory when the subject is walking straight ahead at a steady pace. If this were true, the optical flow would be simple and there would be no labyrinthic information. During locomotion, however, there may of course be movements of the eyes and eye-head system which are not related to the locomotor activity. They can be produced by intentional motor commands (e.g., turning one's head to talk to someone) or unintentional commands (e.g., in vestibular pathology). These movements induce changes in the optical flow, and necessarily have effects on locomotion, which will be discussed later. For now, however, we shall deal only with head movement observed in normal subjects who are walking steadily and straight ahead on flat ground while looking where they are going. In general, the moving body can be considered as a "biodynamic chain" (Bernstein, 1967) which consists of several links connected in such a way that modifying one link will affect the others (Turvey, Fitch, & Tuller, 1982). For the purpose of analysis, the body can be effectively (but very schematically) be divided into two subsystems: the lower limbs and the upper part of the body (80% of the body mass). The motion pattern of the lower limbs affects the rest of the chain, including the trunk and the head, which become involved in complex movements (as shown in Figure 17.5) aimed at minimizing the mechanical energy exchange between the two subsystems (Cappozzo, 1981). Head movements during locomotion can be divided into translational and rotational components. The main characteristics

(especially their frequency) are highly correlated to leg movement. The trajectories of the head and pelvis in the sagittal, transverse, and frontal planes, each related to a system of reference endowed with progressional motion, are shown in Figure 17.5. This same phenomenon could also be directly observed in treadmill walking because subjects do not progress through space.

These curves show how the pattern of motion is modified from the pelvis up to the head. Several points should be stressed here. First, a vertical displacement occurs at the motion frequency of the legs; it is transmitted without decreasing to the head; its amplitude varies with walking speed. Secondly, the anterior-posterior oscillation of the upper body occurs at the same frequency, but its amplitude is lower at the head than at the pelvis; furthermore it decreases with walking speed. Finally, lateral displacement occurs at a frequency half that of the above movements, which corresponds to the motion frequency of one leg; its amplitude decreases with walking speed.

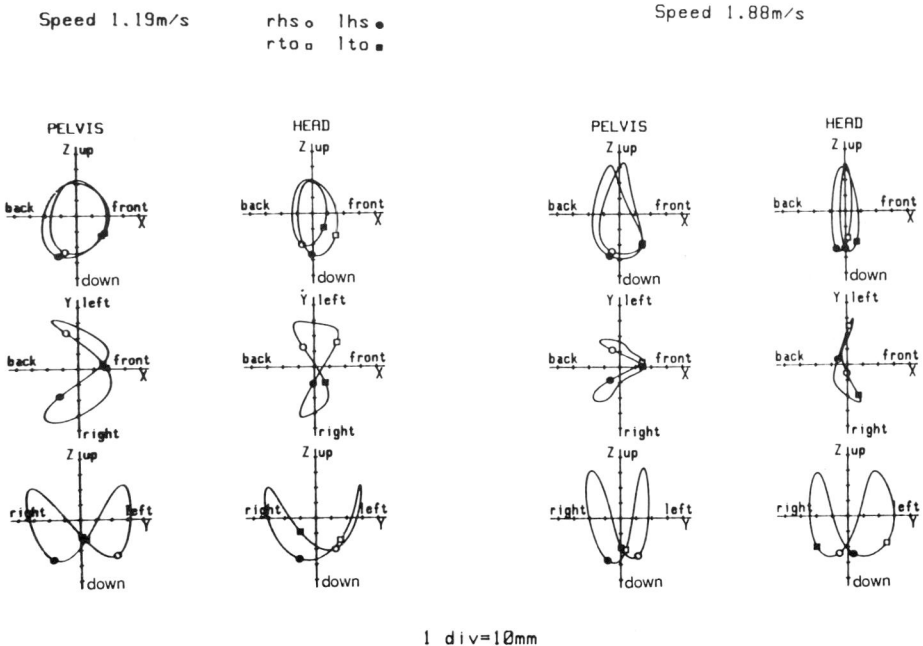

Figure 17.5. Lissajous' figures of the displacement of head- and pelvis-point in the sagittal, transverse and frontal planes during walking at 1.19 and 1.88 meters per second.(lhs: left heel strike; lto: left toe off; rhs: right heel strike; rto: right toe off. (Reproduced with permission from Cappozzo, 1981).

During the transmission from the pelvis up to the head, notice that the translational movements also have rotational components. In the case of "walking in place", Grossman, Leigh, Abel, Lanska, and Thurston (1988) describes rotations in the horizontal plane (yaw) and in the vertical plane (pitch). Horizontal rotations occur at the motion frequency of one leg (like lateral translations); their mean amplitude is 6 degrees, and their maximal angular velocity is 36 degrees per second. Vertical rotations have a double frequency (like vertical translation movements); their mean amplitude is 3 degrees and their maximal angular velocity is 32 degrees per second. In order to better understand the actual displacement of the head through space during locomotion, the longitudinal component of progression through space, whose velocity is equal to the subject's displacement speed, must be added to all of the above components. Figure 17.6a shows the longitudinal displacement of each foot, the head, and the waist during walking.

The longitudinal displacement of the head and the waist is continuous while that of the foot is, of course, discontinuous. In Figure 17.6b the instantaneous velocities of these body parts are plotted against time. It appears that the back-front displacement mentioned above creates a sinusoidal component around the progression velocity. This is not trivial: for a mean walking speed of 1.7 m/s, the velocity of the head oscillates between 1.5 and 2 m/s, depending on the phase of the locomotor cycle. The head velocity is the highest during the double support phase, being equal to the maximal velocity of the waist with a slight phase lag (about 40 degrees). Of course, the maximal velocity of the waist is out of phase with that of the legs. Thus it appears that during ordinary locomotion, the organizing principle underlying head movement is based entirely on leg motion: the head moves rhythmically with a frequency and amplitude that are precisely correlated to leg movement.

From a sensory point of view, these head movements are effective stimulation for the labyrinthine structures, the eye, and the neck muscle receptors, thus producing a very complex proprioceptive flow. From the visual point of view, without taking into account the autonomous rotational movements of the eye in the orbit, the optical flow induced by steady walking in a straight line is complex, being composed of velocities that are rhythmically modulated as a function of gait phase, in addition to rotational components. Grossman, Leigh, Bruce, Huebner, and Lanska (1989) showed that one can compensate for rotational head movements by reflex counter-rotation of the eye. Indeed, these authors found a VOR gain between 0.96 and 0.98 in subjects walking in place at different frequencies while fixating a visual target. Faced with the complexity of the optical flow induced by locomotion, a great number of birds (pigeons, doves, sand pipers, starlings, etc.) have developed strategies for stabilizing the head in space for 63% of their walking time (Frost, 1978).

Figure 17.7 shows the displacement of the breast, head, and one foot of a walking pigeon.

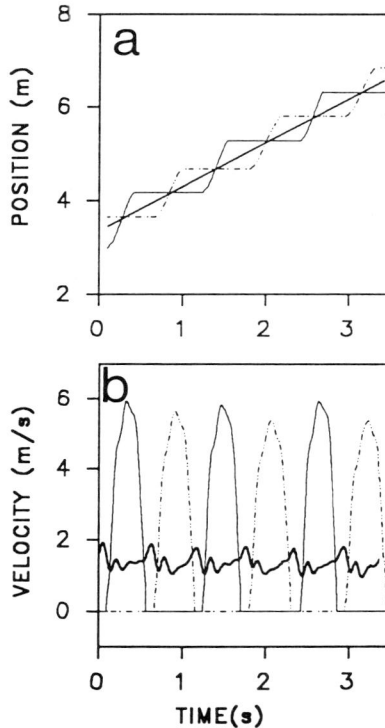

Figure 17.6. (a) Human head position (bold line), and position of foot 1 (solid line) and foot 2 (dotted line) plotted against time. (b) Instantaneous velocity for the same points.

As in humans (Figure 17.6), the displacement of the breast is continuous, while that of the foot is discontinuous. Head displacement is discontinuous, however. In fact, the progression of the head consists of two phases: one where the head slowly moves in the opposite direction, but at the same speed as the forward moving body (the head is "locked" in space), which is followed by another phase where the head is thrust forward in a ballistic way (about 120 ms) to a new position. As emphasized by Frost (1978), the slow displacement of the head with respect to the body during the "locked" phase (which causes the retinal slippage of the visual scene to drop down to between 2.5 and 3.6 m/s) and during the "thrust" phase (when a new spatial position is fixated) reminds us of the two components (slow and fast) of the optokinetic nystagmus. Furthermore, Frost demonstrated that this "head-bobbing" of the pigeon does not occur for biomechanical reasons (to maintain balance) since it disappears when the animal walks on a treadmill. When the visual world of the animal is stabilized in this way, the relative movement of the head with respect to the body is no longer observed. This shows that the counter-translational movement mentioned above was performed to stabilize the visual world. In humans, such stabilization is not observed; whatever the counter-rotation of the eye or the eye-head system, the visual scene can never be stabilized on the retina: it is always slipping. The fact that head and leg movements are synergic suggests that head movements are a biomechanical consequence of leg movement, occurring to produce harmonious displacement of the "biodynamical chain", which in turn minimizes the mechanical energy exchange between the various body parts.

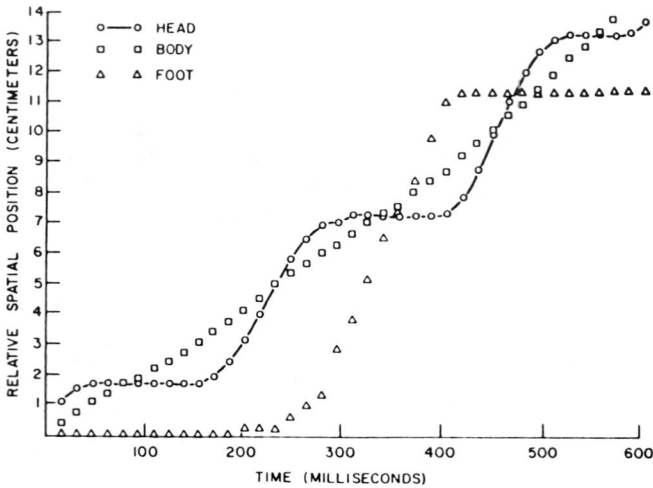

Figure 17.7. Pigeon head, breast and foot position as a function of time. Note the characteristic "hold and thrust" pattern of head movements, where the head is apparently locked in space for a period of time, and then is rapidly moved forward to a new "hold" position. (Reproduced with permission from Frost, 1978).

In conclusion, optical flow is subject to substantial interference from walking. For example, the information it contains about the direction of the displacement is "noisy" due to interference by lateral head movements, just as the information it contains about walking speed is noisy due to interference by anterior-posterior head movements. Ordinary locomotion, which is a gross motor activity, can make do with the visual inaccuracy caused by these perturbations. However, this is not the case when the subject is involved in a visuo-locomotor task (locomotor pointing, walking on a narrow beam). Indeed, in this situation, the motion of the head, which is already biomechanically dampened by its remoteness with respect to the leg and by the transmission of the movement through the flexible spine (except in the vertical plane) and the neck, is also decreased by an active compensation of these "noisy" movements (so important from the energy standpoint) allowing to precisely stabilize the head in space. This stabilization becomes greater as the visual constraints increase (for walking on a beam, see Assaiante and Amblard, in press). Thus, the head behaves like a genuine inertial platform (Berthoz & Pozzo, 1988; Grossman et al., 1989).

Conclusion

In this chapter, we have shown that rhythmical leg movement, which alternates between flexion and extension, is linked with vision in a very specific way. Two main dimensions can be identified. The first deals with problems of maintaining an adapted posture. Vision is important in such postural adaptation, and posture has an immediate effect on the organization of the forces developed by the legs. For example, exerting the same force with the right and left legs produces straight line walking only if the subject is well balanced.

The second dimension deals with the adaptive capabilities of gait. For its automatical aspects, gait appears mainly as a rhythm (walking frequency). For its adaptative aspects, it is mainly a stride amplitude: walking on irregular terrain requires walking with a stride length that fits the spatial characteristics of the terrain. In this case, vision acts "intimately" with locomotor synergy to modulate stride amplitude, especially through intentional intervention during the swing phase of the leg.

Walking is not a "luxurious" behavior; it is simple, robust and adaptive. Two main reasons can be evoked to explain this: (a) The integrating levels which regulate visuo-locomotor coordination are not all cortical, even if by definition they are supra-spinal. The centers integrating sensory information, especially visual, are informed of the locomotor activity; in this way walking is possible even if the optical flow receives interference from head movement, as mentioned above. If a standing human were subject to the same visual perturbations, important balance problems would be observed. (b) For dynamic reasons, maintaining balance is easier during walking than when at rest. It may appear paradoxical that a dynamic equilibrium is simpler to maintain than a static one. To convince yourself of this, try it on a bicycle, or note that balance disorders are often detected well before locomotor disorders.

By displacing the body, and thus the head, locomotion creates visual and labyrinthic stimulation. The legs, like cars, bicycles and so on, are a means of transportation. Whatever the means, a great number of cognitive and perceptive processes are shared. However, it should be noted that due to its fundamental role in the behavioral repertory of most species, including humans, locomotor output is sustained by a very powerful sensorimotor organization at the lower levels of the central nervous system. The relationships linking vision - especially optical flow - and leg movement during walking are different from those linking vision and leg movement (arm movement also!) during driving. In both cases, obstacles always have to be avoided, speed must be controlled, as must direction, but the command of forces (which determines the trajectory and kinematics of the movement) and their control are very different due both to proprioceptive reafferences (direct during walking, highly indirect during driving) and to the specialized nature of the locomotor command. (Learning to drive is not really the same thing as "learning" to walk).

References

Armstrong, D.M. (1988). The supraspinal control of mammalian locomotion. *Journal of Physiology, 405,* 1-37.

Assaiante, C., & Amblard, B. (in press). Head-trunk coordination and locomotor equilibrium in 3 to 8 year old children. In A. Berthoz, W. Graf, & P.P. Vidal (Eds.), *The head-neck sensory-motor system.* Oxford University Press.

Bernstein, N. (1967). *The co-ordination and regulation of movements.* New York: Pergamon Press.

Berthoz, A., & Pozzo, T. (1988). Intermittent head stabilization during postural and locomotory tasks in humans. In B. Amblard, A. Berthoz, & F. Clarac (Eds.), *Posture and gait: Development, adaptation and modulation* (pp. 189-198). Amsterdam: Elsevier.

Bisti, S., Maffei, I., & Piccolino, L. (1974). Visuo-vestibular interactions in the cat superior colliculus. *Journal of Neurophysiology, 22,* 146-155.

Bonnard, M., & Pailhous, J. (1989). Contribution du flux visuel au contrôle locomoteur. [Contribution of optical flow to locomotor control]. *Psychologie française, 34,* 63-70.

Cappozzo, A. (1981). Analysis of the linear displacement of the head and trunk during walking at different speeds. *Journal of Biomechanics, 14,* 411-425.

Clarac, F. (1984). Spatial and temporal coordination during walking in crustacean. *Trends in Neurosciences, 7,* 293-298.

Davis, W.J., & Ayers, J.L. (1972). Locomotion: Control by positive feedback optokinetic responses. *Science, 177,* 183-185.

Drew, T. (1988). Motor cortical discharge during voluntary gait modification. *Brain Research, 457,* 181-187.

Frost, B.J. (1978). The optokinetic basis of head-bobbing in the pigeon. *Journal of Experimental Biology, 74,* 187-195.

Frost, B.J. (1985). Neural mechanisms for detecting object motion and figure ground boundaries, contrasted with self-motion detecting systems. In D.J. Ingle, M. Jeannerod, & D.N. Lee (Eds.), *Brain mechanisms and spatial vision* (pp. 415-451). Dordrecht: Martinus Nijhoff (NATO Asi Series D: Behavioral and Social Sciences n°21).

Gibson, J.J. (1958). Visually controlled locomotion and visual orientation in animals. *British Journal of Psychology, 49,* 182-194.

Göetz, K.G., & Wenking, H. (1973). Visual control of locomotion in the walking fruitfly drosophila. *Journal of Comparative Physiology, 85,* 235-266.

Grillner, S. (1981). Control of locomotion in bipeds, tetrapods and fish. In V.B. Brooks (Ed.), *Handbook of physiology, Section 1: The nervous system, Volume II, Motor control* (pp.1179-1236). Bethesda, MD: American Physiological Society.

Grossman, G.E., Leigh, R.J., Abel, L.A., Lanska, D.J., & Thurston, S.E. (1988). Frequency and velocity of rotational head perturbations during locomotion. *Experimental Brain Research, 70,* 470-476.

Grossman, G.E., Leigh, R.J., Bruce, E.N., Huebner, W.P., & Lanska, D.J. (1989). Performance of the vestibuloocular reflex during locomotion. *Journal of Neurophysiology, 62 ,* 264-272.

Hay, L., & Schoebel, P. (1990). Spatio-temporal invariants in hurdle racing patterns. *Human Movement Science, 9,* 37-54.

Henn, V., Young, L.R., & Finley, C. (1974). Vestibular nucleus units in alert monkeys are also influenced by moving visual fields. *Brain Research, 71,* 144-149.

Kanaya, T., Unno, T., Kawahara, K., & Mori, S. (1985). Functional roles played by Deiters' neurons during controlled locomotion in the mesencephalic cat. In M. Igarashi & F. Black (Eds.), *Vestibular and visual control on posture and locomotor equilibrium* (pp. 193-198). Basel: Karger.

Katz, P.S., & Harris-Warrick, R.M. (1990). Actions of identified neuromodulatory neurons in a simple motor system. *Trends in Neurosciences, 13,* 367-373.

Lee, D.N., & Lishman, J.R. (1977). Visual proprioceptive control of stance. *Journal of Human Movement Studies, 1,* 87-95.

Lee, D.N., Lishman, J.R., & Thomson, J.A. (1982). Regulation of gait in long jumping. *Journal of Experimental Psychology: Human Perception and Performance, 8,* 448-459.

Lestienne, F., Soechting, J.F., & Berthoz, A. (1977). Postural readjustments induced by linear motion of visual scenes. *Experimental Brain Research, 28,* 262-284.

Liddell, E.G., & Phillips, C.G. (1944). Pyramidal section in the cat. *Brain, 67,* 1-9.

Lishman, J.R., & Lee, D.N. (1973). The autonomy of visual kinesthesis. *Perception, 2,* 287-294.

McCloskey, D.I. (1981). Corollary discharges: Motor commands and perception. In V.B. Brooks (Ed.), *Handbook of physiology, Section 1: The nervous system, Volume II, Motor control* (pp. 1391-1415). Bethesda, MD: American Physiological Society.

Mori, S., Matsuyama, K., Takakusaky, K., & Kanaya, T. (1988). The behavior of lateral vestibular neurons during walk, trot and gallop in acute precollicular decerebrate cats. In O. Pompeiano & J.H. Allum (Eds.), *Progress in brain research, 76* (pp. 211-222). Elsevier Science Publishers B.V.

Orlovsky, G.N. (1972). Activity of vestibulospinal neurons during locomotion. *Brain Research, 46,* 85-98.

Orlovsky, G.N., & Pavlova, G.A. (1972). Response of Deiters' neurons to tilt during locomotion. *Brain Research, 42,* 212-214.

Pailhous, J., Ferrandez, A.M., Flückiger, M., & Baumberger, B. (1990). Unintentional modulations of human gait by optical flow. *Behavioral and Brain Research, 38,* 275-281.

Patla, A.E., Robinson, C., Samways, M., & Armstrong, C.J. (1989). Visual control of step length during overground locomotion: Task-specific modulation of the locomotor synergy. *Journal of Experimental Psychology: Human Perception and Performance, 15,* 603-617.

Patla, A.E., Armstrong, C.J., & Silveira, J.M. (1989). Adaptation of the muscle activation patterns to transitory increase in stride length during treadmill locomotion in humans. *Human Movement Science, 8,* 45-66.

Paulus, W.M., Straube, A., & Brandt, T. (1984). Visual stabilization of posture. Physiological stimulus characteristics and clinical aspects. *Brain, 107,* 1143-1163.

Precht, W., Simpson, H., & Llinas, R. (1976). Responses of Purkinje cells in rabbit nodulus and uvula to natural vestibular and visual stimuli. *Pfluegers Archiv Fuer die Gesante Physiologie, 367*, 1-6.

Rock, I., & Smith, D. (1986). The optomotor response and induced motion of the self. *Perception, 15*, 497-502.

Rossignol, S., Lund, J.P., & Drew, T. (1988). The role of sensory inputs in regulating patterns of rhythmical movements in higher vertebrates. In A. Cohen, S. Rossignol, & S. Grillner (Eds.), *Neural control of rhythmic movements in vertebrates* (pp. 201-283). New York: Wiley.

Shik, M.L., & Orlovsky, G.N. (1976). Neurophysiology of locomotor automatism. *Physiological Reviews, 56*, 465-501.

Stein, P.S. (1978). Motor systems, with specific reference to the control of locomotion. *Annual Review of Neuroscience, 1*, 61-81.

Stoffregen, T.A., Schmückler, M.A., & Gibson, E.J. (1987). Use of central and peripheral optic flow in stance and locomotion in young walkers. *Perception, 16*, 121-133.

Talbott, R.E., & Brookhart, J.M. (1980). A predictive model study of the visual contribution to canine postural control. *American Journal of Physiology, 239*, R80-R92.

Taub, E. (1976). Motor behavior following deafferentation in the developing and motorically mature monkey. In R.M. Herman, S. Grillner, P.S.G. Stein, & D.G. Stuart (Eds.), *Neural control of locomotion* (pp. 675-705). London: Plenum Press.

Thelen, E., Ulrich, B.D., & Niles, D. (1987). Bilateral coordination in human infants: Stepping on a split-belt treadmill. *Journal of Experimental Psychology: Human Perception and Performance, 13*, 405-410.

Turvey, M.T., Fitch, H.L., & Tuller, B. (1982). The Bernstein perspective: 1. The problems of degrees of freedom and context-conditioned variability. In J.A.S. Kelso (Ed.), *Human motor behavior: An introduction* (pp. 239-253). Hillsdale, NJ: Erlbaum.

Waespe, W., & Henn, V. (1981). Visual-vestibular interaction in the flocculus of the alert monkey. II. Purkinje cells activity. *Experimental Brain Research, 43*, 349-360.

Warren, W.H., Young, D.S., & Lee, D.N. (1986). Visual control of step length during running over irregular terrain. *Journal of Experimental Psychology: Human Perception and Performance, 12*, 259-266.

Warren, W.H. (1988). Action modes and laws of control for the visual guidance of action. In O.G. Meijer & K. Roth (Eds.), *Complex movement behaviour: 'The' motor-action controversy* (pp. 339-380). Amsterdam: North-Holland.

Young, D.S. (1988). Describing the information for action. In O.G. Meijer & K. Roth (Eds.), *Complex movement behaviour: 'The' motor action controversy* (pp. 419-437). Amsterdam: North-Holland.

VISION AND MOTOR CONTROL
L. Proteau and D. Elliott (Editors)
© 1992 Elsevier Science Publishers B.V. All rights reserved.

CHAPTER 18

VISUAL PERCEPTION OF SELF-MOTION

Daniel R. MESTRE

Université d'Aix-Marseille II, Faculté de médecine
Cognition & Mouvement, URA CNRS 1166
IBHOP, Traverse Charles Susini, 13388 Marseille, France

The Optical Flow Field

Every time we move (our eyes, head or whole-body) we produce visual stimulations that consist in modifying the light pattern (optic array) projected onto the entire retina. For instance, as we move forward the objects we are approaching become larger in our visual field. We do not usually perceive a changing world, but rather our own motion in an otherwise stable environment. From this simple observation we might conclude that the optical motion resulting from our own movements is simply not perceived. A simple "suppression" mechanism could prevent the perception of environmental motion when we are moving, and the perception of self-motion would arise from proprioceptive information. That this is not the case can be demonstrated using visual rearrangement procedures. Specifically, when wearing an inverting lens, an observer perceives environmental (visual) motion as he/she turns the head (Wallach, 1987). If environmental motion arising from head motion were simply not perceived, it should not matter if it is inverted. Visual and proprioceptive information is clearly integrated for the control of self-motion (cf. Pailhous & Bonnard, this volume, also Dichgans & Brandt, 1978) and more generally for the control of movement (see also Proteau, this volume). Herein we will concentrate on the role of visual information in the perception of self-motion.

To our knowledge, Von Helmholtz (1866/1925) and Mach (1875) were among the first scientists to understand that the deformation of the retinal image due to egomotion was not just a superfluous nuisance, but indeed a rich source of information concerning the surrounding environment. Von Helmholtz (1866/1925) noted that one efficient way to estimate the distance to environmental elements involves comparing successive images of an object, when viewed from successive viewpoints. This comparison can be achieved through stereopsis (each eye receives different images of the same object) or monocularly, when the head is moving (e.g., Regan & Beverley, 1979). Von Helmholtz provided an initial qualitative analysis of visual scene transformations during egomotion. He noted that "In walking along, the objects that are at rest (...) appear to glide past us in our field of view in the opposite direction to that in which we are advancing. (...) The apparent angular velocities of objects in the field of vision will be inversely proportional to their real distances; and, consequently, safe conclusions can be drawn as to the real distance of the body from its apparent angular velocity." (Von Helmholtz, 1866/1925, p. 295).

These preliminary observations led to the conception of motion perception as the basis of spatial perception. It was observed that accurate perception of a three-dimensional (3D) object can be derived from the presentation of its two-dimensional (2D) projection, provided it is brought into motion; this was named the "Kinetic Depth Effect" (Wallach & O'Connell, 1953; see also Braunstein, 1976; Doner, Lappin, & Perfetto, 1984; Johansson, 1975, 1982; Rogers & Graham, 1979; Todd, 1982). Johansson (1975) suggested that "(...) The visual system abstracts relational invariances in the visual motion and constructs percepts of rigid objects moving in three-dimensional space." (Johansson, 1975, p. 80). For example, he noted that a rigid square, advancing and receding in relation to the subject, is perceived when the stimulus consists of a square which shrinks and expands on a vertical screen.

Regan and Beverley (1978) argued for the existence of "looming" (changing size) detectors in visual pathways. In addition, experiments by Yonas and his colleagues suggest that the sensitivity to kinetic visual information (specifying objects and events in three-dimensional space, such as impending collision) develops no later than 3 months of age (Yonas & Granrud, 1985; Yonas, Pettersen, & Lockman, 1979). These studies suggest that sensitivity to transformations of the visual scene is a functional sensorial ability which seems to depend on early processing skills (Mather, 1989). The study of this perception remains essential for the understanding of the regulation of spatial behavior.

Gibson (1947, 1950) applied this conception of motion perception to the case of a moving observer, that is to the perception and control of self-motion. He started from the basic conception of a continuously changing ambient optic array (what he called *optical flow*) produced by a continuously moving point of observation. Gibson, Ollum, and Rosenblatt (1955) proposed that "the fundamental visual perception is that of approach to a surface. This perception always has a subjective component as well as an objective component, i.e. it specifies [the Observer's] position, movement, and direction as much as it specifies the location, slant, and shape of the surface." (Gibson et al., 1955, p. 383). They proposed that egomotion through a stable environment produces an optical flow which specifies the properties of the environment and of the observer's trajectory. Within this flow, it was suggested that two types of information are available simultaneously: exteroceptive information about the three-dimensional structure of environmental elements, and proprioceptive information about the movements of the observer.

This fundamental idea that vision plays a proprioceptive role in the control of self-motion was in close agreement with Bernstein's (1967) original idea that all kinds of receptors (including visual receptors) contribute to "functional proprioception" during the control of action. Lee (1980) extended this conception further. He noted that what is needed for the visual control of action is information relative to the environment. He started from a basic "ecological" (Gibson, 1979) principle that any living organism is in constant interaction with its environment. An organism therefore does not need information about the position, orientation and movements of its body in absolute terms, it needs information about the environment-body relationships. Lee (1980) proposed the term "exproprioception" to qualify this necessary relativistic information. The optical flow pattern was then proposed as a means to access predictive information, enabling a subject to control his/her locomotion through a cluttered environment, to avoid obstacles and more generally to adapt his/her action to his/her action capabilities in a given environment.

These observations and early studies suggested the role that optical flow might play in the guidance of action. We will next review some experimental studies which prove that optical flow is actually a major source of information for self-motion perception and postural control.

The Functional Significance of Optical Flow

The proposition that optical flow contributes to self-spatial orientation is deeply grounded in early observations. For instance, Mach (1886/1959) noted that "if we take our stand over a bridge, and look fixedly at the water flowing beneath, we shall generally have the sensation of being at rest, whilst the water will seem in motion. Prolonged gazing, however, almost invariably results in the sensation that suddenly the bridge, with the observer and his whole environment begins to move in the direction opposite to the water, while the water assumes the appearance of being at rest" (Mach, 1886/1959, p. 68).

This accurate early report of visually induced self-motion (often called *vection* nowadays) suggested the overwhelming influence of visual input on an observer's evaluation of his/her spatial orientation. In this example, self-motion perception occurs even though the observer is stationary and is perfectly aware of the fact. It is obvious, however, that vision is not the sole source of information for the perception of self-motion. The visual, vestibular and kinesthetic senses all contribute to the perception and control of one's orientation and movement (see Dichgans & Brandt, 1978, for a review). In particular, the perception of motion of one's own body during its acceleration depends mainly on vestibular inputs. Under natural circumstances however, with eyes open and during motion at constant velocity, the sensation of self-motion appears to be maintained by visual input (Brandt, Dichgans, & Koenig, 1973). Moreover, when placed in conflict with vestibular information, visual information appears to be dominant in producing perception of self-motion. This led Lishman and Lee (1973) to consider vision as an autonomous kinesthetic sense (see also Gibson, 1979).

Mach (1886/1959) was the first to investigate this phenomenon in the laboratory: A subject was seated inside a drum which was covered with vertical stripes on its inner surface. When the drum was rotating, the subject felt him/herself to be rotating inside a stationary environment. Much research has since been conducted on visually induced rotation and translation, in an attempt to determine which characteristics of the visual stimulation produce vection (see Andersen, 1986, for a review).

One hypothesis is that, because self-motion generates a continuous optical flow over the whole retina (global optical flow), visual motion of the whole environment would trigger visually induced self-motion. On the contrary, object motion produces local flows restricted to certain parts of the visual field, which seem to identify a static observer inside a non-stable environment. Observations, however, do not follow this hypothesis: the "bridge over (troubled) water" illusion reported by Mach (see above) or the fact that we have all experienced self-motion when looking through a train window, we see another train leave on an adjacent track. This indicates that we can experience vection when only a limited part of the visual field is stimulated by (a local) optical flow.

A more conclusive demonstration was carried out by Brandt et al. (1973). Using Mach's "rotating drum" paradigm, they showed that circular motion of the entire environment always leads to a strong sensation of self-rotation. Yet, when different parts of the observer's visual field were masked, they found that circular vection occurred only when the stimulation extended beyond a 30-degree circular area of the central visual field. This result indicated that the sensation of vection does not require stimulation of the entire visual field. It also suggested that stimulation of the peripheral visual field induces perception of self-motion, whereas stimulation of the foveal and perifoveal visual field induces the perception of object motion.

This and other works studying visually induced sensations of self-rotation (Held, Dichgans, & Bauer, 1975) or translation (Berthoz, Pavard, & Young, 1975; Johansson, 1977) have been used to argue in favor of a distinction between two modes of processing of visual information, which have physiological foundations (i.e., Held, 1970; Schneider, 1967). The first mode, a focal mode of processing, concerned with object discrimination and identification, would depend on stimulation of the central visual field. Whereas the second, or ambient mode of processing visual information, would be concerned with spatial orientation and would depend on stimulation of the peripheral visual field.

However, Andersen and Braunstein (1985) gave experimental evidence against this hypothesis by inducing self-motion in central vision, with visual displays simulating forward motion through a three-dimensional cloud of dots. They concluded that perception of self-motion (and ambient processing) occurs in central vision, and that it is therefore related to depth perception. The central visual field seems to be engaged in the perception of complex optical flow patterns with the task of detecting not only the direction of self-motion (proprioceptive information), but also potential obstacles lying in the way. The role of depth information was also reported by Brandt, Wist, and Dichgans (1975) and more recently by Ohmi, Howard, and Landolt (1987), suggesting that foreground/background relationships influence self-motion perception. Vection is favored when static references are located in the foreground and the background is moving (see also Delorme & Martin, 1986; Lishman & Lee, 1973).

Other evidence suggesting the role of optical flow in spatial orientation comes from studies of postural control (the maintenance of balance). When comparing open-eye and closed-eye conditions, it appears that visual input attenuates self-generated body sway by 50 per cent (Edwards, 1946; Travis, 1945). By masking different parts of the visual field, Paulus, Straube, and Brandt (1984) showed that the central area of the visual field contributes to the control of fore-aft and lateral sway.

Lishman and Lee (1973) investigated the influence of optical motion on postural regulation more thoroughly. They put their subjects inside a moveable room. When this room was set into horizontal oscillatory motion, the subjects tended to sway approximately in phase and with an amplitude equal to the displacement of the room ("The subjects were like puppets visually hooked to their surroundings and were usually unaware of their disturbance" [Lee, 1980, p. 173]). This effect is so strong that it suggests that the perceptual system takes the optical motion of the surroundings as if it resulted from self-instability, the result being to sway so as to compensate for this disturbance.

Stoffregen (1985, 1986) also used a moveable room to systematically investigate the influence on postural readjustments of the structure and retinal position of the optical flow. He found that, while stimulation of the entire visual field produced the strongest compensatory postural swaying, stimulation of the central visual field resulted in small but consistent postural swaying (see also Andersen & Dyre, 1989). Moreover, he found that when the subject stared at the side wall of the fore-aft swinging room stimulation of the retinal periphery failed to produce postural readjustments. This result is at odds with the thesis that retinal periphery is specialized in spatial orientation. Stoffregen (1985) concluded that the retinal position of the optical flow could not fully explain these results and that flow structure also had to be taken into account.

The Information in Optical Flow

At this point, understanding the visual basis of self-motion perception requires a careful analysis of the structure and characteristics of optical flow patterns. This is a *sine qua non* condition to test the sensitivity of observers to these properties. In the original analysis of optical flow carried out by Gibson et al. (1955) and by Gordon (1966), the simple laws of linear perspective were used to derive the changing optical array of a moving observer. Since then, optical flow has classically been described as a two-dimensional *velocity field*, in which each vector represents the optical motion of an environmental element. Figure 18.1 represents optical flow resulting for pure observer's translation through space, which was the only case envisaged in early analyses. It is noteworthy that, independent of the structure of the environment (Figure 18.1), the optical flow vectors radiate outward from a common "focus of expansion" which corresponds precisely to the direction of self-motion. A preliminary conclusion that can be drawn from this analysis is that the direction of the vectors is determined entirely by that of the observer, thus forming a purely proprioceptive source of information (Gibson et al., 1955). However, the local magnitudes of the vectors depend on the environmental layout (Koenderink, 1986; Nakayama & Loomis, 1974; Prazdny, 1980). It was hypothesized that this "velocity gradient" in the optical flow constitutes an exteroceptive source of information concerning the environmental layout. In this respect, translation and rotation constitute two completely different situations. In the case of observer rotation, the whole environment translates at constant angular velocity (that of the observer's head), regardless of how far away the elements are (Cutting, 1986; Koenderink, 1986). A "rotational flow" contains only proprioceptive and no depth information. These differences between the structure and the informational content of "translational" and "rotational" optical flows should be kept in mind, especially because they might help to reconcile certain divergent conclusions reached from studies of circular and linear vection. Given that these two sensations obviously depend on

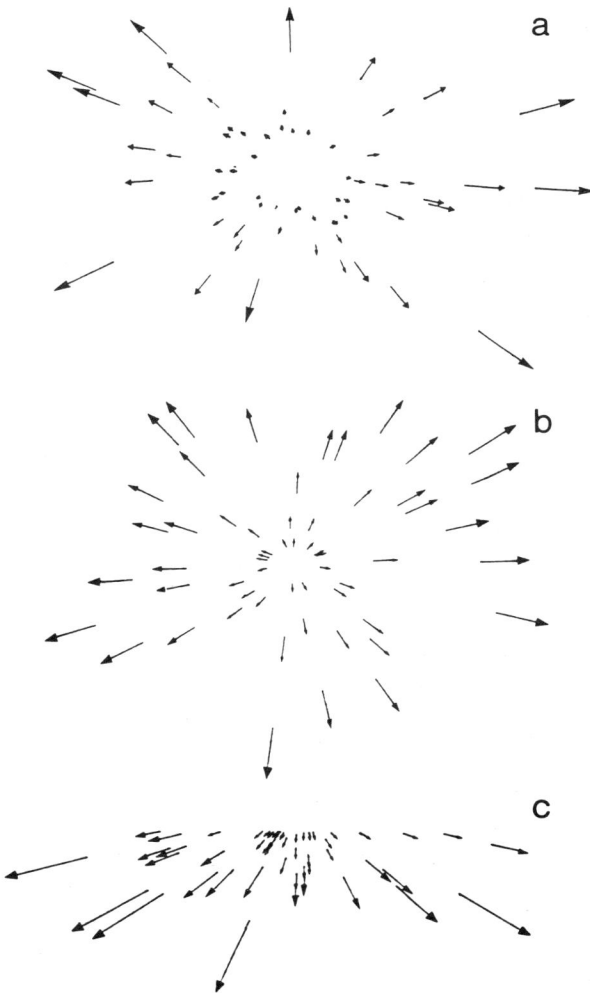

Figure 18.1. Vectorial representation of optical flows generated by simulating forward motion of the observer relative to (a) a "random-dot" cloud of dots, (b) a wall surface and (c) a ground surface. Note that in all cases the radial structure of the optical flow is the same, and that the vectors radiate outward from a common origin, corresponding to the observer's heading direction. The distribution of local vector amplitudes (velocity gradient) depends however on the spatial layout of the environment.

different optical stimulations, they might well be processed and integrated differently by the visuo-motor system.

Following these formal studies, it remained to be determined whether observers could indeed perceive their direction of self-motion (*heading* direction) from optical flow. One intriguing question was whether the "focus of expansion" could be perceived at all by a moving subject, which would give him/her a direct estimation of his/her heading direction. In this context a number of studies in the domain of computational and machine vision have demonstrated that one can build algorithms which compute the focus of expansion by locating the common origin of the vectors (Lawton, 1983 for example). However, these studies did not prove that human observers were able to perceive this focus of expansion, nor did they demonstrate how humans processed the information in optical flows.

Translational Motion

Warren, Morris, and Kalish (1988) conducted a series of experiments examining heading judgments from optical flow during translational motion. This study sought to answer two questions. First, did the perception of heading depend on the location of the focus of expansion (a fixed point) in the optical flow? Secondly, what was the accuracy of the perception of heading?

These questions were related to several prior experimental studies on the question, in which heading judgments were found to be quite inaccurate. Cutting (1986) calculated that an accuracy of 1 degree was necessary to avoid collision with obstacles during locomotion. However, in tasks where subjects had to point in the direction of the focus of expansion during visual simulations of forward motion, experimental data showed heading errors on the order of 5 to 10 degrees (Johnston, White, & Cumming, 1973; Llewellyn, 1971; R. Warren, 1976). While these results led many researchers to doubt the usefulness of the optical flow in the control of locomotion, Warren et al. (1988) suggested that these poor heading judgments might have been due to methodological difficulties, including the observer's task. They devised a new experimental protocol in which observers were presented with displays simulating forward motion either parallel or perpendicular to a plane represented by random dots. After seeing the presentation, the observers had to decide (on a forced-choice basis) whether it looked as if they were heading to the left or to the right of a target line located on the surface (Figure 18.2).

The accuracy of observer judgments was evaluated by varying the angular distance between the target and the heading direction. This enabled the computation of a perceptual threshold, or the angular distance for which more than 75% of the responses were correct. Using this procedure, it was found that heading accuracy was quite high, with thresholds on the average order of 1.2 degrees. However, when manipulating the number of dots represented on the surface, they found that thresholds rose significantly when the display showed only two dots, as opposed to several dots.

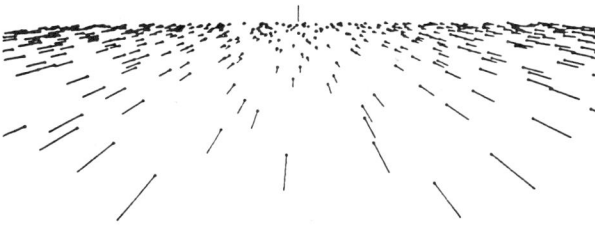

Figure 18.2. The instantaneous velocity field generated by translation parallel to a plane surface. In this figure, the target on the horizon lies in the direction of translation. In the experiment, it could be placed at various angular distances from this heading direction.

They concluded that heading accuracy derived from optical flow information is quite high and that it is not based on the perception of the relative motion between two dots. This would disprove the computational solution presented above. On the other hand, as Gibson first hypothesized, heading perception seems to rely on the perception of the global optical flow. The high accuracy level found when using 10 dots randomly positioned in a display confirms this hypothesis showing that subjects do not need to locate a fixed point in the display, corresponding to the focus of expansion.

Curvilinear Motion

In another series of experiments, the perception of the future path of self-motion on circular paths was examined (circular heading, cf. Warren, Mestre, Blackwell, & Morris, 1991). One basic motivation was that natural locomotion is often more complex than pure translational motion and often includes curvilinear components. Moreover, the perception of circular heading corresponds to predictive information, which certainly plays an important role in enabling a subject to find his/her way through a cluttered environment. Furthermore, curvilinear motion generates a "curved" optical flow in which there is no focus of expansion (Figure 18.3). If observers are able to perceive circular heading, this would reinforce the original hypothesis stating that the perception of the direction of self-motion relies on the global optical flow pattern.

Figure 18.3. The instantaneous velocity field generated by circular movement of the observer parallel to a plane surface. The target line lies on the future path of self-motion. In experimental conditions, this line could be positioned at various angular distances from the path. An identical velocity field would be produced by forward translation accompanied by eye rotation about a vertical axis.

The question was thus to determine not only the accuracy with which observers can perceive circular heading, but also what information they use to do so. From the beginning, a problem arose. If optical flow is represented as an instantaneous velocity field, one cannot distinguish between curvilinear motion and the combination of translational and rotational motions (Chasles' theorem, cf. Whittaker, 1944), as would occur for a driver travelling on a rectilinear path with a pursuit eye movement to track a traffic sign (Mestre & Pailhous, 1988). How then might circular heading be perceived from optical flow? One possibility (as suggested by Rieger, 1983) was that the velocity field might be an incomplete description of the optical flow. One must consider the dots' trajectories over time (what Koenderink, 1986, called streamlines) rather than the velocity vectors defined by two successive positions of the dots. Over time, circular movement with constant curvature produces a "steady" flow, with streamlines "rotating" around the center of curvature (Figure 18.4a), whereas a combination of translation and rotation generates an unsteady flow with streamlines "rotating" around different centers (Figure 18.4b).

Visual displays simulated the optical flow that would occur during observer's motion on a circular path relative to "random dot" environments. The motion could be simulated so that it would appear parallel to a ground surface, perpendicular to a wall surface or through a three-dimensional cloud of dots. Observers had to judge whether they would pass to the right or to the left of a target, if they continued on their current path.

The main result drawn from these experiments was that observers can perceive circular heading under most conditions with an accuracy on the order of 1 or 2 degrees, which is sufficient for the control of locomotion. Observers make use of the changes in direction of the dots' motion over time, which is consistent with the hypothesis of the insufficiency of the "velocity field" description of optical flow patterns. However, using small radii of curvature resulted in significant decreases in accuracy, with thresholds rising to 6 degrees under some circumstances, associated with systematic perceptual biases. With small radii (sharp curves) there was a systematic tendency to underestimate the path curvature on the ground, and inversely to overestimate the curvature when facing a wall. These effects might be due to those properties of the flow structure that are dependent on the environmental layout (Warren et al., 1991). One additional factor that remains to be evaluated is the size and configuration of the displays in these experiments. The simulations occupied 40 degrees of visual angle in central vision, and it could be that under these conditions important information in peripheral vision was missing.

Figure 18.4. Optical trajectories traced over time by dots on the surface for (a) circular movement of the observer and (b) translation accompanied by eye rotation.

At this point, it can be concluded that in a wide range of trajectories and environments the direction of self-motion is perceived from global flow patterns with sufficient accuracy to control locomotion. However, it remains to be explained why poor accuracy was found in previous studies. Aside from data recording and analysis problems, there is one procedural difference which seems worth discussing. In most previous studies, the subjects had to move a pointer toward the estimated heading direction. In the experiments reported above, subjects had to judge their heading relative to an object in the environment. This appears to be a major difference, suggesting that the information concerns the movement of an animal's body relative to its environment. It thus seems reasonable to suppose that absolute information is less important than information relative to actor-environment relationships, or what Lee (1980) named exproprioceptive information.

Furthermore, there are good reasons to believe that relative information is "directly" available in the optical flow. Lee (1976) demonstrated that the time it takes a moving observer to reach an obstacle in the environment (time-to-contact) is optically specified, whereas "absolute" variables such as the speed of self-motion or the distance to the obstacle are only specified within a scale-factor. A number of more recent experimental studies have suggested that optically-specified time-to-contact can actually be used, notably in grasping moving objects (e.g. Savelsbergh, Whiting, & Bootsma, 1991, see also Liebermann & Goodman, this volume).

Perspectives

Where are we? Initially the pattern of optical flow that is generated by our movements through the environment was thought to be a precise source of information for the control of self-motion. Behavioral evidence, such as visually-induced postural readjustments and self-motion perception, support the notion of the functional significance of optical flow. Moreover, experimental data show that the direction of self-motion can be accurately perceived from optical flow.

We should pay tribute here to the great insight of James Gibson, who first described the optical flow as a plausible source of information, leading us to simulate this specific type of visual movement and to test the ability of human observers to detect certain properties (such as the direction of movement), which were geometrically available. One direct outcome of this approach was the discovery by Regan and his colleagues (Regan, 1986; Regan & Beverley, 1978, 1979, 1984) of possible visual mechanisms sensitive to looming, divergence and curl components of optical flow. Another outcome has been the recent physiological evidence for the presence in visual areas of cells which seem to be selectively sensitive to radially expanding, translating or rotating patterns (i.e., Saito, Masao, Tanaka, Hisosaka, Fukada, & Iwai, 1986; Judge, 1990).

Now that optical flow is recognized as a major source of information in the perception of self-motion, one necessary orientation for future research will be to truly understand which variables in the flow pattern are relied upon by observers. For example, the velocity field description of optical flows is obviously insufficient. Warren et al.'s results (1991) suggest that observers have to rely on temporally-extended perception of the dots' trajectories in order to determine their precise direction of heading during curvilinear motion. Whether or not this corresponds to a perceptual ability to detect temporal derivatives of the velocity field (acceleration components), as suggested by Rieger (1983) is a question which refers to an "old" problem in the more general field of visual science : Is acceleration perceived *per se* or is it the result of successive velocity measurements? (cf. review in Regan, Kaufman, & Lincoln, 1986; see also Calderone & Kaiser, 1989).

Directly related to this question is that of the detectability of velocity gradients in optical flow patterns. In the experiments presented herein, motion through rigid three-dimensional environments was simulated. That subjects perceived motion through space does not fully explain which properties of the flow were used. An optical flow corresponding to motion through an environment contains velocity gradients (variations in local optical velocities, refer to the figures), which are assumed to be a source of information about the environmental layout. Simply simulating motion through a rigid environment presents its own limitations, since velocity gradients are inherent to the environment. Thus, a different approach is needed in order to study the role of these gradients. For instance, a radial optical flow can be generated in which the dots move at a constant speed from their central origin (the focus of expansion) to the periphery of the display. The results are quite surprising. Notably, the observer perceives a three-dimensional deformation with dots moving faster at the center of the display than at the periphery. De Bruyn and Orban (1989) systematically measured this phenomenon and concluded that velocity gradients are necessary for the perception of rigid motion in three-dimensional space. This type of experimental manipulation of optical flow components may prove essential in understanding the role of the optical flow structure in self-motion perception and in postural readjustments.

From these comments, the reader might conclude that the only remaining problem, with regards to the visual bases in the control of self-motion, is how to set up the right "visual" experiments to test the observer's sensitivity to significant variables in optical flow patterns. This is clearly not the case, principally because the optical flows we are interested in are generated by an observer's motion through the environment. For that reason the observers themselves are part of a perception-action cycle.

It would be worthy of interest to set up experiments to demonstrate not only that an active subject actually uses a given source of information, but also how he/she uses it in a perception/action cycle. This, however, is the challenge we are confronted with if we want to understand the role of vision during the control of locomotion. Lee and his colleagues conducted extensive work in this context (see Lee & Thomson, 1982, for a review). Lee (1976) proposed that the timing of motor acts could be visually specified. Lee, Lishman, and Thomson (1982) analyzed long jumping. They showed that the last three strides were visually adjusted in order to zero in on the board. Laurent and Thomson (1988) confirmed this conclusion. Furthermore, they suggested that well-coordinated visual regulation of the locomotor act does not require continuous visual guidance (see also Elliott, 1990, for a review). Intermittent information can be efficient, provided it is available at appropriate times in the action sequence (for instance, when the foot hits the ground).

There is, however, more to motor acts than timing. Warren, Young, and Lee (1986) investigated how propulsive forces could be visually regulated during locomotion (see also Patla, Robinson, Samways, & Armstrong, 1989). This is a promising line of research in which correlations between (kinematic) spatio-temporal variables in the optical flow and the dynamics of action are studied.

References

Andersen, G.J. (1986). Perception of self-motion: Psychophysical and computational approaches. *Psychological Bulletin, 99,* 52-65.

Andersen, G.J., & Braunstein, M.L. (1985). Induced self-motion in central vision. *Journal of Experimental Psychology: Human Perception and Performance, 11,* 122-132.

Andersen, G.J., & Dyre, B.P. (1989). Spatial orientation from optic flow in the central visual field. *Perception & Psychophysics, 45,* 453-458.

Bernstein, N. (1967). *The co-ordination and regulation of movements.* Oxford: Pergamon Press.

Berthoz, A., Pavard, B., & Young, L.R. (1975). Perception of linear horizontal self-motion induced by peripheral vision (linear vection): Basic characteristics and visual-vestibular interactions. *Experimental Brain Research, 16,* 476-489.

Brandt, T., Dichgans, J., & Koenig, E. (1973). Differential effects of central versus peripheral vision on egocentric and exocentric motion perception. *Experimental Brain Research, 23,* 471-489.

Brandt, T., Wist, E.R., & Dichgans, J. (1975). Foreground and background in dynamic spatial orientation. *Perception & Psychophysics, 17,* 497-503.

Braunstein, M.L. (1976). *Depth perception through motion.* London: Academic Press.

Calderone, J.B., & Kaiser, M.K. (1989). Visual acceleration detection: Effect of sign and motion orientation. *Perception & Psychophysics, 45,* 391-394.

Cutting, J.E. (1986). *Perception with an eye for motion.* Cambridge: MIT Press.

De Bruyn, B., & Orban, G.A. (1990). The importance of velocity gradients in the perception of three-dimensional rigidity. *Perception, 19,* 21-27.

Delorme, A., & Martin, C. (1986). Roles of retinal periphery and depth periphery in linear vection and visual control of standing in humans. *Canadian Journal of Psychology, 40,* 176-187.

Dichgans, J., & Brandt, T. (1978). Visual-vestibular interactions and motion perception. In R. Held, H.W. Leibowitz, & H.L. Teuber (Eds.), *Handbook of sensory physiology, Volume VIII: Perception* (pp. 755-804). Berlin: Springer Verlag.

Doner, J., Lappin, J.S., & Perfetto, G. (1984). Detection of three-dimensional structure in moving optical patterns. *Journal of Experimental Psychology: Human Perception and Performance, 10,* 1-11.

Edwards, A.S. (1946). Body sway and vision. *Journal of Experimental Psychology, 36,* 526-535.

Elliott, D. (1990). Intermittent visual pickup and goal directed movement: A review. *Human Movement Science, 9,* 531-548.

Gibson, J.J. (1947). *Motion picture testing and research.* A.A.F. Aviation Psychology Research Report N. 7. Washington D.C.: Government Printing Office.

Gibson, J.J. (1950). *The perception of the visual world.* Boston: Houghton Mifflin.

Gibson, J.J. (1979). *The ecological approach to visual perception.* Boston: Houghton Mifflin.

Gibson, J.J., Ollum, P., & Rosenblatt, F. (1955). Parallax and perspective during aircraft landings. *American Journal of Psychology, 68,* 372-385.

Gordon, D.A. (1966). Perceptual mechanisms in vehicular guidance. *Public Roads, 34,* 53-68.

Held, R. (1970). Two modes of processing spatially distributed visual stimulation. In F.O. Schmidt (Ed.), *The neurosciences: Second study* (pp. 317-323). New-York: Rockefeller University Press.

Held, R., Dichgans, J., & Bauer, J.A. (1975). Characteristics of moving visual scenes influencing spatial orientation. *Vision Research, 15,* 337-365.

Johansson, G. (1975). Visual motion perception. *Scientific American, 232,* 76-88.

Johansson, G. (1977). Studies on visual perception of locomotion. *Perception, 6,* 365-376.

Johansson, G. (1982). Visual space perception through motion. In A.H. Wertheim, W.A. Wagenaar, & H.W. Leibowitz (Eds.), *Tutorials in motion perception* (pp. 19-39). New York: Plenum Press.

Johnston, I.R., White, G.R., & Cumming, R.W. (1973). The role of optical expansion patterns in locomotor control. *American Journal of Psychology, 86*, 311-324.

Judge, S.J. (1990). Knowing where you're going. *Nature, 348*, 115.

Koenderink, J.J. (1986). Optic flow. *Vision Research, 26*, 161-180.

Laurent, M., & Thomson, J.A. (1988). The role of visual information in control of a constrained locomotor task. *Journal of Motor Behavior, 20*, 17-37.

Lawton, D.T. (1983). Processing translational motion sequences. *Computer Vision, Graphics and Image Processing, 22*, 116-144.

Lee, D.N. (1976). A theory of visual control of braking based on information about time to collision. *Perception, 5*, 437-459.

Lee, D.N. (1980). The optic flow field: The foundation of vision. *Philosophical Transactions of the Royal Society of London, Series B 290*, 169-179.

Lee, D.N., Lishman, J.R., & Thomson, J.A. (1982). Regulation of gait in long jumping. *Journal of Experimental Psychology: Human Perception and Performance, 8*, 448-459.

Lee, D.N, & Thomson, J.A. (1982). Vision in action: The control of locomotion. In D.J. Ingle, M.A. Goodale, & R.J.W. Mansfield (Eds.), *Analysis of visual behavior* (pp. 411-433). Cambridge: MIT Press.

Lishman, J.R., & Lee, D.N. (1973). The autonomy of visual kinaesthesis. *Perception, 2*, 287-294.

Llewellyn, K.R. (1971). Visual guidance of locomotion. *Journal of Experimental Psychology, 18*, 224-230.

Mach, E. (1875). *Grundlinien der Lehre von den Bewegungs-empfindungen.* [Basic principles for the study of motion perception.] Leipzig: Engelmann.

Mach, E. (1959). *The analysis of sensations*, (C.M. Williams, Trans.) New-York: Dover. (Original work published in 1886).

Mather, G. (1989). Early motion processes and the kinetic depth effect. *The Quarterly Journal of Experimental Psychology, 41A*, 183-198.

Mestre, D.R., & Pailhous, J. (1988). Rotation and translation of vehicles: Some aspects of their dissociation. In A.G. Gale, M.H. Freeman, G.M. Haslegrave, P. Smith, & S. Taylor (Eds.), *Vision in vehicles II* (pp. 35-44). Amsterdam: North-Holland.

Nakayama, K., & Loomis, J.M. (1974). Optical velocity patterns, velocity sensitive neurons, and space perception: A hypothesis. *Perception, 3*, 63-80.

Ohmi, M., Howard, I.P., & Landolt, J.P. (1987). Circular vection as a function of foreground-background relationships. *Perception, 16,* 17-22.

Patla, A.E., Robinson, C., Samways, M., & Armstrong, C.J. (1989). Visual control of step length during overground locomotion: Task-specific modulation of the locomotor synergy. *Journal of Experimental Psychology: Human Perception and Performance, 15,* 603-617.

Paulus, W.M., Straube, A., & Brandt, T. (1984). Visual stabilization of posture (Physiological stimulus characteristics and clinical aspects). *Brain, 107,* 1143-1163.

Prazdny, K. (1980). Egomotion and relative depth map from optical flow. *Biological Cybernetics, 36,* 87-102.

Regan, D. (1986). Visual processing of four kinds of relative motion. *Vision Research, 26,* 127-145.

Regan, D., & Beverley, K.I. (1978). Looming detectors in the human visual pathways. *Vision Research, 18,* 415-421.

Regan, D., & Beverley, K.I. (1979). Visually guided locomotion: Psychophysical evidence for a neural mechanism sensitive to flow patterns. *Science, 205,* 311-313.

Regan, D., & Beverley, K.I. (1984). Psychophysics of visual flow patterns and motion in depth. In L. Spillman & B.R. Wooten (Eds.), *Sensory experience, adaptation and perception: Festschrift for Ivo Kohler* (pp. 215-240). Hillsdale, NJ: Erlbaum.

Regan, D., Kaufman, L., & Lincoln, J. (1986). Motion in depth and visual acceleration. In K.R. Boff, L. Kaufman, & J.P. Thomas (Eds.), *Handbook of perception and human performance* (pp. 19.1-19.46). New-York: Wiley.

Rieger, J.H. (1983). Information in optical flows induced by curved paths of observation. *Journal of the Optical Society of America, A2,* 354-360.

Rogers, B., & Graham, M. (1979). Motion parallax as an independent cue for depth perception. *Perception, 8,* 125-134.

Saito, H., Masao, Y., Tanaka, K., Hisosaka, K., Fukada, Y., & Iwai, E. (1986). Integration of direction signals of image motion in the superior temporal sulcus of the Macaque monkey. *Journal of Neuroscience, 6,* 145-157.

Savelsbergh, G.J.P., Whiting, H.T.A., & Bootsma, R.J. (1991). 'Grasping' tau! *Journal of Experimental Psychology: Human Perception and Performance, 17,* 315-322.

Schneider, G.E. (1967). Contrasting visuomotor functions of tectum and cortex in the golden hamster. *Psychologische Forschung, 81,* 52-62.

Stoffregen, T.A. (1985). Flow structure versus retinal location in the optical control of stance. *Journal of Experimental Psychology: Human Perception and Performance, 11,* 554-565.

Stoffregen, T.A. (1986). The role of optical velocity in the control of stance. *Perception & Psychophysics, 39*, 355-360.

Todd, J.T. (1982). Visual information about rigid and nonrigid motion: A geometric analysis. *Journal of Experimental Psychology: Human Perception and Performance, 8*, 238-252.

Travis, R.C. (1945). An experimental analysis of dynamic and static equilibrium. *Journal of Experimental Psychology, 35*, 216-234.

Von Helmholtz, H. (1962). *Treatise on physiological optics*. (J.P.C. Southall, Trans.). New-York: Dover. (Original work published in 1866).

Wallach, H. (1987). Perceiving a stable environment when one moves. *Annual Review of Psychology, 38*, 1-27.

Wallach, H., & O'Connell, D.N. (1953). The kinetic depth effect. *Journal of Experimental Psychology, 52*, 571-578.

Warren, R. (1976). The perception of egomotion. *Journal of Experimental Psychology: Human Perception and Performance, 2*, 448-456.

Warren, W.H., Young, D.S., & Lee, D.N. (1986). Visual control of step length during running over irregular terrain. *Journal of Experimental Psychology: Human Perception and Performance, 12*, 259-266.

Warren, W.H., Morris, M.W., & Kalish, M. (1988). Perception of translational heading from optical flow. *Journal of Experimental Psychology: Human Perception and Performance, 14*, 646-660.

Warren, W.H., Mestre, D.R., Blackwell, A.W., & Morris, M.W. (1991). Perception of circular heading from optical flow. *Journal of Experimental Psychology: Human Perception and Performance, 17*, 28-43.

Whittaker, E.T. (1944). *A treatise on the analytical dynamics of particles and rigid bodies*. New York: Dover.

Yonas, A., Pettersen, L., & Lockman, J.J. (1979). Young infants' sensitivity to optical information for collision. *Canadian Journal of Psychology, 33*, 268-276.

Yonas, A., & Granrud, C.E. (1985). The development of sensitivity to kinetic, binocular and pictorial depth information in human infants. In D.J. Ingle, M. Jeannerod, & D.N. Lee (Eds.), *Brain mechanisms and spatial vision* (pp. 113-145). Dordrecht: Martinius Nijhoff .

Acknowledgment

The author wishes to thank William H. Warren, Jr. for the enjoyable and productive time spent in Providence analyzing and testing optical flow patterns.

Epilogue

VISION AND MOTOR CONTROL
L. Proteau and D. Elliott (Editors)
© 1992 Elsevier Science Publishers B.V. All rights reserved.

CHAPTER 19

THE VISUAL CONTROL OF MOVEMENT

DANIEL J. WEEKS* and ROBERT W. PROCTOR**

*School of Physical Education and Department of Psychology,
Lakehead University, Thunder Bay, Ontario, P7B 5E1*

**Department of Psychological Sciences,
Purdue University, West Lafayette, IN, 47907-1364*

Almost 100 years have passed since R. S. Woodworth expressed an interest in the role of visual feedback in the control of movements. As documented by Carlton (Chapter 1), there is irony in the fact that the basic research paradigm has changed little since Woodworth's pioneering work. Nevertheless, as the present volume demonstrates, the body of knowledge on the issue has grown substantially.

Research examining vision and movement has expanded to consider a wider variety of task situations. Moreover, significant progress is being made toward the development of detailed theoretical accounts of the interplay between vision and movement. Several research programs with diverse foci have converged on new insights into the functional mechanisms that underlie the use of visual information to control movement. For the purpose of this volume, the research has been categorized as examinations of vision in manual aiming, prehension and gesturing, spatial-temporal anticipation, and posture and locomotion.

Manual Aiming

Studies investigating the role of visual information in manual aiming tasks have focused primarily on temporal processing delays. Carlton (Chapter 1) emphasizes that even "mundane activities such as pedestrian and automobile travel, sport performance, and catching living prey in animals, [are] influenced by visual processing delays." Historically, one goal has been to determine the shortest movement time at which vision could benefit movement accuracy.

This time would reflect the minimum duration required for the processing and use of visual feedback. Through refinements in procedures and instrumentation, this minimum duration has diminished from an initial estimate by Woodworth of 450 ms to recent estimates as low as 100 ms. However, Carlton notes that the estimate of minimum processing time varies as a function of task demands. Consequently, he concludes that there may be no single minimum latency for the use of visual feedback in motor control. Rather, the limit for rapid visual processing involves a subtle interaction between the available stimulus information and the response correction requirements incurred by the task demands.

Elliott (Chapter 2) has extended the traditional question regarding temporal processing delays to the continued processing of visual information following the removal of input. He postulates the existence of a visual representation that can, under some conditions, provide an adequate substitute for direct visual input. In fact, Elliott argues in favour of two mechanisms, a brief "iconic-like" representation and a longer-lasting representation that may decay over 10 seconds or longer (cf. Steenhuis & Goodale, 1988). This suggestion is in accord with proposed subdivisions of iconic memory in the human perception literature, such as the distinction between visible persistence and informational persistence (Coltheart, 1980).

Recent investigations into the nature of informational persistence have produced results consistent with Elliott's analysis. Irwin and Yeomans (1986) argue that informational persistence has both a brief analogue component and a more enduring, abstract component that can maintain coarse spatial and structural codes able to mediate performance. These subdivisions of iconic representation are proposed to be task sensitive, consistent with Carlton's argument that task demands influence the time required to process visual feedback. Thus, the important point is that the subdivisions are proposed to differ in terms of both content and temporal delay. Their respective roles in the control of movement are likely candidates for future research.

Carson (Chapter 3) discusses research attempting to relate manual asymmetries in performance to hemispheric differences in the processing of visual feedback. He laments that the traditional dichotomies used to describe hemispheric asymmetries have impeded the study of manual asymmetries as a means of understanding the nature of movement regulation. Indeed, Carson strikes a note that is becoming increasingly apparent in cognitive psychology. During the past half-century, the right cerebral hemisphere in humans has been described as specialized for nonverbal, holistic, parallel processing, whereas the left hemisphere has been characterized as a verbal, analytic, serial information processor. None of these dichotomies (or any others) can capture the entire family of hemispheric asymmetries, because the dichotomies label processing systems that are necessarily comprised of many subsystems. As a consequence, the relative superiorities of the subsystems can cut across the dichotomies (Allen, 1983). Thus, the quest for neurological correlates of the broad dichotomies is unlikely to be successful. A viable alternative to this approach might be to seek out the neurological correlates of the subsystems that can be inferred from a computational assessment of response dynamics.

Proteau (Chapter 4) considers the control changes that occur as a function of learning. In contrast to the traditional position that the efficacy of sensory feedback diminishes with practice (e.g., Schmidt, 1975), he argues that practiced subjects actually rely more on sensory information (with a clear emphasis on visual information) to control their movement. Proteau proposes that a shift occurs from intramodal sensory comparison to reliance on an integrated intermodal sensory representation. Consequently, after extensive practice, the addition of visual feedback hinders performance when it has not been available previously, as does its removal when it has been available.

In Proteau's research subjects were trained to particular levels of expertise. However, a different approach might be to use a novice-expert paradigm in which individuals of different skill levels in sport or work-related perceptual-motor tasks are tested with and without the benefit of vision. For example, Proteau's model makes the prediction that an expert dart player would suffer greater performance deterioration from the withdrawal of visual feedback than would a novice player.

Redding and Wallace (Chapter 5) distinguish between perceptual-motor skill acquisition, as studied by Proteau, and perceptual adaptation, which is their primary concern. According to them, skill learning involves tuning of the relation between progressive stimulus changes and corresponding motor responses, whereas perceptual adaptation consists of learning the transformation to distorted perceptual input. On the basis of extensive research using prism adaptation, Redding and Wallace have developed a model of eye-hand coordination in which adaptive spatial alignment of the perceptual-motor system is mediated by various subsystems. When a discrepancy is introduced between the cooperative functions of the eye-head and eye-hand subsystems, adaptation occurs for the one that is subordinate in that particular task situation.

Abrams (Chapter 6) considers the retinal and extraretinal information offered by the eye during limb movements. Specifically, he proposes that different types of retinal and extraretinal information are important at the different component phases of an aimed movement. A wrist-rotation experiment showed that the eye and hand movements begin at about the same time, with the eye arriving at the target region in advance of the hand. Most important, the nature of the eye movement (saccadic or smooth pursuit) affects the initial impulse, whereas eye position affects error corrections. Extraretinal information about eye position seems to aid the assessment of target location when an eye movement is made to that location.

Finally, Cohen and Welch (Chapter 7) present work concerned with manual aiming conducted under quite different environmental constraints. Given that coordinative action has emerged as a result of earth-bound constraints placed on the nervous system, it is of interest to consider the operation of the perceptuomotor system in different gravitational environments. Cohen notes that reaching responses in new gravitational environments are initially subject to opposing errors from muscle unloading/loading and the "elevator illusion" that cancel each other. The muscle loading/unloading effects show rapid adaptation through proprioceptive feedback. Such results have considerable implication for training human performance in extra-terrestrial environments.

Prehension and Gesturing

Other researchers have related the manual aiming literature to a similar but different visuomotor behavior -- grasping. This work also investigates the interactive nature of a number of subsystems that cooperate to achieve the skill of prehension. Carnahan (Chapter 8) takes an approach similar to Abrams of considering extraretinal information. She presents evidence that pointing accuracy is enhanced by proprioceptive information from the positioning of the head. Moreover, the temporal organization of eye and head movements is altered by the inclusion of limb movements, and the relation between all three depends on whether the goal is pointing or grasping.

Jeannerod and Marteniuk (Chapter 9) summarize current knowledge regarding the transport and grasp components of prehension. For the transport component, they distinguish between directional coding, which orients the arm toward the target, and amplitude coding, which generates the proper dynamics for the chosen movement. Directional coding has been examined by introducing perturbations in the target location. According to Jeannerod and Marteniuk, movements can be corrected quickly if the perturbation occurs during the movement, with longer delays required if the perturbation occurs during movement preparation. They argue that the quick changes reflect automatic readjustments based on comparison between the afferent and efferent signals, whereas the slow reactions reflect reprogramming of movement direction. The grasp component is shown to be influenced by cognitive representation. Jeannerod and Marteniuk argue that evidence for parallel control of the two components, rapid information processing, and motor equivalence converge to promote neuromorphic computation as a viable theoretical tool for modeling prehension.

Sivak and MacKenzie (Chapter 10) consider the consequences for prehension of the structural and functional differences between the central and peripheral retinal regions. They conclude that peripheral vision provides poor information about the object to be grasped. Consequently, both transport and grasping are slowed when only peripheral vision is available. Central vision provides the details necessary for the grasp component; thus, only the transport component is affected by limiting vision to the central region.

Similar to Carson (Chapter 3), Roy and Hall (Chapter 11) address the interaction of hemispheric subsystems in the execution of coordinated movement. However, the emphasis is on individuals with a particular neurological disorder. The authors note that, relative to normal adults, apraxic individuals tend to exhibit one of two typical patterns of performance when cued by verbal command or imitative gestures to pantomime movements. Specifically, for one pattern performance suffers with both types of cue, whereas for the other pattern performance is substantially better with the imitative cue. This suggests that vision plays a mediating role in pantomiming movements. Because other evidence links the left hemisphere damage characteristic of the apraxic syndrome with sequencing problems for movements performed from memory, Roy and Hall suggest that a visual image generation process (cf. Kosslyn, 1980) is the relevant subsystem affected. The resulting model proposed by Roy and Hall is comprised of a number of components that interact to account for apraxic performance. However, the model is sufficiently global so as to provide a useful theoretical tool to address broader issues in motor behavior.

Spatial-Temporal Anticipation

The authors of the present section seek to identify the control parameters that underlie motor tasks requiring spatial and temporal coincidence. Bootsma and Peper (Chapter 12) attempt to determine candidate action parameters that can be deduced from the optical flow (see also Mestre, Chapter 17, for a similar approach). They conclude that the critical information for effective action in hitting and throwing tasks (i.e., "when things will be where") is available from the optic array. Thus, the environment can provide sufficient information to allow successful prediction of temporal events that determine the spatial unfolding of actions.

Fleury, Bard, Gagnon, and Teasdale (Chapter 13) continue the efforts to advance an ecological approach to perception-action. In particular, they address the interaction of perceptual and motor systems in the interception of moving objects. Not only do Fleury et al. consider a range of stimulus and perceptual characteristics that influence the effective interface between perception and action, but they include discussion of more traditional movement issues, such as practice, knowledge of results, and aging. Fleury et al. argue that systematic manipulation of perceptual parameters, as well as task complexities, is a most promising means of determining the nature of perceptual-motor interactions in the execution of interceptive actions.

Goodman and Liebermann (Chapter 14) further consider the functional importance of optical flow variables in determining effective action for whole body "landing tasks." Of particular interest is their finding that subjects who were not blindfolded prior to release performed similarly to subjects who were blindfolded. Goodman and Liebermann's analysis leaves room for a functional role for cognitive mechanisms and strategies not typically included in the ecological approach to action. Moreover, the cognitive representation of the environment that they propose is consistent with that proposed by Elliott (Chapter 2). This additional evidence for control via the intermittent pickup of visual information suggests that some form of iconic representation may mediate action.

Posture and Locomotion

In the previous sections the various authors isolated local interactions in which a stimulus event occurs and the organism responds. However, as behavior unfolds in the course of our daily lives, each of these local interactions is embedded in a sequence that is subsumed within the context of the behavior that follows. In the final section the authors consider the larger system that guides humans in their complex interactions with the environment.

The role of vision in the control of posture is considered by Starkes, Riach, and Clarke (Chapter 15). They argue that age-related improvement in postural control, as well as control deficits exhibited by Parkinson patients, may be attributable to changes in control strategies that differ in their emphasis on sensory guidance. Specifically, Starkes et al. propose that the lower sway velocity, characteristic of normal adults, may reflect a closed-loop strategy of postural control.

Corlett (Chapter 16) addresses the nature of the mental representation of environmental information that would serve to mediate effective navigation in that environment. He reconsiders the issue of intermittent versus continuous visual sampling of the environment discussed by Elliott (Chapter 2). Corlett concludes that the evidence favours intermittent sampling and that the issue has largely been replaced by the work attempting to define the time-course constraints placed on such sampling. Corlett also discusses work attempting to isolate candidate environmental cues that constitute the mental representations of movement space. He raises important issues regarding the representational format for this information (cf. Kosslyn, 1980).

Mestre (Chapter 17) describes the functional importance of sensory guidance through a specific analysis of information in the optical flow. He reports a number of experiments that suggest that the direction of self-motion, as well as change in speed, can be deduced from optical flow. Because optical flow is determined by the observing individual, Mestre suggests that the critical parameters may vary as a function of action characteristics. Mestre suggests that such changes could explain the shifts in control strategies postulated by Starkes et al. (Chapter 15).

Finally, Pailhous and Bonnard (Chapter 18) continue the discussion of locomotive behavior by noting that it emerges from an interactive cooperation of higher and lower levels of the central nervous system. As a consequence, they argue that the liaison between vision and locomotion is functionally specific. In particular, vision plays an important mediating function in the maintenance of posture as well as the adaptive characteristics of gait.

Summary
The role of visual information in the control of human movement, as summarized in the chapters of this book, is indeed complex. The work presented here emanates from distinct research programs within the realms of manual aiming, grasping and gesturing, spatial-temporal anticipation, and posture and locomotion. As a consequence, different issues are stressed in the respective chapters. Nevertheless, considerable consensus regarding the role of vision emerges. Following the tone set by Carlton (Chapter 1), virtually all of the authors stress that the role of vision in motor control is multidimensional, with the specific effects on behavior varying in subtle ways as a function of the task requirements.

One implication of the complex interaction between vision and motor control is that perception and action should not be studied in isolation. The way perceptual information is translated into action, and the manner in which action influences perception, constitute primary problems to be resolved by cognitive science. This point is evident in Newell's (1990) efforts to consolidate information processing subsystems into a unified theory of cognition that can capture the broad range of human behavior. In the next section, we discuss some of the salient points of contact that the research on visual control of movement makes with the larger enterprise of cognitive science.

Theoretical Contact with Cognitive Science
It has been suggested that two distinct camps have arisen in the study of movement control (Fisk & Goodale, 1989). On the one hand, there are those committed to the principles of engineering (e.g., Atkeson & Hollerbach, 1985) to uncover the functional constraints on motor control. On the other, there are

those who approach the control of action from the perspective of cognitive psychology (e.g., Proctor & Reeve, 1990; Zanone & Haurt, 1987). Although this dichotomy is apparent in the present volume, we are struck by the degree of consensus among authors about the necessity for conducting systematic analyses of the information provided by the visual system and the role played by that information in the performance of specific movement tasks. Those researchers from the cognitive tradition tend to examine information processing associated with visual input, whereas those from the ecological tradition look at dynamic and environmental constraints. As in all sciences, the cooperative competition among these approaches is needed in the development of unified theories of behavior (Proctor & Weeks, 1990; Proctor, Van Zandt, Lu, & Weeks, in press).

The research strategy adopted by many of the contributing authors is to examine variations of a prototypical task, with the goal of determining the nature of the processes that underlie performance. As a consequence, many of the emerging points of interest overlap with other domains concerned with the understanding, prediction, and control of human behavior.

Representation

One of the more salient points to emerge from this volume on visual information and movement control is the recent concern with specification of the representational format for that information. For example, consistent with the historic emphasis on brief iconic memories of the type postulated by Neisser (1967) and Sperling (1960), Elliott (Chapter 2), as well as Goodman and Liebermann (Chapter 14), has argued in favour of an "iconic-like" veridical representation for temporary storage of visual input relevant for movement control. However, the use of an analog (iconic) representation has not gone uncriticized in cognitive psychology. For example, Pylyshyn (1973) has argued that analog representation is a restricted medium for higher-level conceptual abstractions. Whether this restriction will impede the theoretical unification of cognition and motor behavior remains an empirical issue.

An alternative to iconic representation is a propositional format. Basically, a propositional representation is a description of the relation between informational items (see Kosslyn, 1980, for a detailed presentation of the characteristics of propositional representations). A propositional representation for the visual information crucial to successful movement control might include relations pertaining to size, shape, spatial location and texture, among others. In fact, other researchers (e.g. Norman & Rumelhart, 1975) have argued that high-level propositional representations can be decomposed to reveal root propositions that are essentially equivalent to an analogue representation. As a consequence, propositional representations can demonstrate operations such as mental rotation (Cooper & Shepard 1973) and stimulus-response compatibility (John & Newell, 1990).

Given that many of the authors of the present volume report cognitive behavior occurring at less than 100 ms, we think that the relative merits of alternative representational formats must be considered when addressing motor behavior. Such consideration likely will be a necessary prerequisite for the emergence of theories that unify the perceptual, cognitive and motoric contributions to human behavior.

Connectionism

Another representational format considered in the present volume is neuromorphic computation. The "connectionist uprising" has the possibility of unifying cognition and neuroscience. It is interesting to note that one of the earliest attempts at developing a neural network model was applied to visual information processing in the retina (Hartline & Ratliff, 1958). Recent efforts indicate that neuromorphic computation could provide important insight into the functional constraints associated with movement control (Jordan, 1990). In particular, the massive parallelism that is characteristic of neural network models can effectively deal with the rapid reorganization of perturbed movement, whereas a strictly serial program cannot explain how humans perform in such situations. A similar version of this point was raised by Feldman (Feldman & Ballard, 1982), one of the leading figures in the connectionist movement. Feldman argued that in getting a serial computer to exhibit "brain-like" behavior there exists a real-time constraint of 100 machine instructions. Basically, the constraint is imposed by the number of operations that can fit between the activation of neural circuitry and the emergence of cognitive behavior. Because it is impossible to accomplish even simple tasks under such constraints, this argument has been used to dismiss symbolic processing models in favor of connectionist models. However, as Newell (1990) definitively demonstrates, "Though widespread, the opinion that symbolic-systems models of cognition are somehow so serial that they cannot be made to jibe with the brain's organization and speed of operations is simply misguided" (p. 486). In other words, symbolic systems with parallel aspects are sufficient models.

As noted by Estes (1988), connectionist models in cognitive science have been most successful when applied to lower level behavior (e.g., speech perception and production). It is not suprising, then, that the most successful models to date for acquisition of cognitive skill have been symbolic, production system models (Anderson, 1983; Newell, 1990). Estes argues that the convergence of neuromorphic and symbolic models may provide a means of establishing comprehensive accounts of behavior. As applied to motor control, this suggests that symbolic information-processing models as well as neural network models need to be explored. If network models can be successful at capturing aspects of movement control it may provide an opportunity for interfacing such models with symbolic processing models of motor learning and performance.

Discrete *vs.* Continuous Processing
The question of discrete versus continuous sampling of visual information is discussed in several chapters, with Corlett (Chapter 16) concluding that the evidence argues for discrete sampling. He holds that since "light itself can be thought of as existing in quantum packages and [given] that the retina consists of discrete receptor cells, each sending electrical signals ultimately to the brain in the form of individual action potentials, any visually-based phenomenon is inevitably intermittent when taken to its neurological level." Although the question of visual sampling may ultimately favour an intermittent system, the broader question of discrete versus continuous information processing must be answered at a behavioral rather than neurological level. Thus, the choice of what type of description is appropriate for a particular system is largely pragmatic, with the goal being to facilitate the prediction of behavior.

The discrete/continuous issue is one that has been prominent more generally in the study of human performance (Miller, 1988; Schweickert, in press). Although this issue has been difficult to resolve, it is clear that across a variety of reaction tasks the relation between reaction times and errors can be explained readily by random-walk and related models. These models are characterized by the gradual accumulation of information to a criterion. The criteria settings determine the quality of the information on which the response is based. One possible relation to the evidence for discrete sampling of visual input in motor control is that a series of continuous accumulation operations are performed.

Summary
This volume attests to the diversity of research activity examining the role of vision in movement control. Despite this diversity, the various research programs are converging on some common control mechanisms and theoretical explanations. The almost 100 years of research on this topic has produced a substantial body of data. The present volume indicates that a comprehensive theory of vision and motor behavior is a promising possibility. This promise is most likely to be realized if the current attempts to integrate local theories in motor behavior with global theories in cognitive science and neuroscience continue to be developed.

References
Allen, M. (1983). Models of hemisphere specialization. *Psychological Bulletin, 93,* 73-104.

Anderson, J.R. (1983). *The architecture of cognition.* Cambridge, MA: Harvard University Press.

Atkeson, C.G., & Hollerbach, J.M. (1985). Kinematic features of unrestrained vertical arm movements. *The Journal of Neuroscience, 5,* 2318-2330.

Coltheart, M. (1980). Iconic memory and visual persistence. *Perception & Psychophysics, 27,* 183-228.

Cooper, L.A., & Shepard, R.N. (1973). Chronometric studies of the rotation of mental images. In W.G. Chase (Ed.), *Visual information processing* (pp. 75-176). New York: Academic Press.

Estes, W.K. (1988). Toward a framework for combining connectionist and symbol-processing models. *Journal of Memory and Language, 27,* 196-212.

Feldman, J.A., & Ballard, D. (1982). Connectionist models and their properties. *Cognitive Science, 6,* 205-254.

Fisk, J.D., & Goodale, M.A. (1989). The effect of instructions to subjects on the programming of visually directed reaching movements. *Journal of Motor Behavior, 21,* 5-19.

Hartline, H.K., & Ratliff, F. (1958). Spatial summation of inhibitory influences in the eye of Limulus, and the mutual interaction of receptor units. *Journal of General Physiology, 41,* 1049-1066.

Irwin, D.E., & Yeomans, J.M. (1986). Sensory registration and information persistence. *Journal of Experimental Psychology: Human Perception and Performance, 12,* 343-360.

John, B.E., & Newell, A. (1990). Toward an engineering model of stimulus-response compatibility. In R.W. Proctor & T.G. Reeve (Eds.), *Stimulus-response compatibility: An integrated perspective* (pp. 427-479). Amsterdam: North Holland.

Jordan, M.I. (1990). Motor learning and the degrees of freedom problem. In M. Jeannerod (Ed.), *Attention and performance XIII* (pp. 796-836). Hillsdale, NJ: Erlbaum.

Kosslyn, S.M. (1980). *Image and mind.* Cambridge, MA: Harvard University Press.

Miller, J. (1988). Discrete and continuous models of human information processing: Theoretical distinctions and empirical results. *Acta Psychologica, 67,* 191-251.

Neisser, U. (1967). *Cognitive psychology.* New York: Appleton-Century-Crofts.

Newell, A. (1990). *Unified theories of cognition.* Cambridge, MA: Harvard University Press.

Norman, D.A., & Rumelhart, D.E. (1975). *Explorations in cognition.* San Francisco, CA: W. H. Freeman.

Proctor, R.W., & Reeve, T.G. (Eds.). (1990). *Stimulus-response compatibility: An integrated perspective.* Amsterdam: North Holland.

Proctor, R.W., & Weeks, D.J. (1990). *The goal of B. F. Skinner and behavior analysis.* New York: Springer-Verlag.

Proctor, R.W., Van Zandt, T., Lu, C.-H., & Weeks, D.J. (in press). Stimulus-response compatibility for moving stimuli: Perception of affordances or directional coding?. *Journal of Experimental Psychology: Human Perception and Performance.*

Pylyshyn, Z.W. (1973). What the mind's eye tells the mind's brain: A critique of mental imagery. *Psychological Bulletin, 80,* 1-24.

Schmidt, R.A. (1975). A schema theory of discrete motor skill learning. *Psychological Review, 82,* 225-260.

Schweickert, R. (in press). Information, time, and the structure of mental events: A twenty-five year review. In D.E. Meyer & S. Kornblum (Eds.), *Attention and performance XIV.* Hillsdale, NJ: Erlbaum.

Sperling, G. (1960). The information available in brief visual presentations. *Psychological Monographs, 11,* whole No. 498.

Steenhuis, R., & Goodale, M.A. (1988). The effects of time and distance on accuracy of target-directed locomotion: Does an accurate short-term memory for spatial location exist? *Journal of Motor Behavior, 20,* 399-415.

Zanone, P., & Haurt, C. (1987). For a cognitive conception of motor processes: A provocative standpoint. *European Bulletin of Cognitive Psychology, 7,* 109-129.

Acknowledgement

The first author would like to acknowledge the support of the Natural Sciences and Engineering Research Council of Canada.

Author index

Abbs, J.H., 67, 98, 206, 221, 227, 228
Abel, L.A., 412, 417
Abel, M.R., 111, 127
Abernethy, B., 290, 310
Abrams, R.A., 7, 13, 30, 43, 44, 47, 52, 53, 56, 58, 62, 63, 71, 98, 100, 129, 131, 132, 133, 134, 138, 143, 147, 148, 149, 162, 173, 207, 230
Acredolo, L.P., 387, 394
Adams, J.A., 13, 28, 67, 68, 79, 80, 81, 91, 97, 98
Adams, S., 261, 262, 263, 265, 268, 275, 277, 279, 282
Agarwal, G.C., 61, 63
Albus, K., 316, 328
Alderson, G.J.K., 296, 297, 310, 322, 328
Allard, F., 13, 18, 19, 29, 34, 38, 46, 74, 99, 102, 187, 194, 206, 228, 236, 237, 256
Allen, M., 61, 62, 443, 450
Allum, J.J., 355, 370
Alstermark, B., 21, 28
Alston, W., 132, 149
Amblard, B., 83, 100, 238, 254, 258, 415, 416
Andersen, G.J. 424, 425, 426, 434,
Anderson, D.J., 154, 173
Anderson, J.R.,449, 450
Anderson, M., 295, 310
Anderson, M.E., 179, 194
Angell, R.W., 132, 149
Annett, J., 53, 59, 62, 77, 78, 98
Annett, M., 53, 59, 62
Anstis, T., 111, 124
Anton, J., 387, 394
Arbib, M.A., 206, 212, 216, 227, 228, 229, 274, 278
Armstrong, C.J., 376, 396, 404, 418, 434, 437

Armstrong, D.M., 402, 416
Armstrong, W., 132, 151
Arndt, D.R., 387, 395
Aronson, E., 356, 371
Arrott, A.P., 170, 173
Ashamead, D.H., 180, 195
Ashby, A.A., 317, 319, 333
Assaiante, C., 415, 416
Athenes, S., 188, 195, 208, 209, 210, 211, 212, 214, 217, 218, 227, 230, 234, 240, 249, 257, 296, 300, 312
Atkeson, C.G., 239, 256, 447, 451
Aume, N.M., 162, 170
Ayers, J.L., 403, 405, 417

Bahill, T.A., 319, 328
Baker, E., 273, 279
Baker, J.T., 153, 170
Bakker, F.C., 306, 308, 310
Ball, C.T., 317, 328
Ballard, D., 449, 451
Ballinger, E.R., 161, 171
Bard, C., 14, 28, 74, 83, 89, 98, 102, 237, 238, 254, 256, 317, 318, 320, 322, 323, 324, 326, 327, 328, 329, 330
Barlow, H.B., 316, 328
Barnes, G.R., 179, 193
Barr, C.C., 111, 123
Bauer, J.A., 71, 90, 96, 100, 425, 435
Baumberger, B., 405, 407, 418
Bawa, P., 356, 370
Beaubaton, D., 13, 19, 20, 28, 74, 84, 89, 90, 96, 98, 99, 102, 237, 254, 256, 258
Bedford, F., 107, 124
Beek, P.J., 296, 310, 323, 328
Beggs, W.D.A., 9, 12, 28, 34, 42, 46, 72, 100
Bellec, J., 310, 317, 324, 328

Bernard, M.C., 166, 171
Bernstein, N., 216, 227, 401, 410, 416, 423, 434
Berthoz, A., 206, 230, 409, 415, 416, 418, 425, 434
Beuchert, D.E., 364, 370
Beverley, K.I., 290, 301, 312, 422, 432, 437
Biguer, B. ,129, 132, 149, 180, 181, 182, 193, 203
Bingham, G., 212, 229
Bisti, S., 402, 416
Bizzi, E., 148, 151, 179, 180, 193, 194, 224, 226, 230
Black, F.O., 355, 372
Blackwell, A.W., 310, 313, 429, 431, 433, 438
Blouin, J., 83, 89, 102
Bondar, R.L., 160, 174
Bonnard, M., 405, 416
Bootsma, R.J., 25, 26, 28, 294, 295, 296, 297, 298, 300, 304, 305, 306, 308, 310, 311, 321, 323, 325, 328, 341, 346, 348, 432, 437
Borst, A., 316, 317, 328
Bossom, J., 117,124
Bowden, J.M., 42, 45
Bowen, K.F. 18, 19, 31, 34, 47
Bower, T.G.R., 286, 311
Bradshaw, J.L., 60, 61, 62, 63
Brandt, T., 355, 370, 409, 418, 421, 424, 425, 434, 435, 437
Braunstein, M.L., 422, 425, 434
Bressan, P., 305, 313
Bridgeman, B., 203, 227
Broadbent, D.E. 9, 29
Brookhart, J.M., 409, 419
Brooks, V.B., 42, 45, 57, 62
Broughton, J.M., 286, 311
Brown, J.F., 318, 328
Brown, P., 262, 281
Brown, R.H., 154, 172

Bruce, E.N., 412, 415, 417
Brunt, D., 185, 195
Buckolz, E., 266, 279, 387, 395
Buekers, M.J.A., 315, 329
Bullock, T.H., 61, 62
Burgess-Limerick, R., 290, 310
Burton, H.W., 71, 98
Butterworth, G., 356, 370
Byblow, W., 391, 395

Calderone, J.B., 433, 435
Calne, D.B., 364, 370
Calvert, R. 34, 36, 37, 38, 39, 40, 41, 46
Campbell, F.W., 316, 331
Canfield, A.A., 156, 157, 164, 171
Canon, L.K. ,110, 117, 124, 126, 139, 147, 150, 160, 174
Cappozzo, A., 410, 411, 417
Carel, W.L., 286, 311
Carello, C. ,304, 310, 336, 348
Carlton, L.G., 3, 7, 14, 15, 17, 19, 22, 23, 24, 26, 27, 28, 34, 36, 42, 45, 85, 86, 98, 129, 130, 131, 134, 137, 139, 143, 149, 150, 162, 171, 187, 194, 236, 256, 381, 396
Carlton, M.J., 7, 14, 17, 19, 22, 23, 24, 26, 27, 29, 381, 396
Carmon, A., 60, 63
Carnahan, H., 180, 183, 184, 185, 186, 187, 188, 189, 191, 194
Carrière, L., 317, 324, 328
Carson, R.G., 42, 43, 44, 45, 46, 55, 56, 57, 58, 59, 60, 62
Castiello, U., 206, 210, 212, 217, 227, 228
Cavallo, V., 287, 290, 311, 323, 327, 329
Caviness, J.A., 285, 313
Cermak, S., 269, 272, 277, 278

Chaffin, D.B., 42, 47, 139, 151
Chamberlin, C.J., 326, 330, 331
Chambers, D., 148, 151
Chan, C.W.Y., 364, 367, 370, 372
Charlton, J., 277, 278
Chew, R.A., 80, 98, 100
Choe, C.S., 109, 123, 127
Christina, R.W., 19, 29
Chua, R., 42, 43, 44, 45, 46, 55, 56, 57, 58, 59, 62
Churchland, P.M., 223, 227
Churchland, P.S., 123, 124
Cisneros, J., 49, 52, 64
Clarac, F., 401, 417
Clark, B., 154, 155, 172
Clark, S.E., 111, 155, 122, 125
Clayton, T.M.H., 25, 30, 293, 298, 312, 323, 331, 341, 345, 348
Cohen, G., 51, 60, 62
Cohen, L., 182, 194
Cohen, M.M., 154, 155, 159, 160, 162, 164, 165, 167, 171, 174
Cole, K.J., 67, 78, 101, 221, 228
Coltheart, M., 38, 46, 277, 278, 442, 451
Comrey, A.L., 156, 157, 164, 171
Cone, S.L., 91, 101
Conti, P., 84, 98
Cook, E., 110, 111, 125
Cooper, L.A., 448, 451
Corballis, M.C., 61, 62
Corcos, D.M., 61, 63
Cordo, P.J., 15, 29, 71, 98, 99
Coren, S., 111, 124
Corlett, J.T., 383, 387, 388, 389, 390, 391, 392, 394, 395
Correia, M.J., 155, 164, 171
Cournoyer, J., 36, 47, 84, 85, 86, 101, 236, 258
Cox, R.H., 71, 99
Craik, K.J.W., 9, 29
Craske, B., 112, 124

Crossman, E.R.F.W., 8, 14, 29, 42, 46, 89, 99, 131, 132, 150
Cumming, R.W., 428, 436
Cutting, J.E., 426, 428, 435
Cynader, M., 301, 312

Davids, K.W., 73, 96, 99
Davis, G., 342, 343, 344, 348
Davis, R., 9, 20, 29
Davis, W.J., 403, 405, 417
Davson, H., 234, 256
De Bruyn, B., 433, 435
De Graff, B., 319, 329
De Renzi, E., 261, 262, 263, 268, 273, 275, 278
de Rijke, W., 354, 355, 365, 372
Dean, J., 226, 228
Decety, J., 214, 229, 249, 257, 267, 278
Del Rey, P., 324, 329
Delorme, A., 425, 435
Deming, D., 356, 371
deMonasterio, F.M., 235, 256
Denier Van der Gon, J.J., 67, 102, 200, 231
Detwiler, M.L., 287, 313
Dichgans, J., 355, 370, 421, 424, 425, 434, 435
Dickinson, J., 346, 348
Diggles, V.A., 73, 99
Dinh Phung, R., 290, 294, 311, 341, 347
Dixon, W.J., 307, 308, 311
Doane, T., 49, 50, 51, 52, 53, 60, 63, 64
Dobkin, R.S., 129, 131, 143, 149
Doherty, S., 214, 229
Donaldson, I.M.L., 182, 193
Donders, F.C., 4, 29
Doner, J., 422, 435
Doolittle, J.H., 155, 171
Dorfman, P.W., 322, 324, 329

Drew, T., 399, 400, 402, 417, 419
Dreyfus, H.L., 61, 63
Drotman, M., 147, 149
Dugas, C., 80, 84, 101, 188, 195,
    208, 209, 210, 218, 230, 239,
    257, 300, 312, 323, 332
Dunham, P., 317, 329
Dyhre-Poulsen, P., 345, 346
Dyre, B.P., 426, 434

Earl, M., 271, 279
Echallier, J.E., 74, 84, 86, 90, 96,
    101, 131, 132, 152
Echallier, J.F., 59, 64, 67, 84, 86,
    90, 96, 100, 129, 130, 132,
    152, 203, 231, 236, 237, 238,
    258
Eddy, J.K., 276, 279
Edwards, A.S., 425, 435
Egelhaaf, M., 316, 317, 328
Eickmeier, B., 239, 257, 300, 312
Ekman, G., 215, 228
Elliott, D. ,13, 18, 19, 29, 33, 34,
    35, 36, 37, 38, 39, 40, 41,
    42, 43, 44, 45, 46, 47, 49,
    53, 54, 55, 56, 57, 58, 59,
    60, 62, 63, 73, 74, 84, 85,
    86, 97, 99, 129, 143, 150,
    187, 194, 206, 228, 236, 237,
    238, 248, 253, 256, 257, 265,
    279, 345, 346, 378, 382, 383,
    384, 385, 395, 396, 434, 435
Enroth-Cugell, C., 235, 257
Epstein, W., 110, 111, 125
Estes, W.K., 449, 451
Evarts, E.V., 206, 228, 364, 370

Faber, C.M., 297, 313
Factor, S.A., 364, 370
Faglioni, P., 273, 278
Farah, M.J., 266, 267, 279, 280

Favilla, M., 199, 228
Fel'dman, A.G., 271, 279
Feldman, J.A., 449, 451
Fernandez-Pone, E., 153, 172
Ferrandez, A.M., 405, 407, 418
Festinger, L. ,139, 147, 150
Finley, C., 402, 417
Firoozmand, F., 355, 371
Fischman, M.G., 73, 99, 296, 311
Fishburne, G., 266, 279
Fisher, J., 361, 372
Fisk, J.D., 132, 151, 447, 451
Fitch, H.L., 294, 301, 311, 410,
    419
Fitts, P.M., 7, 9, 14, 29, 54, 63,
    69, 73, 99, 208, 218, 228
Flaherty, D., 262, 275, 279
Flaherty, J.F., 364, 371
Flanders, M., 71, 98, 99, 102,
    167, 174, 181, 196, 251, 258
Fleishman, E.A., 71, 95, 99
Fleury, M., 14, 28, 74, 83, 89, 98,
    102, 237, 238, 254, 256, 296,
    311, 317, 318, 320, 322, 323,
    324, 326, 327, 328, 329, 330
Flowers, K., 49, 50, 51, 52, 54,
    56, 58, 59, 63, 364, 371
Flückiger, M., 405, 407, 418
Fodor, J.A. 111, 124, 390, 395
Folkerts, N., 355, 371
Forssberg, H., 356, 371
Foulke, J.A., 42, 47, 134, 151
Fox, C.R., 155, 173
Frank, J.S. 52, 53, 64, 147, 152,
    162, 164, 174, 271, 279
Franks, I.M., 325, 329
Fraser, C., 210, 212, 218, 232,
    242, 259
Friesen, H., 262, 263, 265, 275,
    279, 282
Frost, B.J., 405, 412, 413, 414,
    417
Fukada, Y., 432, 437

Fukuda, H., 346
Fukuoka, M., 346
Funk, C.J., 179, 194

Gagnon, M., 317, 318, 322, 323, 324, 326, 327, 328, 329, 330
Garhammer, J., 73, 99
Garland, H., 132, 149
Garneau, M., 160, 174
Garrett, J.B., 109, 126
Gauer, O.H., 153, 171
Gazenko, O.G., 153, 171
Gazzaniga, M.S., 181, 194, 266, 267, 279, 280
Geffen, G., 60, 63
Gentilucci, M., 210, 212, 217, 228
Gentner, D.R., 296, 311
Georgopoulos, A.P., 20, 21, 22, 29, 187, 190, 194, 200, 201, 207, 228
Gerathewohl, S.J., 155, 158, 164, 172
Ghez, C., 199, 228
Gibson, E.J., 71, 95, 97, 99, 321, 323, 330, 408, 419
Gibson, J.J., 285, 313, 321, 323, 330, 335, 336, 346, 386, 390, 395, 401, 417, 422, 423, 424, 426, 435
Gielen, C.C.A.M., 67, 102, 200, 206, 231, 232
Gill, E.B., 10, 11, 12, 31, 320, 334
Gilson, R.D., 153, 172
Girouard, Y., 80, 84, 101, 323, 332
Gladwin, T., 377, 395
Glass, A.L., 276, 279
Glass, B., 134, 150
Glass, D., 42, 47
Glencross, D., 317, 328
Glenn, J.H., 153, 172

Goetz, E.T., 79, 80, 81, 91, 98
Goldberg, I.A., 164, 174
Gollhofer, A., 346, 347
Goodale, M.A., 20, 30, 44, 46, 47, 56, 63, 130, 131, 132, 137, 143, 151, 152, 180, 187, 188, 189, 194, 195, 196, 201, 202, 203, 204, 228, 230, 383, 384, 385, 393, 396, 447, 451
Goodeve, P.J., 8, 14, 29, 42, 46, 89, 99, 131, 132, 150
Goodglass, H., 273, 277, 279
Goodman, D., 42, 43, 44, 45, 46, 55, 56, 57, 58, 59, 60, 62, 345, 348
Goodwin, G.M., 182, 194
Gooskens, R.H.J.M., 67, 102
Gopher, D., 79, 81, 97, 98
Gordon, D.A., 426, 435
Gordon, I.E., 335, 347
Gorska, T., 21, 28
Gottlieb, G.L., 61, 63
Gottlieb, N., 158, 162, 172
Gottsdanker, R., 200, 228
Göetz, K.G., 403, 417
Grabiner, M.D., 73, 99
Gracco, V.L., 67, 98, 206, 227
Graham, M., 422, 437
Granrud, C.E., 422, 438
Gray, S., 38, 46
Graybiel, A., 153, 154, 155, 164, 170, 172, 173
Green, K.J., 326, 330
Greenwald, G., 317, 330
Greenwood, R., 337, 338, 339, 342, 347
Gregor, R.J., 342, 348
Gregoric, M., 369, 371
Gregory, R.L., 339, 347
Grillner, S., 190, 194, 399, 400, 417
Grimm, R.J., 355, 372
Grose, E.J., 324, 330

Grossberg, S., 123, 124
Grossman, G.E., 412, 415, 417
Guedry, F.E., 153, 172
Guirao, M., 215, 231
Guitton, D., 179, 180, 194
Gurnee, H., 337, 347

Haaland, K.Y., 262, 275, 279
Haggard, P., 219, 228
Hakkinen, S., 318, 330
Hall, C., 266, 279, 387, 395
Hamilton, C.R., 110, 124
Hancock, P.A., 7, 30
Hansen, R.M., 130, 137, 148, 150
Hardt, M.E., 108, 124
Harris, C.S. 109, 127
Harris-Warrick, R.M., 399, 418
Harrison, J.S., 23, 30
Harshman, R.A., 387, 395
Hartline, H.K., 449, 451
Hasher, L., 115, 124
Haslwanter, T., 180, 196
Hattori, K., 369, 371
Haurt, C., 447, 452
Hawkins, B., 5, 7, 16, 17, 19, 26,
    31, 34, 48, 52, 53, 64, 129,
    131, 137, 147, 152, 162, 164,
    174, 175, 187, 196
Hay, J.C., 107, 109, 124, 166, 173
Hay, L., 13, 14, 19, 20, 28, 74,
    83, 84, 89, 90, 96, 98, 99,
    237, 238, 254, 256, 354, 261,
    371, 401, 417
Hayes, K.D., 354, 355, 356, 371,
    372
Haywood, K.M., 317, 324, 326,
    330, 333
Hazen, N.L., 378, 395
Hécaen, H., 275, 279
Hefti, F., 364, 371

Heilman, K.M., 262, 263, 265,
    267, 268, 275, 276, 277, 279,
    282
Hein, A., 90, 99, 100, 167, 172,
    182, 193
Heinrich, D.R., 109, 123, 127
Held, R., 71, 90, 96, 99, 100, 107,
    108, 112, 124, 158, 162, 167,
    172, 207, 226, 228, 248, 253,
    257, 425, 435
Hendrickson, A., 235, 259
Henn, V., 402, 417, 419
Henning, W., 199, 228
Henry, F.M., 22, 30
Hepp, K., 180, 196
Hepp-Reymond, M.C., 180, 196
Herman, E., 147, 150
Herman, R., 180, 194
Herman, R., 180, 194
Hernandez-Korwo, R., 153, 172
Heuer, F., 148, 151
Heuer, H., 296, 311
Hicks, L., 356, 370
Hill, A.L., 130, 131, 137, 144,
    147, 150
Hill, R.M., 316, 328
Hintzman, D.L., 387, 395
Hisosaka, K., 432, 437
Hixson, W.C., 153, 155, 164, 171,
    172
Hoffler, G.W., 154, 170
Hoffman, J.S., 324, 330
Hogg, S., 261, 265, 268, 272, 277,
    282
Holding, D.H., 35, 46
Hollenberg, M.J., 235, 257
Hollerbach, J.M., 239, 256, 447,
    451
Holmes, C.H., 70, 101
Holt, K.G., 223, 229
Holtzman, J.D., 266, 267, 279,
    280

Homick, J.L., 153, 154, 170, 172, 173
Honda, H., 147, 150
Hooper, P., 272, 280
Hopkins, A., 337, 338, 339, 342, 347
Horak, F.B., 185, 195
Hore, J., 180, 194
Houbiers, M.H.J., 295, 313
Howard, I.P., 108, 109, 110, 111, 123, 124, 125, 126, 181, 195, 425, 437
Howard, R.M., 317, 319, 333
Howarth, C.I., 9, 12, 28, 34, 42, 45, 71, 100
Hoyle, F., 286, 311
Hoyt, D.F., 336, 347
Hu, S., 164, 172
Hubbard, A.W., 294, 311, 325, 331
Hubel, D.H., 316, 331, 377, 395
Hudson, P.T.W., 53, 59, 62
Huebner, W.P., 412, 415, 417
Husak, W.S., 317, 319, 333

Iberall, T., 212, 216, 228, 229
Imwold, C.H., 324, 330
Ingvar, D.N., 267, 278
Inhelder, B., 377, 396
Irwin, D.E., 442, 451
Isaacs, L.D., 317, 331
Iwai, E., 432, 437

Jaeger, M., 34, 36, 37, 38, 39, 40, 46, 92, 97, 99
Jagacinski, R.J., 42, 47, 134, 150
James, W., 266, 279
Jason, G., 263, 264, 277, 279, 280
Jeannerod, M., 14, 20, 21, 22, 26, 27, 30, 43, 47, 56, 63, 67, 74, 84, 86, 90, 96, 100, 101, 129, 130, 131, 132, 149, 151, 152, 180, 181, 182, 187, 188, 193, 195, 198, 199, 201, 202, 203, 204, 205, 206, 207, 208, 209, 210, 211, 212, 213, 214, 215, 216, 218, 219, 220, 224, 226, 227, 229, 230, 231, 233, 234, 236, 237, 238, 240, 245,248, 249, 253, 257, 258, 268, 272, 280, 296, 300, 311, 312, 378, 380, 396
Jensen, B.E., 326, 331
Johansson, G., 321, 331, 422, 425, 435
Johansson, R.S., 206, 216, 229, 245, 257
John, B.E., 448, 451
Johnson, P., 79, 91, 100
Johnson, R.L., 154, 170
Johnson-Laird, P.N., 274, 280
Johnston, I.R., 428, 436
Jolicoeur, P., 276, 280
Jones, B., 68, 100
Jones, R., 34, 36, 37, 38, 39, 40, 46
Jordan, M.I., 198, 223, 224, 226, 229, 449, 451
Judge, S.J., 432, 436
Junge, K., 215, 228

Kaiser, M.K., 433, 435
Kalaska, J.F., 20, 21, 22, 29, 187, 194, 200, 228
Kalil, R.E., 179, 180, 193, 194
Kalish, M., 310, 313, 428, 438
Kanaya, T., 402, 403, 417, 418
Kandel, E.R., 71, 100
Kaplan, E., 277, 279
Katz, P.S., 399, 418
Kaufman, L., 433, 437
Kawahara, K., 402, 403, 417

Keele, S.W., 8, 9, 11, 13, 14, 16, 17, 33, 34, 44, 42, 47, 67, 100, 129, 131, 137, 150, 1897, 195
Kelso, J.A.S., 110, 111, 128, 217, 223, 229, 231, 274, 282, 337, 349
Kerr, B., 131, 150
Kertesz, A., 265, 272, 280
Kieker-Van Ophen, A., 355, 372
Kilbreath, S.L., 365, 368, 371
Kim, K.H. 23, 24, 27, 29
Kimura, D., 60, 63, 277, 280
Kirch, M., 203, 227
Kirshen, A., 356, 371
Kisselburgh, L., 5, 7, 16, 17, 19, 26, 31, 34, 48, 75, 102, 129, 131, 137, 152, 162, 175, 187, 196
Klapp, S.T., 41, 47
Klatzky, R.L., 214, 229, 231
Klein, R.M., 37, 47, 115, 125
Koch, K.L., 164, 172
Koenderink, J.J., 426, 430, 436
Koenig, E., 424, 434
Kohler, I., 107, 125
Kolb, B., 379, 396
Koller, J.A., 324, 330
Komilis, E., 59, 64, 67, 74, 84, 86, 90, 96, 100, 101, 129, 130, 131, 132, 152, 203, 231, 236, 237, 238, 258
Konzag, G., 315, 331
Konzag, I., 315, 331
Koozenkanani, S., 355, 371
Kornblum, S., 7, 13, 30, 43, 44, 47, 52, 53, 56, 58, 62, 63, 71, 98, 100, 129, 131, 132, 133, 134, 138, 147, 149, 151, 162, 173, 207, 230
Kornheiser, A.S., 166, 172

Kosslyn, S.M., 266, 267, 270, 276, 279, 280, 387, 396, 445, 446, 448, 451
Kozlovskaya, I.B., 153, 171
Kozub, S., 387, 394, 395
Krampitz, J.B., 317, 319, 333
Kreydich, Y.V., 153, 172
Kugler, P.N., 223, 224, 225, 229, 336, 349
Kuperstein, M., 123, 124

Laabs, G.J., 390, 396
Lackner, J.R., 153, 154, 164, 170, 173
Lacquaniti, F., 20, 21, 31, 200, 231, 239, 259
Landgraf, J.Z., 148, 149
Landolt, J.P., 425, 437
Langolf, G.D., 42, 47, 134, 151
Lanska, D.J., 412, 415, 417
Lappin, J.S., 422, 435
LaRitz, T., 319, 328
Larsen, B., 266, 281
Larue, J., 83, 98
Lassen, N.A., 266, 281
Laurent, M., 287, 290, 294, 311, 323, 327, 329, 331, 341, 345, 347, 434, 436
Laursen, A.M., 355, 356
Laurutis, V.P., 179, 195
Lavric, A., 369, 371
Lawton, D.T., 428, 436
Leavitt, J.L., 210, 211, 212, 217, 218, 230, 234, 240, 249, 257, 296, 300, 312, 323, 331
Lebedev, V.I., 156, 157, 164, 170, 173
Lee, D.N., 25, 30, 286, 287, 288, 290, 292, 293, 294, 298, 301, 302, 304, 310, 311, 312, 321, 323, 327, 331, 333, 335, 339, 340, 341, 345, 346, 347, 348,

349, 356, 371, 376, 381, 396, 397, 401, 418, 419, 423, 424, 425, 432, 434, 436, 448
Lee, R.G., 206, 229
Leen, T., 356, 371
Legge, G.E., 316, 331
Lehmkuhl, G., 265, 280
Leigh, R.J., 412, 415, 417
Lennie, P., 235, 257
Leonov, A.A., 156, 157, 164, 170, 173
Lestienne, F., 409, 418
Levick, W.R., 316, 328
Lewis, C., 317, 330
Lévesque, L., 92, 94, 101
Lichtenberg, B.K., 153, 170, 173, 175
Liddell, E.G., 402, 418
Liebermann, D.G., 340, 342, 343, 344, 345, 346, 347
Liepmann, H., 262, 263, 280
Lincoln, J., 433, 437
Lintern, G., 79, 81, 97, 98
Lishman, J.R., 294, 312, 341, 348, 381, 396, 401, 418, 424, 425, 436
Liske, D., 239, 257, 300, 312
Llewellyn, K.R., 428, 436
Llinas, R., 402, 419
Lobovits, D., 164, 173
Lockman, J.J., 378, 395, 422, 438
Loomis, J.M., 426, 436
Lough, S., 25, 30, 293, 298, 312, 323, 331, 341, 345, 348
Lu, C.-H., 448, 452
Lucas, D.R., 121, 125
Luchelli, F., 262, 278
Lucia, H.C., 111, 124
Lucy, S.D., 355, 356, 371
Lund, J.P., 399, 400, 419
Lundberg, A., 21, 28
Lynch, K., 378, 396

MacCorquodale, K., 155, 172
Mach, E., 421, 423, 424, 436
Mack, A., 147, 148, 151
Mack, L., 262, 267, 275, 277, 281
MacKenzie, C.L., 21, 22, 26, 27, 30, 188, 195, 204, 205, 206, 208, 209, 210, 211, 212, 213, 215, 217, 218, 219, 220, 222, 226, 227, 230, 234, 238, 239, 240, 243, 249, 257, 258, 268, 277, 278, 282, 296, 300, 312, 323, 331, 378, 396
Madalena, J., 34, 35, 36, 37, 42, 46, 73, 99, 129, 143, 150, 345, 346
Maffei, I., 402, 416
Magill, R.A., 326, 330, 331
Mainor, R., Jr 91, 101
Malpeli, J.G., 235, 258
Malucci, R., 180, 194
Mann, I.C., 235, 257
Marcus, J.T., 156, 173
Mark, L.S., 337, 348
Marks, L.E., 215, 230
Marriott, A.M., 296, 297, 313, 320, 333
Marsden, C.D., 364, 371
Marshall, P.H., 79, 91, 98
Marteniuk, R.G., 21, 22, 26, 27, 30, 80, 84, 87, 92, 94, 101, 180, 181, 182, 183, 184, 185, 186, 187, 188, 189, 194, 195, 204, 205, 206, 208, 209, 211, 212, 217, 218, 219, 220, 222, 226, 227, 230, 233, 239, 240, 249, 257, 277, 278, 296, 300, 312, 323, 331, 332
Martin, C., 425, 435
Martinez-Fernandez, S., 153, 172
Masao, Y., 432, 437
Massey, F.J., 307, 308, 311
Massey, J.T., 20, 21, 22, 29, 187, 194, 200, 201, 228

Massone, L., 2224, 226, 230
Mather, G., 422, 436
Mather, J.A., 132, 151
Matin, L., 155, 173
Matsuyama, K., 402, 418
Matthews, B.H.C., 337, 348
Matthews, P.B.C., 182, 194
Mauritz, K.H., 355, 370
Maxwell, J.C., 61, 63
McCabe, J.F., 69, 101
McClelland, J.L., 198, 221, 223, 224, 225, 231
McCloskey, B.P., 214, 229, 231
McCloskey, D.I., 27, 30, 71, 100, 181, 182, 194, 196, 406, 418
McCollum, G., 185, 195, 356, 371
McGhee, R., 355, 371
McKee, S.P., 316, 331
McKeever, W.F., 61, 63
McKinley, P.A., 342, 348
McLaughlin, C., 323, 332
McLeod, P., 23, 26, 27, 30, 323, 332
McNitt-Gray, J., 342, 343, 344, 348
Mead, B.J., 324, 334
Megaw, E.D., 132, 151, 200, 230
Melvill Jones, G., 337, 342, 344, 348
Mestre, D.R., 310, 313, 429, 430, 431, 433, 436, 438
Metzler, J., 377, 396
Meugens, P., 315, 329
Meyer, D.E., 7, 13, 30,43, 44, 47, 52, 53, 56, 58, 62, 63, 71, 98, 100, 129, 131, 132, 133, 134, 138, 143, 147, 149, 151, 162, 173, 207, 230
Michaels, C.F., 336, 348
Michaud, D., 318, 323, 324, 326, 329
Milgram, P., 45, 47
Milich, M., 206, 232

Miller, E.F. II, 153, 172
Miller, G.A., 274, 280
Miller, J., 450, 451
Miller, J.M., 147, 151
Milner, B., 61, 63
Minenko, V.A., 153, 172
Mishkin, M., 236, 259
Mittelstaedt, M., 319, 333
Miyashita, M., 346
Money, K.E., 153, 160, 174, 175
Moore, M.K., 286, 311
Moore, S., 185, 195
Moran, M.S., 42, 47, 134, 150
Morasso, P., 179, 194
Morenz, C., 326, 331
Morgan, C.L., 130, 131, 144, 147, 151
Mori, S., 402, 403, 417, 418
Morris, M.W., 310, 313, 428, 429, 431, 433, 438
Motti, F., 263, 278
Mountcastle, V.B., 377, 396
Murray, M., 356, 372
Murray, M.A., 73, 102

Nachson, I., 60, 63
Nakayama, K., 316, 331, 332, 426, 436
Nashner, L.M., 185, 195, 206, 230, 355, 356, 371, 372
Nebes, R.D., 60, 63
Neisser, U., 38, 47, 448, 451
Nettleton, N.C., 60, 61, 62, 63
Newell, A., 447, 448, 449, 451
Newell, K.M., 7, 14, 19, 29, 30, 31, 59, 64, 73, 74, 76, 80, 96, 100, 102, 131, 152, 327, 332, 381, 396
Nichelli, P., 263, 278
Nicogossian, A.E., 153, 157, 170, 173
Niles, D., 401, 419

Nimmo-Smith, I., 323, 332
Nissen, M.J., 37, 47
Niven, J.I., 153, 155, 164, 171, 172, 174
Njiokiktjien, C., 354, 355, 365, 371, 372
Norman, D.A., 183, 196, 274, 280, 448, 451
Northam, C., 318, 333

O'Connell, D.N., 422, 438
O'Dell, C.S., 387, 395
Odenrick, P., 356, 372
Ohmi, M., 425, 437
Oldak, R., 287, 289, 313
Ollum, P. ,422, 423, 426, 435
Olson, M.E., 110, 211, 125
Oman, C.M., 153, 175
Orban, G.A., 433, 435
Orlovsky, G.N., 401, 402, 403, 418, 419
Ornitz, E., 356, 372
Oudejans, R.R.D., 298, 300, 305, 310
Owens, P., 147, 149

Pailhous, J., 405, 407, 416, 418, 430, 436
Paillard, J., 83, 98, 100, 138, 143, 151, 198, 230, 237, 238, 254, 257, 258
Paivio, A., 266, 267, 270, 271, 281, 387, 395
Parker, D.E., 153, 170, 173
Parker, J.F., 153, 173
Patla, A.E., 376, 378, 383, 388, 389, 390, 391, 392, 395, 396,401, 418, 434, 437
Paulignan, Y., 21, 22, 26, 27, 30, 204, 205, 206, 211, 212, 213,

217, 220, 222, 226, 227, 230, 234, 258
Paulus, W.M., 409, 418, 425, 437
Pauwels, J., 315, 329
Pavard, B., 425, 434
Pavlova, G.A., 403, 418
Payne, V.G., 318, 332
Pellegrino, J.W., 214, 229, 231
Perenin, M.T., 236, 258
Perfetto, G., 422, 435
Peters, M., 53, 64
Peterson, J.R., 9, 29
Pettersen, L., 422, 438
Pettersson, L.-G., 21, 28
Pew, R.W., 68, 69, 100
Pélisson, D., 20, 21, 27, 30, 44, 46, 47, 56, 63, 130, 131, 132, 137, 143, 151, 152, 187, 188, 195, 201, 202, 203, 204, 207, 228, 230, 231
Phillips, C.G., 402, 418
Piaget, J., 377, 396
Picado, M.E., 326, 331
Piccolino, L., 402, 416
Pick, H.L., 70, 86, 98, 101, 107, 109, 124, 166, 173, 297, 301, 309, 312, 320, 332, 378, 395
Pitcairn, T., 295, 310
Poeck, K., 262, 265, 280, 281
Poizner, H., 277, 281
Polit, A., 148, 151
Pongrac, J., 387, 395
Posner, M.I., 8, 9, 12, 13, 14, 16, 17, 30, 34, 37, 47, 115, 125, 131, 137, 150, 187, 195
Poulton, E.C., 20, 30, 73, 100, 187, 195, 315, 332
Pozzo, T., 415, 416
Prablanc, C., 20, 21, 27, 30, 44, 46, 47, 56, 63, 64, 67, 84, 86, 90, 96 100, 101, 129, 130, 131, 132, 137, 143, 149, 151, 152, 180, 181, 187, 188,

193, 195, 201, 202, 203, 204, 227, 228, 230, 231, 236, 237, 238, 257, 258, 380, 396
Prazdny, K., 426, 437
Precht, W., 402, 419
Proctor, R.W., 447, 448, 451, 452
Proteau, L., 36, 47, 80, 81, 83, 85, 86, 87, 92, 94, 97, 101, 236, 258, 323, 332
Purdy, W.C., 286, 312
Pylyshyn, Z.W., 390, 395, 448, 452

Quick, H., 387, 395
Quinn, J.T., 27, 31, 52, 53, 64, 147, 152, 162, 164, 174

Rader, S.J., 124, 129
Rakhamanov, A.S., 153, 172
Ratliff, F., 449, 451
Raymond, J.E., 319, 332
Raz, T., 346, 348
Redding, G.M., 108, 109, 111, 112, 113, 115, 117, 118, 119, 120, 121, 122, 123, 125, 126
Reddish, P.E., 25, 30, 292, 293, 298, 312, 323, 327, 331, 341, 345, 347, 348
Reed, E.S., 321, 332
Reeve, T.G., 91, 101, 447, 451
Regal, D.M., 180, 195
Regan, D., 290, 301, 312, 422, 432, 433, 437
Reid, D.J., 153, 156, 161, 174
Repperger, D.W., 42, 47, 134, 150
Reschke, M.F., 153, 154, 170, 172, 173
Rex de Palmer, D., 73, 96, 99
Reymond, M.C., 180, 196
Rhodes, R.W., 111, 127

Riach, C., 354, 355, 356, 357, 362, 371, 372
Rich, S., 71, 95, 99
Rieger, J.H., 430, 433, 437
Ripoll, H., 290, 294, 311, 341, 347
Rizzolatti, G., 210, 212, 217, 228
Roberts, L., 223, 231
Robertson, D.G.E., 325, 329
Robinson, C., 376, 396, 401, 418, 434, 437
Robinson, D.A., 111, 112, 123, 126
Robinson, J.A., 179, 195
Robson, J.G., 234, 257
Rock, I.,107, 126, 166, 173, 403, 404, 419
Rodieck, R.W., 235, 258
Rogers, B., 422, 437
Rogers, M.W., 364, 367, 372
Rohr, L., 59, 64
Roland, P.E., 266, 281
Rondot, P., 275, 279
Ronnqvist, L., 217, 231, 240, 249, 259
Rosenbaum, D.A., 183, 196, 198, 229, 236
Rosenblatt, F., 422, 423, 426, 435
Rosengren, K.S., 86, 101, 297, 301, 309, 312, 320, 332
Ross, H.E., 155, 164, 166, 173, 174
Rossignol, S., 399, 400, 419
Rothi, L.J.G., 262, 267, 268, 275, 276, 277, 279, 281
Roy, E.A.,36, 47, 49, 53, 54, 59, 60, 64, 248, 253, 256, 261, 262, 263, 264, 265, 268, 272, 273, 274, 275, 277, 278, 279, 281, 282
Rumelhart, D.E., 183, 196, 198, 223, 224, 225, 231, 448, 451
Runnesson, S., 336, 348

Russell, D.G., 70, 101

Saito, H., 432, 437
Sakitt, B., 148, 152
Sakstein, R. 12, 28
Salapatek, P., 180, 195
Salmoni, A.W., 326, 332
Saltzman, C., 274, 282
Samways, M., 376, 396, 401, 418, 434, 437
Sandstedt, P., 356, 372
Savelsbergh, G.J.P., 73, 86, 96, 99, 101, 294, 297, 298, 301, 309, 312, 313, 341, 348, 432, 437
Scarpa, M., 210, 212, 217, 228
Scharf, M.K., 129, 131, 143, 149
Schiff, W., 285, 286, 287, 289, 312, 313
Schiller, P.H., 235, 258
Schmidt, R.A., 13, 23, 26, 31, 52, 53, 64, 67, 69, 70, 89, 95, 96, 101, 147, 152, 162, 164, 174, 206, 232, 296, 313, 326, 332, 334, 443, 452
Schmitt, H.H., 153, 156, 161, 174
Schmückler, M.A., 408, 419
Schneider, G.E., 425, 437
Schneider, T., 73, 102, 296, 311
Schneider, W., 119, 129
Schoebel, P., 401, 417
Schöne, H., 155, 164, 174
Schultheis, L.W., 111, 123
Schwartz, J.H., 71, 100
Schweickert, R., 450, 452
Scully-Power, P., 160, 174
Seireg, A., 356, 3729
Semmes, J., 60, 64
Seng, C.N., 294, 311, 325, 331
Sepic, A., 356, 372
Shambes, G., 361, 372
Shanks, M.D., 320, 333

Shapiro, K.L., 319, 332
Sharp, R.H., 320, 333, 334
Shea, C.H., 317, 318, 319, 324, 333, 334
Shepard, R.N., 377, 396, 448, 451
Sherwood, D.E., 27, 31
Shiffrin, R.M., 115, 126
Shik, M.L., 401, 419
Shimojo, S., 71, 102
Shumway-Cook, A., 356, 361, 372
Sidaway, B., 342, 343, 344, 348
Silva, J.M., 266, 282
Silveira, J.M., 401, 418
Simpson, H., 402, 419
Sivak, B., 234, 238, 239, 243, 257, 258, 268, 282, 378, 396
Sivak, J., 243, 258
Skavenski, A.A., 130, 137, 148, 150
Skinhoj, E., 266, 281
Slater-Hammel, A.T., 9, 31
Smith, D., 403, 404, 419
Smith, J.E.K., 7, 13, 30, 43, 44, 47, 52, 53, 56, 58, 63, 71, 100, 131, 132, 134, 143, 147, 151, 162, 173, 207, 230
Smith, J.L., 342, 348
Smith, T., 214, 229
Smith, W.M., 18, 19, 31, 34, 47
Smyth, M.M., 73, 78, 102, 296, 297, 313, 320, 333
Snyder, C.R.R., 119, 128
Soechting, J.F., 20, 21, 31, 71, 102, 167, 174, 178, 181, 196, 200, 208, 231, 239, 251, 258, 259, 409, 418
Solomon, H.Y., 336, 348
Solomon, J., 391, 397
Sorgato, P., 273, 278
Spencer, J. ,356, 371
Sperling, A., 203, 227
Sperling, G., 38, 47, 448, 452

Spira, A.W., 235, 257
Square, P.A., 269, 273, 275, 281, 282
Square-Storer, P.A., 261, 262, 263, 264, 265, 268, 272, 277, 278, 279, 281, 282
Stadulis, R.E., 324, 333
Stallings, H.D., 158, 164, 172
Stark, L., 179, 196
Starkes, J., 354, 355, 356, 357, 362, 369, 371, 372
Steenhuis, R.E., 383, 384, 385, 393, 396, 442, 452
Stein, P.S., 339, 419
Stein, R.B., 61, 64
Steinbach, M.J., 108, 124
Stelmach, G.E., 13, 31, 115, 126
Stephenson, J.M., 10, 11, 12, 31, 320, 334
Stern, R.M., 164, 172
Stevens, S.S., 215, 231
Stockwell, C., 355, 371
Stoffregen, T.A., 408, 419, 426, 437, 438
Stoper, A.E., 155, 174
Stratford, R., 73, 99
Stratton, G.M., 107, 126
Straube, A., 409, 418, 425, 437
Strauman, D., 180, 196
Strelow, E.R., 386, 387, 396
Strughold, H. ,158, 164, 192
Stubbs, D.F., 34, 47, 84, 102
Sully, D.J., 296, 297, 310, 322
Sully, H.G., 296, 297, 310, 322

Tagliasco, V., 180, 193
Takahashi, T., 369, 371
Takakusaky, K., 402, 418
Talbott, R.E., 409, 419
Tanaka, K., 432, 437
Tardif, M., 387, 394
Tatton, W.G., 206, 229

Taub, E., 164, 174, 405, 419
Taylor, B. 391, 395
Taylor, C.F., 336, 347
Taylor, J.L., 181, 182, 196
Tdlohreg, C.W., 306, 308, 310
te Linde, J., 267, 281
Teasdale, N., 83, 89, 102
Teghtsoonian, M., 215, 231
Telford, C.W., 20, 31
Templeton, W.B., 109, 110, 111, 125, 126
Teravainen, H.T., 364, 370
Teuber, H.L., 319, 333
Thelen, E., 401, 419
Thirsk, R.B., 160, 174
Thomson, J.A., 34, 35, 36, 48, 294, 312, 341, 345, 347, 348, 379, 381, 382, 383, 384, 385, 386, 388, 389, 393, 396, 397, 404, 418, 434, 436
Thurston, S.E., 412, 417
Thyer, L., 60, 62
Todd, J.T., 287, 300, 301, 304, 305, 313, 422, 438
Todor, J.I., 49, 51, 52, 53, 60, 64
Tolman, E.C., 378, 397
Tolson, H., 317, 319, 333
Torres, J., 316, 333
Townsend, J.T., 60, 64
Travis, R.C., 425, 438
Tresilian, J.R., 290, 298, 301, 309, 313
Trevarthen, C.B., 50, 64, 335, 349
Tritschler, K., 361, 372
Tuller, B., 410, 419
Turner, A., 53, 59, 62
Turton, A., 73, 102, 210, 218, 232, 242, 259
Turvey, M.T., 61, 64, 223, 224, 225, 229, 294, 301, 311, 336, 349, 391, 397, 410, 419

Tyldesley, D.A., 295, 313, 325, 333
Tyler, C.W., 316, 332, 333

Uhlarik, J.J., 111, 117, 126, 130, 160, 174
Ullman, S., 339, 349
Ulrich, B.D., 401, 419
Umilta, C., 210, 212, 217, 228
Ungerleider, L.G., 236, 259
Unno, T., 402, 403, 417

Valenstein, E., 268, 272, 279
Van den Biggelaar, J., 239, 257
Van der Horst, A.R.A., 294, 213
Van der Meulen, J.H.P.,67, 102
Van Holten, C.R., 156, 173
Van Parys, J., 354, 372
Van Rossum, J.H.A., 89, 102
Van Snippenberg, F.J., 306, 308, 310
Van Sonderen, J.F., 200, 231
Van Wieringen, P.C.W., 25, 26, 28, 294, 295, 296, 300, 310, 321, 323, 328, 341, 346
Van Zandt, T., 448, 452
Vaughn, W.J., 206, 228
Velay, J.-L., 84, 102
Verfaellie, M., 261, 277, 281
Vighetto, A., 236, 258
Villardi, K., 148, 151
Villas, T., 180, 194
Vince, M.A., 6, 7, 8, 12, 13, 31, 36, 48, 128, 152
Vogele, D., 337, 348
Volle, M., 179, 180, 194
Von Beckh, H.J.A., 157, 158, 161, 164, 170, 174
Von Helmholtz, H., 107, 123, 155, 174, 421, 422, 438

Von Hofsten, C., 86, 101, 217, 231, 240, 249, 259, 297, 301, 309, 312, 320, 321, 322, 323, 324, 332, 333
Von Holst, E., 319, 333
Voohoeve-Coebergh, O., 355, 372

Wade, M.G. ,324, 334
Waespe, W., 402, 419
Wagner, H., 341, 349
Walkuski, J.J., 71, 99
Wall, C., 355, 372
Wallace, B., 109, 111, 113, 115, 117, 128, 119, 122, 123, 125, 126
Wallace, S.A., 19, 31, 59, 64, 73, 76, 96, 102, 131, 152, 210, 215, 217, 218, 231
Wallach, H., 421, 422, 438
Walter, C.B., 326, 332
Ward, S.L., 42, 47, 134, 150
Warren, R., 428, 438
Warren, W.H., 294, 310, 313, 336, 337, 341, 349, 376, 397, 401, 419, 428, 429, 431, 433, 434, 438
Watson, R.T., 267, 281
Watt, D.G.D., 153, 160, 174, 175, 337, 342, 344, 348
Weeks, D.J., 265, 269, 448, 452
Weeks, D.L., 210, 215, 217, 218, 231, 317, 334
Weicker, D. ,325, 329
Weiner, W.J., 364, 370, 371
Welch, R.B., 107, 109, 111, 123, 127, 155, 159, 162, 164, 166, 167, 171, 174
Welford, A.T., 9, 31, 187, 196
Wenking, H., 403, 417
Wertheim, A.H., 319, 329, 334
Westling, G., 206, 216, 229, 245, 257

Whishaw, I.Q., 379, 396
White, G.R., 428, 436
Whitehurst, M., 324, 329
Whiteside, T.C.D., 155, 158, 162, 174, 337, 348
Whiting, H.T.A., 10, 11, 12, 31, 86, 101, 294, 295, 296, 297, 298, 310, 312, 313, 320, 321, 325, 327, 333, 334, 341, 348, 432, 437
Whitney, R.J., 360, 372
Whittaker, E.T., 430, 438
Wiesel, T.N., 316, 331, 377, 395
Wihelm, K., 67, 102
Wilkinson, D.A., 109, 111, 126, 127
Williams, H., 361, 372
Williams, J.G. ,383, 388, 389, 395
Willmes, K., 265, 280
Wilson, R.C., 156, 157, 164, 171
Wing, A.M. 210, 212, 214, 218, 219, 227, 228, 232, 242, 359
Winstein, C.J., 326, 334
Wist, E.R., 425, 434
Wood, C., 164, 172
Woodworth, R.S., 3, 4, 6, 7, 8, 12, 13, 31, 34, 48, 89, 102, 129, 131, 152, 378, 397
Woollacott, M., 356, 361, 372
Wright, C.E., 5, 13, 29, 43, 44, 47, 52, 53, 56, 58, 63, 71, 100, 131, 132, 134, 143, 147, 151, 162, 173, 207, 230
Wrisberg, C.A., 324, 334
Wughalter, E.H., 324, 329

Yeomans, J.M., 442, 451
Yilmaz, E.H., 292, 314
Yonas, A., 422, 438
Young, D.E., 325, 326, 327, 334
Young, D.S., 25, 30, 290, 293, 294, 298, 301, 312, 313, 323, 331, 335, 339, 341, 345, 347, 348, 349, 376, 397, 401, 408, 419, 434, 438,
Young, L.R., 153, 175, 402, 417, 425, 434
Yuodelis, C., 235, 259

Zacks, R.T., 115, 124
Zangmeister, W.H., 179, 196
Zangwill, D.L., 61, 65
Zanone, P., 447, 452
Zarriello, J.J., 154, 172
Zeffren, M., 147, 149
Zelaznik, H.N., 5, 7, 16, 17, 18, 19, 26, 31, 34, 48, 52, 53, 64, 75, 102, 129, 131, 137, 147, 152, 162, 164, 174, 175, 187, 196, 206, 232
Zimmerman, W.S., 157, 171
Zuidema, G.D., 153, 171

Subject index

absolute information, 432
adaptation, 68, 105, 106, 107,
    108, 109, 110, 111, 112, 113,
    114, 115, 116, 117, 119, 120,
    121, 122, 123, 154, 156, 157,
    159, 160, 161, 163, 164, 166,
    167, 168, 169, 170, 296, 327,
    336, 399, 415, 443, 444
affordance, 214, 274
aftereffect, 154, 157, 159, 160
ambient, 318, 422, 425
aperture, 210, 211, 212, 214, 217,
    218, 219, 220, 221, 238, 239,
    240, 241, 242, 243, 244, 245,
    246, 247, 249, 250, 251, 252,
    253, 255
apraxia, 262, 263, 265, 266, 267,
    268, 269, 270, 272, 273, 275,
    276, 277, 278, 445
automatic processing, 115, 122,
    323
autonomic motor system, 236

balance, 335, 355, 356, 364, 399,
    408, 409, 414, 415, 416, 425
batting, 294
blind locomotion, 387

catching, 3, 10, 11, 12, 68, 71, 73,
    76, 86, 285, 294, 296, 297,
    298, 309, 320, 323, 324, 327,
    341, 379
center of expansion, 301
central pattern generator, 401
central planning, 67
central vision, 83, 96, 187, 234,
    236, 237, 238, 239, 240, 241,
    242, 243, 248, 251, 252, 253,
    254, 255, 425, 431, 445
centrifugation, 156, 158, 159, 160

cerebral hemisphere, 49, 50, 51,
    60, 442
closed-loop, 13, 34, 43, 69, 70,
    83, 91, 94, 400
cognitive, 444, 446, 449
cognitive load, 115, 121, 122
cognitive processes, 108, 112, 344
coincidence-anticipation, 315,
    316, 320, 321, 322, 324, 326
complexity, 322
comprehension, 262, 263, 265,
    272, 277
computational assessment, 443
cones, 234, 235, 316
connectionism, 449
connectionist model, 449
contact lenses, 243
context, 33, 84, 86, 90, 93, 160,
    208, 226, 261, 265, 270, 271,
    274, 275, 277, 317, 318, 391
continuous control, 44
controlled processing, 122, 123
coordinative linkage, 111, 112,
    114, 115, 116, 117, 118, 119,
    120, 121, 122
corollary discharge, 207, 226, 319
cross-modal, 68
curvilinear motion, 429, 430, 433

deafferentation, 166, 400
discordance, 108, 109, 111, 113,
    114, 116, 122, 123
discrete adjustment, 42, 43, 44
discrete sampling, 450
displacement, 69, 71, 80, 106,
    108, 109, 112, 155, 156, 160,
    164, 180, 181, 192, 218, 219,
    309, 315, 316, 317, 319, 320,
    322, 325, 399, 401, 404, 405,
    411, 412, 413, 414, 415, 425
distortions, 112, 167
distribution of practice, 164, 169

Down's syndrome, 263
driving, 287, 301, 318, 323, 327, 416
dynamic visual information, 84, 86

egocentric viewing, 376
elevator illusion, 155, 156, 157, 158, 160, 161, 162, 163, 164, 165, 166, 167, 168, 169, 170
EMG, 181, 337, 338, 339, 340, 343, 345, 346, 361, 402, 403
episodic memory, 267
exteroceptive, 71, 226, 336, 401, 423, 426
extraretinal information, 130, 131, 132, 138, 139, 140, 142, 143, 145, 146, 147, 148, 149, 182, 443, 444
eye movement, 443, 444
eye movements, 45, 129, 130, 132, 134, 135, 136, 137, 138, 139, 140, 142, 143, 144, 145, 146, 147, 148, 149, 181, 185, 203, 320, 394
eye-hand coordination, 107, 108, 109, 110, 111, 112, 113, 115, 146, 164, 443

feedback, 4, 5, 6, 7, 8, 9, 10, 12, 13, 14, 15, 16, 17, 18, 19, 20, 25, 34, 44, 49, 50, 51, 52, 53, 54, 55, 56, 57, 58, 59, 67, 68, 69, 70, 71, 73, 76, 77, 78, 79, 84, 90, 95, 96, 111, 114, 115, 117, 118, 119, 120, 121, 122, 129, 130, 132, 137, 138, 139, 140, 144, 157, 158, 159, 160, 161, 163, 166, 167, 168, 169, 170, 193, 206, 214, 215, 237, 238, 253, 254, 266, 272, 277, 309, 326, 327, 357, 361, 362, 363, 367, 368, 381, 383, 384, 393, 441, 442, 443, 444
feedforward, 20, 34, 43, 44, 89, 90, 96
Fitts' law, 7, 8, 9, 14, 50, 51, 52, 59, 69, 73, 208, 218
fixate/fixation, 133, 138, 139, 140, 142, 143, 144, 147, 180, 182, 183, 185, 187, 191, 192, 234, 237, 240, 307, 412, 414
fovea, 83, 156, 234, 235, 236, 239, 240, 255, 256, 259, 269, 424
free fall landings, 339

gait, 153, 336, 364, 365, 367, 376, 391, 392, 394, 402, 405, 412, 415
ganglion cells, 235
gaze, 148, 180, 183, 342
gestures, 261, 262, 263, 264, 265, 266, 267, 268, 269, 270, 271, 272, 273, 274, 275, 276, 445
grasping, 21, 26, 109, 179, 188, 189, 190, 191, 192, 193, 197, 198, 204, 208, 209, 210, 212, 214, 215, 216, 217, 218, 219, 233, 234, 235, 236, 238, 239, 240, 241, 242, 243, 245, 246, 248, 249, 250, 251, 253, 254, 255, 256, 269, 277, 296, 297, 298, 323, 432, 441, 444, 445, 447
gravity, 153, 154, 155, 156, 157, 158, 159, 161, 165, 166, 167, 168, 169, 170, 185, 303, 355, 357
grip accuracy, 214

head movement, 179, 180, 181, 182, 183, 185, 193, 402, 404, 410, 412, 414, 415, 444
head orientation, 164, 169
heading direction, 427, 428, 429, 432
high resolution vision, 243, 246, 248
hitting, 12, 71, 208, 285, 294, 296, 297, 310, 323, 325, 445
homing-in, 83, 238
hypergravity, 156, 158, 159, 167, 168, 169
hypogravity, 158, 167, 168

iconic memory, 38, 40, 41, 442
illumination, 10, 11, 21, 53, 54, 55, 234
illusion, 155, 156, 163, 169, 253, 405, 424
image generation, 262, 264, 267, 269, 270, 271, 276, 445
imagery, 266, 267, 270, 276, 377, 386, 387, 388, 394
imitation, 261, 262, 263, 265, 268, 269, 270, 272, 273, 276, 278
impulse variability, 53
index of difficulty, 14, 15, 51, 59, 69, 70, 73
initial impulse, 131, 132, 135, 136, 141, 142, 146, 147, 148, 149, 444
interception, 315, 321, 323, 325, 445
intermittency, 9, 33, 381, 382, 383, 385, 393, 394
intermodal, 97, 269
intersensory integration, 277
invariance, 217, 239, 294, 296, 327, 422

just noticeable difference, 318, 319

kinematics, 13, 14, 21, 22, 42, 52, 58, 119, 188, 199, 204, 207, 208, 217, 220, 221, 222, 239, 242, 297, 394, 416
kinesthetic, 8, 18, 27, 34, 43, 44, 68, 71, 72, 78, 79, 84, 95, 159, 163, 168, 266, 271, 272, 273, 276, 390
kinetic depth effect, 422
knowledge, 214, 223, 224, 225, 273, 274, 317, 387, 401
knowledge of results/KR, 8, 19, 78, 79, 80, 81, 82, 87, 88, 89, 91, 92, 93, 95, 326, 327, 445

landings, 337, 342
landmarks, 217, 221, 388, 389
language, 263, 265, 273
lateral geniculate body, 235, 236
learning, 13, 67, 68, 70, 71, 72, 76, 77, 78, 80, 90, 91, 92, 93, 94, 95, 96, 97, 105, 106, 107, 108, 116, 223, 224, 266, 267, 323, 325, 326, 327, 443
long jump, 341, 381, 434
looming, 285, 290, 422, 432

manual asymmetry, 49, 50, 52, 59, 61, 62
mental practice, 266
modules, 111, 112, 116, 123
moments of force, 355, 361
motor equivalence, 198, 221, 226, 227, 444
motor programming, 13, 89, 148, 167, 215, 296, 381, 385, 386

movement detection, 316
movement goal, 20, 22, 96, 188, 199, 226, 227
movement planning, 96, 223
moving objects, 190, 341, 432, 445
moving targets, 189, 239
multiple correction, 42
muscle activity, 44, 341, 342, 344, 361, 399
muscle loading, 155, 156, 157, 160, 161, 162, 163, 164, 165, 166, 167, 168, 169, 170

neural network, 198, 224, 226, 227, 317, 449

object characteristics, 233, 239, 248, 249, 253, 254, 255
object motion, 239, 255, 424
object size, 198, 210, 211, 212, 213, 215, 218, 219, 221, 222, 239, 249, 251, 300
observation point, 323
oculomotor commands, 130, 138
open-loop, 68, 69, 71, 94, 109, 169, 206, 400
optic flow, 289, 323, 339, 341
optical flow, 341, 399, 401, 402, 403, 404, 405, 406, 407, 409, 410, 412, 415, 416, 421, 422, 423, 424, 425, 426, 427, 428, 429, 430, 431, 432, 433, 434, 445, 446, 447
optomotor responses, 402
otolith-vestibular, 337, 339, 342

pantomime, 261, 264, 265, 268, 269, 270, 273, 274, 275, 276, 278, 445

parallel distributed process, 198
parallel processing, 50, 225, 442
parameterization, 49
Parkinson's disease, 354, 355, 363, 364, 365, 366, 367, 368, 369, 370, 446
peripheral vision, 50, 233, 234, 236, 237, 238, 239, 240, 241, 242, 243, 244, 245, 246, 247, 248, 249, 250, 251, 254, 255, 269
persistence, 36, 442
perturbations, 26, 27, 148, 155, 199, 201, 206, 207, 213, 218, 219, 361, 364, 405, 406, 410, 415, 416, 444
postural control adjustments, 185, 206, 361, 363
postural control, 354, 355, 356, 357, 361, 362, 363, 365, 368, 369, 370, 409, 425, 446
practice, 8, 23, 26, 36, 37, 67, 69, 70, 72, 76, 77, 78, 79, 80, 81, 83, 84, 85, 86, 87, 88, 89, 90, 91, 92, 93, 94, 95, 96, 97, 159, 160, 163, 164, 316, 317, 323, 324, 325, 326, 384
prediction of speed, 317
predictive spatial information, 297, 301, 304, 306
predictive temporal information, 285
prehension, 26, 71, 197, 198, 199, 204, 205, 207, 211, 212, 213, 214, 216, 217, 218, 219, 220, 221, 222, 223, 224, 226, 233, 234, 236, 238, 243, 296, 300, 444, 445
preprogramming, 70, 188, 189
propositional, 274, 387, 388, 448
proprioception, 71, 72, 79, 90, 111, 159, 182, 185, 423

proprioceptive information, 130,
163, 181, 182, 185, 192, 238,
251, 421, 423, 425, 444
proprioceptive shift, 108, 110,
118, 119, 120, 122
pursuit eye movement, 138, 139,
140, 141, 142, 143, 146, 147,
148, 319, 320, 430, 443

rate of dilation, 25, 286, 287, 288,
298, 300, 341, 345
recalibration, 108, 109, 113, 114
reflex, 153, 155, 319, 320, 337,
338, 339, 342, 344, 356, 364,
368, 370, 375, 399, 400, 406,
412
relative information, 432
representation, 18, 33, 34, 36, 37,
38, 39, 40, 41, 43, 44, 45,
69, 89, 96, 97, 113, 123,
131, 148, 167, 187, 207, 208,
214, 223, 224, 236, 268, 270,
273, 275, 345, 377, 378, 383,
385, 386, 387, 394, 399, 442,
443, 444, 446, 448, 449
response complexity, 316, 324
retina, 20, 25, 45, 130, 131, 132,
137, 139, 142, 143, 144, 145,
146, 149, 155, 156, 215, 233,
234, 235, 236, 239, 242, 243,
248, 249, 319, 320, 323, 324,
336, 341, 345, 393, 405, 406,
414, 421, 424, 426, 443, 445,
449, 450
retinal information, 130, 131,
138, 139, 142, 143, 145
retinotopic information, 236
rods, 234
Romberg test, 354, 363, 369
rotation, 80, 132, 133, 134, 135,
136, 137, 138, 139, 140, 141,
142, 143, 154, 180, 182, 185,

201, 212, 221, 267, 406, 410,
412, 414, 424, 425, 426, 430,
431, 443, 448

saccadic, 134, 137, 139, 140, 145,
148, 179, 319, 443
secondary movement, 14, 203
self-motion, 404, 421, 422, 423,
424, 425, 426, 428, 429, 430,
431, 432, 433, 447
self-organizing principles, 224,
227
semantic memory, 267
sensorimotor store, 67, 71
sensory-motor rearrangement,
159
sequencing, 262, 263, 264, 265,
266, 277, 445
serial processing, 60
single correction, 42
skill, 8, 69, 105, 106, 108, 116,
197, 198, 266, 322, 327, 361,
378, 386, 387, 422, 443, 449
spatial ability, 377, 378, 394
spatial alignment, 107, 114, 116,
123, 443
specificity, 90, 91, 92, 93, 95, 97,
167, 322, 323
speed-accuracy tradeoff, 5, 151
stability, 217, 240, 354, 355, 356,
357, 358, 359, 360, 361, 362,
365, 366, 367, 368, 369, 370
static visual information, 87
step cycle, 403
stimulus-response compatibility,
448
strategies, 11, 16, 20, 37, 56, 57,
69, 115, 119, 120, 183, 185,
206, 210, 212, 240, 245, 246,
247, 249, 266, 293, 294, 295,
302, 318, 320, 325, 341, 342,
344, 345, 354, 361, 368, 369,

370, 381, 386, 388, 412, 446, 447
submovements, 8, 13, 14, 42, 134, 140, 223, 253
superior colliculus, 235, 236
sway velocity, 354, 356, 362, 363, 365, 367, 368, 369, 370, 446
synergy, 185, 221, 223, 356, 401, 415

tactile contact, 251, 253
target uncertainty, 40, 122, 123
tau, 299, 300, 339, 340, 341, 343
tau-margin, 290, 293, 294, 295, 297, 298, 300
time to contact, 288, 290, 292, 293, 294, 298, 323, 335, 339, 340, 341, 344, 345
total shift (TS), 116, 123
trajectory, 10, 12, 16, 21, 23, 42, 43, 44, 56, 58, 83, 162, 187, 188, 189, 192, 199, 200, 201, 204, 213, 219, 222, 237, 238, 239, 242, 287, 298, 304, 320, 324, 410, 416, 423
translation, 113, 123, 180, 185, 313, 406, 410, 412, 424, 425, 426, 429, 430, 431
translational motion, 428, 429
two-hand coordination, 71

vection, 423, 424, 425, 426
vestibular information, 369, 424
visual acuity, 234, 235, 238, 245, 248, 249
visual field, 18, 50, 106, 116, 155, 156, 157, 163, 169, 234, 242, 243, 248, 254, 315, 318, 323, 324, 361, 401, 404, 405, 421, 424, 425, 426

visual processing time, 4, 7, 8, 9, 13, 14, 16, 17, 19, 25, 187
visual shift, 108, 110, 118, 119, 120, 122
visuomotor channels, 198, 224, 240, 248, 253, 255

W-like cells, 235
WATSMART, 42, 57, 239

Y-like, 235

zero-crossings, 42, 43, 57, 58, 134